# analysis of biological data
a soft computing approach

# SCIENCE, ENGINEERING, AND BIOLOGY INFORMATICS

**Series Editor:** Jason T. L. Wang
*(New Jersey Institute of Technology, USA)*

---

*Published:*

Vol. 1: Advanced Analysis of Gene Expression Microarray Data
*(Aidong Zhang)*

Vol. 2: Life Science Data Mining
*(Stephen TC Wong & Chung-Sheng Li)*

Vol. 3: Analysis of Biological Data: A Soft Computing Approach
*(Sanghamitra Bandyopadhyay, Ujjwal Maulik & Jason T. L. Wang)*

# analysis of biological data
## a soft computing approach

*editors*

### Sanghamitra Bandyopadhyay
*Indian Statistical Institute, India*

### Ujjwal Maulik
*Jadavpur University, India*

### Jason T L Wang
*New Jersey Institute of Technology, USA*

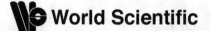
**World Scientific**

NEW JERSEY · LONDON · SINGAPORE · BEIJING · SHANGHAI · HONG KONG · TAIPEI · CHENNAI

*Published by*

World Scientific Publishing Co. Pte. Ltd.
5 Toh Tuck Link, Singapore 596224
*USA office:* 27 Warren Street, Suite 401-402, Hackensack, NJ 07601
*UK office:* 57 Shelton Street, Covent Garden, London WC2H 9HE

**British Library Cataloguing-in-Publication Data**
A catalogue record for this book is available from the British Library.

Science, Engineering, and Biology Informatics — Vol. 3
**ANALYSIS OF BIOLOGICAL DATA**
**A Soft Computing Approach**

Copyright © 2007 by World Scientific Publishing Co. Pte. Ltd.

*All rights reserved. This book, or parts thereof, may not be reproduced in any form or by any means, electronic or mechanical, including photocopying, recording or any information storage and retrieval system now known or to be invented, without written permission from the Publisher.*

For photocopying of material in this volume, please pay a copying fee through the Copyright Clearance Center, Inc., 222 Rosewood Drive, Danvers, MA 01923, USA. In this case permission to photocopy is not required from the publisher.

ISBN-13 978-981-270-780-2
ISBN-10 981-270-780-8

Editor: Tjan Kwang Wei

Typeset by Stallion Press
Email: enquiries@stallionpress.com

Printed in Singapore.

To my parents Satyendra Nath Banerjee
and Bandan Banerjee and son Utsav
*Sanghamitra Bandyopadhyay*

To my parents Manoj Kumar Maulik
and Gauri Maulik
*Ujjwal Maulik*

To my wife Lynn and daughter Tiffany
*Jason T. L. Wang*

# PREFACE

Over the past few decades, major advances in the field of molecular biology, coupled with advances in genomic technologies, have led to an explosive growth in the biological information generated by the scientific community. This deluge of genomic information has, in turn, led to a requirement for computerized databases to store, organize and to index the data, and there is pressure for specialized tools to view and analyze the data. Bioinformatics, a field devoted to the interpretation and analysis of biological data using computational techniques, has evolved in response to this need. It is an interdisciplinary field involving biology, computer science, mathematics and statistics to analyze biological sequence data, genome content & arrangement, and to predict the function and structure of macromolecules. The ultimate goal of the field is to enable the discovery of new biological insights as well as to create a global perspective from which unifying principles in biology can be derived.

Soft computing is a consortium of methodologies that work synergistically and provides, in one form or another, flexible information processing capabilities for handling real life ambiguous situations. Its aim, unlike conventional (hard) computing, is to exploit the tolerance for imprecision, uncertainty, approximate reasoning and partial truth in order to achieve tractability, robustness, low solution cost, and close resemblance with human like decision making. At this juncture, fuzzy sets (FS), artificial neural networks (ANN), evolutionary algorithms (EAs) (including genetic algorithms (GAs), genetic programming (GP), evolutionary strategies (ES)), support vector machines (SVM), wavelets, rough sets (RS), simulated annealing (SA), particle swarm optimization (PSO), memetic algorithms (MA), ant colony optimization (ACO), tabu search (TS) and chaos theory are the major components of soft computing.

The different tasks involved in the analysis of biological data include sequence alignment, genomics, proteomics, DNA and protein structure prediction, gene/promoter identification, phylogenetic analysis, analysis of gene expression data, metabolic pathways, gene regulatory networks, protein folding, docking, and molecule and drug design. Data analysis tools used earlier in bioinformatics were mainly based on statistical techniques like regression and estimation. Role of soft computing in bioinformatics gained significance with the need of handling large, complex, inherently uncertain, data sets in biology in a robust and computationally efficient manner.

Much of the biological data are inherently uncertain and noisy, thus making fuzzy sets a natural framework for analyzing them. The learning capability of neural nets both in the supervised and unsupervised domains can be utilized effectively when extracting patterns from large datasets. This is particularly true in data-rich environments as in the case of biological data. Most of the bioinformatic tasks involve search and optimization of different criteria (like energy, alignment score, overlap strength), while requiring robust, fast and close approximate solutions. Evolutionary and other search algorithms like TS, SA, ACO, PSO etc. provide powerful searching methods to explore huge and multi-modal solution spaces. Moreover, since many of the problems involve multiple conflicting objectives, application of multi-objective optimization algorithms like multi-objective genetic algorithms appears to be natural and appropriate.

This book is an attempt to bring together research articles by eminent scientists and active practitioners reporting recent advances in integrating soft computing techniques, either individually or in an hybridized manner, for analyzing biological data in order to extract more and more meaningful information and insights from them. Biological data to be considered for analysis include sequence data, structure data, and microarray data. These data types are typically complex in nature, and require advanced methods to deal with them. Characteristics of the methods and algorithms reported here include the use of domain-specific knowledge for reducing the search space, dealing with uncertainty, partial truth and imprecision, efficient linear and/or sub-linear scalability, incremental approaches to knowledge discovery, and increased level and intelligence of interactivity with human experts and decision makers.

The book has three parts. The first part provides an overview of the areas of bioinformatics and soft computing. The second part deals with applications of different soft computing techniques for sequence and structure analysis. The last part deals with different studies involving gene expression data.

In Chapter 1, Tang and Kim provide an overview of bioinformatics with special reference to the task of mining massive amount of data from high throughput genomic experiments. They discuss about the classical tasks in bioinformatics and their recent developments. These include sequence alignment, genome sequencing and fragment assembly, gene annotation, RNA folding, motif finding and protein structure prediction. Among the emerging topics resulting from new genome technologies, they discuss about comparative genomics, pathway reconstruction, microarray analysis, proteomics and protein-protein interaction.

In Chapter 2, Konar and Das present a lucid overview of the soft computing paradigm. They discuss in detail about the scope of soft computing to overcome the limitations of traditional artificial intelligence techniques. Some of the major soft computing tools, such as fuzzy logic, neural networks, evolutionary algorithms and probabilistic reasoning, are introduced. Merits of hybridizing these techniques are mentioned along with a discussion on some popular hybridizations namely, neuro-fuzzy, neuro-genetic, fuzzy-genetic etc. Finally some emerging areas of soft computing like artificial life, particle swarm optimization, artificial immune system, rough sets and granular computing, chaos theory and ant colony systems are discussed.

Gallardo, Cotta and Fernández consider the problem of inferring a phylogenetic tree given genomic, proteomic, or even morphological data in Chapter 3. Although the classical approaches for solving this problem are inherently limited, given the computational hardness of this problem, they can be useful when used in combination with other meta heuristic search techniques. Such a model, hybridizing memetic algorithms and branch-and-bound techniques, is described in this chapter.

In Chapter 4, Wang and Wu tackle the problem of RNA classification using support vector machines. First, a review of the recent advances in this field is presented. Thereafter, they present a new kernel that takes advantage of both global and local structural information in RNAs and uses the information to classify RNAs. A part of the kernel is based on recurring substrings from RNA molecules while the other part is based on counting bi-grams in RNA molecules. Experimental results using nine families of non-coding RNA sequences taken from the Rfam database demonstrate the good performance of the new kernel and show that it outperforms existing kernels when applied to classifying non-coding RNA sequences.

Wavelet transform is a powerful signal processing tool that can analyze the data in multiple resolutions. Recently there has been a growing interest in using wavelet transforms in the analysis of biological sequences and biology related signals. In Chapter 5, Krishnan and Li review the applications of wavelet transforms for motif search, sequence comparison, protein secondary structure prediction, detection of transmembrane proteins and hydrophobic cores, mass spectrometry and disease related applications.

Moving from sequence information to surface information, a system for surface motif extraction from protein molecular surface data, called SUMOMO, is proposed by Shrestha and Ohkawa in Chapter 6. Since surface motifs cannot be assigned predetermined shapes and sizes, to extract surface motifs of different sizes, a given set of protein molecular surfaces is divided into several small surfaces collectively called unit surfaces. A filtering process is then applied based on the fact that active sites from proteins having a particular function have similar shape and physical properties. Subsequently, negative instances of active sites are used to further reduce the number of possible active site candidates.

In Chapter 7, Pollastri, Baú and Vullo present a simple and effective scalable architecture called Distill for *ab initio* prediction of protein $C_\alpha$ traces based on predicted structural features. It uses a set of state-of-the-art predictors of protein features based on machine learning techniques and trained on large, non-redundant subsets of the PDB, and a simple and fast 3D reconstruction algorithm guided by a pseudo-energy defined according to these predicted features. The reconstruction algorithm employs simulated annealing in its search phase. Results show that Distill can generate topologically correct predictions for a significant fraction of short proteins with 150 or fewer residues.

In Chapter 8, Bandyopadhyay, Santra, Maulik and Muehlenbein deal with the problem of using evolutionary computation techniques for designing small ligands that can bind to the active site of a target protein, there by inhibiting its function. The proposed method uses a variable string length genetic algorithm to encode a tree-shaped ligand constructed using functional groups from a given library. The size of the tree is kept variable. Results on four proteins demonstrate the superiority of the method as compared with some earlier attempts in this direction.

The following five chapters deal with applications of soft computing to different problems related to microarray data analysis. In Chapter 9, Noman and Iba tackle the task of reconstructing genetic network from expression profile by using an improved evolutionary algorithm. The method is tested on simulated data, and is also used to analyze microarray data for predicting the interaction among the genes in SOS DNA repair network in Escherichia coli.

In Chapter 10, Deb, Reddy and Chaudhuri model the task of classifying gene expression data by identifying a relevant subset of the genes as one of multi-objective optimization. The minimizing criteria are the classifier size and the number of misclassified instances in training and test samples. A multi-objective evolutionary algorithm (EAs) is applied as the underlying optimization technique. The standard weighted voting method is used to design a unified procedure for handling two and multi-class problems. The use of multi-objective EAs here is unique in finding multiple high-performing classifiers in a single simulation run. The designed classifier is used to classify three two-class cancer data sets, Leukemia, Lymphoma, and Colon. Using the multi-objective genetic algorithm NSGA-II for this task, the authors report much higher accuracies on several data sets as compared to previous studies.

A similar study is reported in Chapter 11, where Gupta, Jayaraman and Kulkarni employ ant colony optimization for performing feature selection for classification of microarray data. Support vector machine is used as the underlying classification method. Results again demonstrate the effectiveness of the application of soft computing techniques to this problem.

It is observed in Chapter 12 that since microarray data can be noisy and incomplete, selected features with feature selection methods can be incomplete. Moreover, no one classification algorithm can be perfect for several data sets. To solve this problem, Cho and Park propose an ensemble of three methods for classifying gene expression data. The first method uses negatively correlated gene subsets and combines their results with Bayesian approach. The second one uses combinatorial ensemble approach based on elementary single classifiers, and the last one searches the optimal pair of feature-classifier ensemble with genetic algorithm. Results are demonstrated on lymphoma and colon data sets.

Finally, Chapter 13 deals with the task of clustering microarray data using a fuzzy partitioning method. Here, Mukhopadhyay, Maulik and Bandyopadhyay use a novel multi-objective clustering algorithm for this purpose. The clustering problem is posed as one of optimization of two different fuzzy cluster validity indices, namely, the Xie-Beni index and the FCM-index $J_m$. NSGA-II is used as the underlying multi-objective optimization strategy. Comparison with a single objective version of the problem, and the widely used K-means, K-medoids, fuzzy C-means and hierarchical clustering methods demonstrate the superiority of the proposed method.

In summary, the chapters on the applications of soft computing techniques for analyzing biological data provide a representative selection of the available methods and their evaluation in real domains. While the field is rapidly evolving with the availability of new data and new tools, these chapters clearly indicate the importance and potential benefit of synergetically combining the potentials of classical and soft computing methods for facing the newer challenges in biological data mining. The book will be useful to graduate students and researchers in computer science, bioinformatics, computational and molecular biology, electrical engineering, system science, and information technology both as text and reference book for some parts of the curriculum. The researchers and practitioners in industry and R & D laboratories will also be benefited.

We take this opportunity to thank all the authors for contributing chapters related to their current research work that provide the state of the art in advanced methods for analyzing biological data. We are grateful to Ms. Yubing Zhai of World Scientific Publishing Co. Pte. Ltd. for her initiative and constant support.

*Sanghamitra Bandyopadhyay*
*Ujjwal Maulik*
*Jason T. L. Wang*

August, 2006

# CONTENTS

**Preface**     vii

**Part I    OVERVIEW**     1

**Chapter 1    Bioinformatics: Mining the Massive Data from High Throughput Genomics Experiments**
*Haixu Tang and Sun Kim*

| | | |
|---|---|---|
| 1 | Introduction | 3 |
| 2 | Recent Development of Classical Topics | 5 |
| | 2.1   Sequence alignment | 5 |
| | 2.2   Genome sequencing and fragment assembly | 8 |
| | 2.3   Gene annotation | 9 |
| | 2.4   RNA folding | 11 |
| | 2.5   Motif finding | 12 |
| | 2.6   Protein structure prediction | 13 |
| 3 | Emerging Topics from New Genome Technologies | 14 |
| | 3.1   Comparative genomics: beyond genome comparison | 15 |
| | 3.2   Pathway reconstruction | 16 |
| | 3.3   Microarray analysis | 17 |
| | 3.4   Proteomics | 18 |
| | 3.5   Protein-protein interaction | 20 |
| 4 | Conclusion | 20 |

**Chapter 2    An Introduction to Soft Computing**
*Amit Konar and Swagatam Das*

| | | |
|---|---|---|
| 1 | Classical AI and its Pitfalls | 25 |
| 2 | What is Soft Computing? | 27 |
| 3 | Fundamental Components of Soft Computing | 28 |
| | 3.1   Fuzzy sets and fuzzy logic | 28 |

|     |                                               |    |
| --- | --------------------------------------------- | -- |
| 3.2 | Neural networks                               | 31 |
| 3.3 | Genetic algorithms                            | 36 |
| 3.4 | Belief networks                               | 39 |

4 Synergism in Soft Computing 44
   4.1 Neuro-fuzzy synergism 44
   4.2 Neuro-GA synergism 44
   4.3 Fuzzy-GA synergism 45
   4.4 Neuro-belief network synergism 45
   4.5 GA-belief network synergism 45
   4.6 Neuro-fuzzy-GA synergism 46
5 Some Emerging Areas of Soft Computing 46
   5.1 Artificial life 46
   5.2 Particle swarm optimization (PSO) 47
   5.3 Artificial immune system 48
   5.4 Rough sets and granular computing 49
   5.5 Chaos theory 50
   5.6 Ant colony systems (ACS) 51
6 Summary 52

## Part II  BIOLOGICAL SEQUENCE AND STRUCTURE ANALYSIS     57

### Chapter 3  Reconstructing Phylogenies with Memetic Algorithms and Branch-and-Bound

*José E. Gallardo, Carlos Cotta and Antonio J. Fernández*

1 Introduction 59
2 A Crash Introduction to Phylogenetic Inference 60
3 Evolutionary Algorithms for the Phylogeny Problem 65
4 A BnB Algorithm for Phylogenetic Inference 66
5 A Memetic Algorithm for Phylogenetic Inference 69
6 A Hybrid Algorithm 73
7 Experimental Results 75
   7.1 Experimental setting 76
   7.2 Sensitivity analysis on the hybrid algorithm 76
   7.3 Analysis of results 77
8 Conclusions 80

## Chapter 4   Classification of RNA Sequences with Support Vector Machines

*Jason T. L. Wang and Xiaoming Wu*

| | | |
|---|---|---|
| 1 | Introduction | 85 |
| 2 | Count Kernels and Marginalized Count Kernels | 88 |
| | 2.1   RNA sequences with known secondary structures | 88 |
| | 2.2   RNA sequences with unknown secondary structures | 92 |
| 3 | Kernel Based on Labeled Dual Graphs | 94 |
| | 3.1   Labeled dual graphs | 94 |
| | 3.2   Marginalized kernel for labeled dual graphs | 95 |
| 4 | A New Kernel | 97 |
| | 4.1   Extracting features for global structural information | 98 |
| | 4.2   Extracting features for local structural information | 100 |
| 5 | Experiments and Results | 102 |
| | 5.1   Data and parameters | 102 |
| | 5.2   Results | 104 |
| 6 | Conclusion | 106 |

## Chapter 5   Beyond String Algorithms: Protein Sequence Analysis using Wavelet Transforms

*Arun Krishnan and Kuo-Bin Li*

| | | |
|---|---|---|
| 1 | Introduction | 109 |
| | 1.1   String algorithms | 110 |
| | 1.2   Sequence analysis | 110 |
| | 1.3   Wavelet transform | 111 |
| 2 | Motif Searching | 114 |
| | 2.1   Introduction | 114 |
| | 2.2   Methods | 115 |
| | 2.3   Results | 116 |
| | 2.4   Allergenicity prediction | 118 |
| 3 | Transmembrane Helix Region (HTM) Prediction | 121 |
| 4 | Hydrophobic Cores | 122 |
| 5 | Protein Repeat Motifs | 122 |
| 6 | Sequence Comparison | 123 |
| 7 | Prediction of Protein Secondary Structures | 125 |

| | | |
|---|---|---|
| 8 | Disease Related Studies | 126 |
| 9 | Other Functional Prediction | 126 |
| 10 | Conclusion | 126 |

### Chapter 6 Filtering Protein Surface Motifs Using Negative Instances of Active Sites Candidates

*Nripendra L. Shrestha and Takenao Ohkawa*

| | | |
|---|---|---|
| 1 | Introduction | 133 |
| 2 | Protein Structural Data and Surface Motifs | 135 |
| | 2.1 Protein structural data | 135 |
| | 2.2 Protein molecular surface data | 136 |
| | 2.3 Functions of a protein and structural motifs | 137 |
| 3 | Overview of SUMOMO | 138 |
| | 3.1 Surface motif extraction | 139 |
| | 3.2 Filtering using similarity between local surfaces | 140 |
| | 3.3 Problems with SUMOMO | 142 |
| 4 | Filtering Surface Motifs using Negative Instances of Protein Active Sites Candidates | 142 |
| | 4.1 Survey on the features to distinguish real active sites from the active sites candidates | 143 |
| | 4.2 Ranking active sites candidates | 147 |
| 5 | Evaluations | 148 |
| 6 | Conclusions and Future Works | 151 |

### Chapter 7 Distill: A Machine Learning Approach to *Ab Initio* Protein Structure Prediction

*Gianluca Pollastri, Davide Baú and Alessandro Vullo*

| | | |
|---|---|---|
| 1 | Introduction | 153 |
| 2 | Structural Features | 155 |
| | 2.1 One-dimensional structural features | 155 |
| | 2.2 Two-dimensional structural features | 157 |
| 3 | Review of Statistical Learning Methods Applied | 159 |
| | 3.1 RNNs for undirected graphs | 159 |
| | 3.2 1D DAG-RNN | 161 |
| | 3.3 2D DAG-RNN | 163 |

| | | |
|---|---|---|
| 4 | Predictive Architecture | 164 |
| | 4.1 Data set generation | 165 |
| | 4.2 Training protocols | 165 |
| | 4.3 One-dimensional feature predictors | 166 |
| | 4.4 Two-dimensional feature predictors | 168 |
| 5 | Modeling Protein Backbones | 169 |
| | 5.1 Protein representation | 170 |
| | 5.2 Constraints-based pseudo energy | 170 |
| | 5.3 Optimization algorithm | 171 |
| 6 | Reconstruction Results | 173 |
| 7 | Conclusions | 178 |

## Chapter 8  In Silico Design of Ligands using Properties of Target Active Sites

*Sanghamitra Bandyopadhyay, Santanu Santra, Ujjwal Maulik and Heinz Muehlenbein*

| | | |
|---|---|---|
| 1 | Introduction | 184 |
| 2 | Relevance of Genetic Algorithm for Drug Design | 186 |
| 3 | Basic Issues | 187 |
| | 3.1 Core formation | 187 |
| | 3.2 Chromosome representation | 190 |
| | 3.3 Fitness computation | 191 |
| 4 | Main Algorithm | 192 |
| 5 | Experimental Results | 193 |
| 6 | Discussion | 199 |

## Part III  GENE EXPRESSION AND MICROARRAY DATA ANALYSIS  203

## Chapter 9  Inferring Regulations in a Genomic Network from Gene Expression Profiles

*Nasimul Noman and Hitoshi Iba*

| | | |
|---|---|---|
| 1 | Introduction | 205 |
| 2 | Modeling Gene Regulatory Networks by S-system | 208 |
| | 2.1 Canonical model description | 208 |

|     |                                                                                          |     |
| --- | ---------------------------------------------------------------------------------------- | --- |
|     | 2.2 Genetic network inference problem by S-system                                         | 209 |
|     | 2.3 Decoupled S-system model                                                              | 210 |
|     | 2.4 Fitness function for skeletal network structure                                       | 211 |
| 3   | Inference Method                                                                          | 212 |
|     | 3.1 Trigonometric Differential Evolution (TDE)                                            | 213 |
|     | 3.2 Proposed algorithm                                                                    | 214 |
|     | 3.3 Local search procedure                                                                | 217 |
| 4   | Simulated Experiment                                                                      | 217 |
|     | 4.1 Experiment 1: inferring small scale network in noise free environment                 | 217 |
|     | 4.2 Experiment 2: inferring small scale network in noisy environment                      | 219 |
|     | 4.3 Experiment 3: inferring medium scale network in noisy environment                     | 220 |
| 5   | Analysis of Real Gene Expression Data                                                     | 222 |
|     | 5.1 Experimental data set                                                                 | 224 |
| 6   | Discussion                                                                                | 226 |
| 7   | Conclusion                                                                                | 227 |

## Chapter 10   A Reliable Classification of Gene Clusters for Cancer Samples Using a Hybrid Multi-Objective Evolutionary Procedure

*Kalyanmoy Deb, A. Raji Reddy and Shamik Chaudhuri*

|     |                                                                |     |
| --- | -------------------------------------------------------------- | --- |
| 1   | Introduction                                                   | 232 |
| 2   | Class Prediction Procedure                                     | 233 |
|     | 2.1 Two-class classification                                   | 234 |
|     | 2.2 Multi-class classification                                 | 235 |
| 3   | Evolutionary Gene Selection Procedure                          | 236 |
|     | 3.1 The optimization problem                                   | 237 |
|     | 3.2 A multi-objective evolutionary algorithm                   | 237 |
|     | 3.3 A multi-modal NSGA-II                                      | 238 |
|     | 3.4 Genetic operators and modified domination operator         | 240 |
|     | 3.5 NSGA-II search using a fixed classifier size               | 241 |
|     | 3.6 Overall procedure                                          | 241 |
| 4   | Simulation Results                                             | 242 |
|     | 4.1 Complete leukemia study                                    | 244 |
|     | 4.2 Diffuse large B-cell lymphoma dataset                      | 248 |

|   |   |   |
|---|---|---|
| 4.3 | Colon cancer dataset | 250 |
| 4.4 | NCI60 multi-class tumor dataset | 252 |
| 5 | Conclusions | 255 |

### Chapter 11  Feature Selection for Cancer Classification using Ant Colony Optimization and Support Vector Machines

*A. Gupta, V. K. Jayaraman and B. D. Kulkarni*

|   |   |   |
|---|---|---|
| 1 | Introduction | 259 |
| 2 | Ant Colony Optimization | 262 |
| 3 | Support Vector Machines | 263 |
| 4 | Proposed Ant Algorithm | 266 |
|   | 4.1  State transition rules | 266 |
|   | 4.2  Evaluation procedure | 268 |
|   | 4.3  Global updating rule | 268 |
|   | 4.4  Local updating rule | 269 |
| 5 | Algorithm Outline | 269 |
| 6 | Experiments | 270 |
|   | 6.1  Datasets | 270 |
|   | 6.2  Preprocessing | 272 |
|   | 6.3  Experimental setup | 273 |
| 7 | Results and Discussion | 274 |
| 8 | Conclusions | 277 |

### Chapter 12  Sophisticated Methods for Cancer Classification using Microarray Data

*Sung-Bae Cho and Han-Saem Park*

|   |   |   |
|---|---|---|
| 1 | Introduction | 281 |
| 2 | Backgrounds | 282 |
|   | 2.1  DNA microarray | 282 |
|   | 2.2  Feature selection methods | 283 |
|   | 2.3  Base classifiers | 284 |
|   | 2.4  Classifier ensemble methods | 285 |
| 3 | Sophisticated Methods for Cancer Classification | 286 |
|   | 3.1  Ensemble with negatively correlated features | 286 |
|   | 3.2  Combinatorial ensemble | 289 |
|   | 3.3  Searching optimal ensemble with GA | 291 |

| 4 | Experiments | 293 |
|---|---|---|
| | 4.1 Datasets | 293 |
| | 4.2 Ensemble with negative correlated features | 294 |
| | 4.3 Combinatorial ensemble | 296 |
| | 4.4 Optimal ensemble with GA | 298 |
| 5 | Conclusions | 300 |

## Chapter 13 Multiobjective Evolutionary Approach to Fuzzy Clustering of Microarray Data

*Anirban Mukhopadhyay, Ujjwal Maulik and Sanghamitra Bandyopadhyay*

| 1 | Introduction | 304 |
|---|---|---|
| 2 | Structure of Gene Expression Data Sets | 306 |
| 3 | Cluster Analysis | 306 |
| | 3.1 K-means | 307 |
| | 3.2 K-medoids | 308 |
| | 3.3 Fuzzy C-means | 308 |
| | 3.4 Hierarchical agglomerative clustering | 309 |
| 4 | Multiobjective Genetic Algorithms | 310 |
| 5 | The Multiobjective Fuzzy Clustering Technique | 312 |
| | 5.1 Chromosome representation and population initialization | 312 |
| | 5.2 Computation of objective functions | 312 |
| | 5.3 Selection, crossover and mutation | 313 |
| | 5.4 Choice of objectives | 314 |
| | 5.5 Distance measures | 314 |
| 6 | Experimental Results | 315 |
| | 6.1 Yeast sporulation data | 315 |
| | 6.2 Human fibroblasts serum data | 316 |
| | 6.3 Performance validation | 316 |
| | 6.4 Input parameter values | 318 |
| | 6.5 Quantitative assessments | 318 |
| | 6.6 Visualization of results | 320 |
| | 6.7 Biological interpretation | 322 |
| 7 | Conclusions and Discussions | 326 |

**Index**     **329**

# I.
# OVERVIEW

OVERVIEW

# CHAPTER 1

# BIOINFORMATICS: MINING THE MASSIVE DATA FROM HIGH THROUGHPUT GENOMICS EXPERIMENTS

Haixu Tang* and Sun Kim[†]

*School of Informatics*
*Center for Genomics and Bioinformatics*
*Indiana University, Bloomington, IN 47408, USA*
*\*hatang@indiana.edu*
*†sunkim2@indiana.edu*

The recent accomplishment of Human Genome Project (HGP) has revolutionized the life sciences and the medical sciences in many ways. Consequently, the field of bioinformatics has emerged as a new multidisciplinary field that attempts to solve biological and medical problems by analyzing data from high throughput experiments with computational methods. In this chapter we will review the recent development of classical and emerging topics in bioinformatics.

## 1. Introduction

Human Genome Project (HGP), in which researchers across the world collaborated to determine the whole genetic information in human body, i.e. the human genome, has revolutionized the life and medical sciences in many ways.[1] Among them, an emerging shift of the paradigm in biological research is probably most influencing. Conventionally, biology knowledge was accumulated mainly through a *hypothesis-driven* approach, in which biologists conceive theory for a particular biological problem and then carry out an experiment to test it. In a *hypothesis-driven* approach, experiments are intentionally designed to collect data only relevant to the to-be-tested hypothesis. As a result, these "intentional" data collection often requires several complementary experimental platforms, but produces only a small amount of data. The "hypothesis-driven" approach have been extremely successful and resulted in many critical discoveries in life sciences.

Human genome project, however, has demonstrated a different model of successful biology research. First, the anticipation of this project is great but rather non-specific, and, more importantly, it started without a clear hypothetical theory! Instead, biologists anticipate the generation of a large amount of data (i.e. genome sequences) that may verify or disprove some old hypothesis, and inspire many valuable new theories. Second, this project of "blind" data collection centers on a single technique platform, i.e. DNA sequencing. Indeed, one of the HGP's goals was to advance the DNA sequencing technology itself for a more efficient data collection. Finally, for the first time in biology research, many biological laboratories across the world work closely and collaboratively on the same project. They carefully planned the project, split the efforts, and share the technologies and results openly with the entire biology community.

Now, the model of HGP is often named as *technology-driven* or *data driven* approach. As HGP has shown, this approach has several distinct features in comparing with the conventional "hypothesis-driven" approach: a high throughput technique platform, a blind collection of large amount of data, and a plan of free data sharing to the community. The success of this model has been copied to several other biological projects, such as the sequencing-based population genetics (HapMap),[2] microarray-based transcriptomics [3] and mass-spectrometry-based proteomics,[4] thus has given rise to a new kind of life science, often called *genomics*.[5]

The large amount of data generated by genomics indicate a new pathway for biological findings, through computational analysis instead of laboratory experiments. A new multidisciplinary field (now called bioinformatics) emerges, which combines life sciences, computer science and physical sciences to solve biological and medical problems. Bioinformatics offers a playground for the applications of novel approaches to data analysis and data mining. Surprisingly, in spite of its short history, several core algorithms in bioinformatics were developed long before the formation of the discipline. For example, computer scientists have started developing algorithms for comparing DNA sequences with several megabases long time ago, even before HGP was initiated. Nevertheless, the advancement of genome technology always poses new challenges for bioinformaticists. To provide an overview of the current status in bioinformatics research, in this chapter, we will first review the recent development of several classical topics in bioinformatics, and then introduce a couple of new emerging problems from genome technologies.

## 2. Recent Development of Classical Topics

It is arguable that the origin of bioinformatics history can be traced back to Mendel's discovery of genetic inheritance in 1865. However, bioinformatics research in a real sense started in late 1960s, symbolized by Dayhoff's *atlas of protein sequences*,[6] and the early modeling analysis of protein[7] and RNA[8] structures. In fact, these early works represented two distinct provenances of bioinformatics: evolution and biochemistry, which still largely define the current bioinformatics research topics.

Bioinformatics is in nature strongly linked to the advancement of genome technologies. As some technologies are proved infeasible in practice or replaced by newer ones, the related bioinformatics topics become outdated. Nevertheless, some classical topics remain important. We will review recent progresses of some of these topics.

### 2.1. *Sequence alignment*

The most frequently used computer procedure nowadays in life sciences is sequence alignment, which is also one of the most extensively studied problems in bioinformatics. Important biological molecules, such as nucleic acids (DNAs and RNAs) and proteins, are all linear polymers composed of a limited number of building units (monomers). Hence, they can be often represented as sequences on a small alphabet. For example, a DNA molecule can be represented as a sequence of letters A, C, G and T representing 4 nucleotides {A,C,G,T}, whereas a protein can be represented as a sequence of 20 letters representing 20 different amino acids. To identify similar regions between two sequences, a sequence alignment procedure is applied, in which gaps are inserted and the sequences are shifted accordingly. The following shows a pairwise alignment of two DNA sequences (top and bottom):

ACTT−GACCCTATTAACTTGCATGCTCTC−−ATCAAAA
CCTTTGACCTTAATAACA−−CATCCTCTCGCATCGAAA

The algorithms to obtain the optimal pairwise alignment between two sequences have been well studied in computer science, known as *string pattern matching* algorithms.[9] The early approaches to pairwise sequence alignment problem aimed at aligning two *entire* sequences, now referred to as the *global* alignment problem. A dynamic programming solution to this problem was proposed by Needleman and Wuncsh.[10] However, in many biological applications, global sequence alignment fails to reveal the similarity between two given biological sequences, because the alignment score of the entire sequence is lower than the alignment score between their two subsequences. Smith and Waterman made a small but critical modification of the original dynamic programming algorithm that can solve the *local* alignment problem.[11]

Dynamic programming algorithms for pairwise sequence alignments are *exact*, i.e. they are guaranteed to report the optimal alignment with a given scoring scheme. Although the optimal alignment is not necessarily the *correct* alignment in a biological sense, we hope to obtain an alignment reflecting the evolutionary process by which these two DNA (or protein) sequences evolved from their common ancestor. Scoring schemes used in sequence alignment usually award identical symbols, and penalize substitutions and gaps based on different evolution models. As a result, scoring schemes affect the resulting alignment in ways as important as the alignment algorithms.[12]

The most important application of pairwise sequence alignment is to find similar sequences of a newly sequenced gene (or protein) in a collection of previously known genes (or proteins), i.e. *gene (or protein) databases*. Thanks to the genome sequencing projects, the size of sequence databases (e.g. Genbank) increases dramatically in the past few years, now achieving $10^{11}$ bases. Hence, searching such huge databases using dynamic programming algorithm, which takes a quadratic amount of time in relation to the size of query and match sequences, is still too slow. It is not until the invention of the rapid database searching programs (e.g. FASTA[13] and BLAST[14]) that sequence similarity comparison became a popular exercise in molecular biology. Nevertheless, the suggestion of *k-tuple filtering* to speed up sequence comparison, which is essential for FASTA and BLAST, goes back to earlier time.[15] This is not the only example of the foresight of bioinformatics researchers on the increasing algorithmic needs for data analysis in genomics. Even long before the first complete genome was sequenced, computer scientists started thinking of algorithms for comparing megabase-long genome sequences.[16] Many novel programs developed to align genomic sequences adopt the same *seed-and-extend* strategy, which first identify near-exact matches, then filter them based on various criteria into a reliable subset, called *anchors*, and finally chain them into long pairwise alignments by filling in the gaps between anchors using classical global and local pairwise alignment algorithms.[17] Novel seeding and filtering techniques, such as maximal unique matches (MUMs),[18] maximal exact matches (MEMs),[19] and gapped seeds,[20] were applied to acquire accurate genome alignments more rapidly and memory-efficiently. With more and more complete genomes, especially mammalian genomes like mouse genome,[21] was sequenced, large scale genome sequence alignment programs become essential tools[17] for studying the function and evolution of genes in genomics.

Unlike pairwise alignment, the exact multiple alignment algorithm for large amount of sequences is not feasible.[21] A straight forward heuristic is known as *progressive alignment*, initially proposed by Feng and Doolittle.[22] They aligned the most similar pair of sequences and merged them together to create a new pseudo-sequence called *sequence profile*, thus reduced the problem of aligning the original $N$ sequences into the problem of aligning $N-1$ sequences, following the concept of "once a gap, always a gap". After iterating this procedure, a multiple alignment of all original sequences could be built progressively. Similar strategies were also used by many other multiple alignment programs for protein and DNA sequences (e.g. the commonly used program ClustalW[23]). In recent years, progressive multiple alignment was also used in aligning multiple genomic sequences, in which the order of alignment was defined based on the previously known phylogenetic tree of the input genomes.[24]

To improve the accuracy of protein sequence alignment, alternative heuristics for multiple alignment were developed. The divide-and-conquer method applies an empirical rule to divide long sequences into small segments, then uses a dynamic programming algorithm to acquire their multiple alignment and finally combines these small sections of alignments into a long one.[25] Another heuristic for multiple alignment is recently implemented in T-Coffee,[26] which attempts to use empirical rules to combine the library of every optimal pairwise alignments between input sequences into a multiple alignment.

Despite the long history of research, sequence alignment problem remains one of the hottest topics in bioinformatics. The future developments of sequence alignment algorithms will still be focused on two directions: the alignment efficiency, especially for many long genomic sequences; and the alignment accuracy, especially for protein sequences with low similarities.

## 2.2. *Genome sequencing and fragment assembly*

Modern DNA sequencing machine based on Sanger's principle [27] can determine the sequence of a short DNA fragment, typically 500–800 base pairs (bps) long. To sequence a long DNA fragment, biologists usually use a shotgun approach: first break the DNA molecule into short overlapping fragments, then sequence each fragment separately until the enough number of fragments are sequenced (typically 10 times of the target DNA size), and finally assemble these fragment sequences into the complete target DNA sequence on computer.

The first fragment assembly program was developed in the same year when the DNA sequencing method was published,[28] which used a greedy algorithm to merge the fragment sequences with strong overlaps. Most later developed assembly programs followed a similar three step procedure[29]:

- *Overlap*: Identifying overlaps between fragments;
- *Layout*: Determining the order of fragments;
- *Consensus*: Deriving the complete DNA sequence.

Due to the potential sequencing errors from DNA sequencing machine (typically 1%), in the popular assembly programs like Phrap,[30] the overlaps and layout of the fragment are determined based on not only the sequences of the fragments, but also on the reliability of each nucleotide output from the sequencing (*base calling*).

Conventional assembly programs were very successful in sequencing DNA molecules of medium size (about 200 000 bps). However, it encountered a new challenge when moving toward assembling shotgun fragments of whole genomes. Depending on the genome complexity, various portions of a genome may be present in more than one copy, referred to as *repeats*. It turns out that about 25% of human genome are repetitive sequences. To address this issue, *double-barreled sequencing* was suggested, in which two fragments are sequenced from a same relative long DNA clone and paired together in assembly.[31] Many algorithms were developed since then for repeat resolution combining double-barreled data and advanced sequence analysis techniques.[32-34] The resulting new assembly programs were successfully used to assemble many large genomes, such as human and mouse genomes.

Although the conventional sequencing technology has accomplished great success in genomics, it remains an expensive experiment. New technologies, pyrosequencing, following the same principle of "sequence by synthesize",[35] were developed towards more affordable experiment for sequencing a higher diversity of genomes. These experiments, however, produced DNA fragments much shorter than the conventional technologies, typically from 20 to 100 bps. It raises new challenges for fragment assembly since the fragment length limits the size of repeats that can be resolved.[36] As a result, the *de novo* sequencing of even a small bacterium genome using the new technology may result in many gaps (caused by repeats in the genome).[37] The development of new computational methods and tools to overcome this difficulty will be an active research topic in bioinformatics.

## 2.3. Gene annotation

The first type of analysis that a biologist would want to carry out after a new genome is sequenced is to find genes (often referred to as protein coding regions) within it. The first approach in detecting protein coding regions is to recognize *Open Reading Frames* (ORFs), i.e. a long (typically $\geq 50$) sequence of codons (triplets of nucleotides) starting from a Start Codon and ending with a Stop Codon. In addition to its length, protein coding regions have other statistical properties different from the non-coding regions. One of them that is commonly used in current gene finding programs is *codon usage*, which describes the frequencies of 64 possible codons in coding and non-coding regions. High order Markov models are often built using species specific parameters for coding and non-coding regions, respectively.[38,39] Discriminative approaches can be applied to estimate the conditional probabilities of a given DNA sequence to be within coding or non-coding regions.[40]

The discovery of split genes created another complication for gene annotation in eukaryotic genomes. The coding sequence of a single gene is not continuous in the genome, forming a number (up to thousands) of segments whose transcripts are joined together in cells through a process called *splicing*. It turns out that there are some sequence signals ("splicing signal") embedded in the junction between the coding ("exons") and the non-coding segments ("introns") that is used to guide the gene splicing in cells. Successful eukaryotic gene annotation tools like Genscan[41] attempt to combine the properties of the coding regions and splicing signals using more complex statistical models (e.g. hidden Markov models, HMMs) to annotate the gene structure.[42]

Other than the *ab initio* methods mentioned above, gene similarity search can also be used for gene structure prediction. A spliced alignment algorithm, the modified version of the conventional dynamic programming algorithm for pairwise sequence alignment, is proposed to find an assembly of putative exons that is closest to a related protein, thus deriving the gene structure from genomic sequences.[43] A useful extension of spliced alignment algorithm is based on the comparison between genome sequences and *Expressed Sequence Tags* (ESTs), rapidly sequenced fragments from message RNAs.[44] Current gene annotation programs usually integrate both the statistical and similarity searching approaches when annotating genes from a newly sequenced genome, and then provide an option to include putative ESTs from the same organism.[45]

With increasingly closely related genomes being sequenced, a novel approach to gene annotation emerges, based on the comparison of syntenic regions across multiple genome. The concept of this approach is that the coding regions are in general more conserved than non-coding regions in evolution, owing to the selection pressure. Furthermore, the level of conservation in the coding regions is different from one reading frame to another, since the mutations at the third position of synonymous codons do not change the coding amino acids, thus are under lower selection pressure compared to the 1st and 2nd positions. Gene annotation systems now allow the use of more than one genomes for gene structure prediction.[46,47] Results have shown that the incorporation of multiple genomes across a variety of evolutionary distance can significantly improve gene annotation.[48] Gene annotation will remain an important research topic in bioinformatics and its accuracy will be continuously pushed to the limit by newly developed methods as well as the accumulated genomic sequences.

## 2.4. RNA folding

Unlike DNAs, RNAs usually function as single strand molecules. The nucleotides of a single RNA molecule can pair with each other (through hydrogen bonds) and form a stable *secondary structure*. Figure 1 shows the common nomenclature for loops in RNA secondary structures. The stable secondary structure of an RNA molecule is thought to be the one with the lowest free energy, and the problem of finding this stable structure computationally is called RNA folding problem.

RNA secondary structures can be represented by a list of base-pairs in a RNA sequence. An approximate solution to RNA folding problem is to find a secondary structure of a given RNA sequence with the maximal number of base-pairs using a cubic dynamic programming algorithm.[49,50] More realistic thermodynamic models of RNA folding take into consideration of free energy of loops in addition to base-pairs, and were implemented in commonly used programs such as MFOLD[51] and ViennaRNA.[52]

Recently, a surprisingly large number of functional RNA molecules encoded by non-coding RNA genes have been found by large scale experimental screening methods. It shows that RNAs play a more important role in cells than biologists initially imagined.[53] As a result, computational identification of non-coding RNA genes has become a very important problem. Non-coding RNA genes encode functional RNAs instead of proteins. Hence, they have different statistical properties from the protein coding genes, and the computational methods described above for protein coding gene annotation cannot be applied directly to this problem. It has been shown that the folding energy alone is insufficient to distinguish non-coding RNA

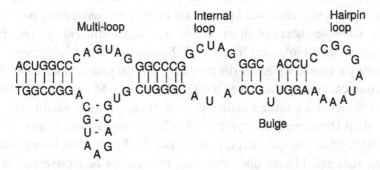

**Fig. 1.** A schematic illustration of an RNA secondary structure and its loop components.

sequences from the other genome sequences.[54] On the other hand, automated methods similar to the one used in the first approach to determining the theoretical secondary structure of tRNAs[8] are developed.[55] When comparing non-coding RNA genes in different species, it is often found that some substitutions occurred at two sides of a base pair such that the base pair retains. The substitutions are referred to as *compensatory mutations*. Since the structures of non-coding RNAs are important for their functions, many more compensatory mutations can be observed in the aligned non-coding RNA genes than other aligned genomic sequences. This property has been implemented in a few non-coding RNA gene finding programs, using two[56] or more[57] aligned RNA sequences from different species. The most significant progress by applying these methods is the discovery of a new class of RNA regulatory elements, *riboswitches*.[58] As a part of an mRNA molecule, riboswitches can directly bind a small target molecule, and regulate (activate or repress) the gene's activity. Nearly all riboswitch elements were found through the computational analysis of multi-aligned mRNAs that are presumably co-regulated. The conserved secondary structure among these mRNAs then can be identified based on the compensatory mutations in the alignment.[59] The discovery of riboswitch demonstrates the power of bioinformatics methods in identifying novel molecular elements in biology.

## 2.5. *Motif finding*

A sequence motif in a nucleic acid or a protein is referred to a conserved sequence pattern that is determined or conjectured to have a biological function. It was noticed long ago that proteins sharing similar functions may not share sequence similarity along their entire sequences, but only one or a few segments of them, which are often sufficient for proteins carrying out their biological functions.[60] Similarly, there are also essential sequence patterns in DNAs. A simple example is the palindromic site of the restriction enzyme that activates the DNA cleavage. More complex DNA sequence motifs are those binding sites of Transcriptional Factors (TFs). These short DNA segments (typically 5 to 20 nucleotides long) can bind to TFs and regulate the gene expression. Since the interactions between the binding sites and TFs are often complex, they cannot be represented by a simple DNA sequence (*word*), but a pattern (*motif*).[61]

The motif finding problem, i.e. finding the most conserved sequence motif among a set of given DNA (or protein) sequences, has been studied as extensively as the sequence alignment problem in bioinformatics. There are several various formulation of this problem that differ in the rigorous definition of a sequence motif (e.g. *consensus, sequence profile* or a set of *words*). The most successful algorithms for solving these problems, however, are not combinatorial algorithms like dynamic programming for solving sequence alignment problem, but probabilistic methods. Gibbs sampling is a procedure to iteratively improve the identified motif, starting from an arbitrarily chosen one.[62] Another popular motif finding program MEME adopts the Expectation-Maximization algorithm to achieve the same goal.[63]

In spite of successful applications of these methods, there is still room for further improvement of motif finding, in particular those *weak* motifs that carry on subtle signals over the random noises.[64] Obviously, an exhaustive searching for all potential motifs will guarantee detection of the motif if there is one in the input sequences. However, there is a tradeoff between the sensitivity and computer time. Advanced probabilistic methods can detect weak motifs within a reasonable computer time,[65] whereas sophisticated data structure can further speed up the searching process.[66]

### 2.6. *Protein structure prediction*

All classical topics we discussed so far are about analyzing the sequences of biomolecules. There is a second source of bioinformatics research coming from the modern biochemistry. During the 1960s, soon after Sanger designed the experimental method for determining the sequence of a protein, Anfinsen concluded from his protein refolding experiments that the native structure of a protein can be determined from its sequence. Anfinsen's theory set one of the most important and difficult goals in bioinformatics, known as the *protein structure prediction problem*.

The early approaches to protein structure prediction were based on the free energy optimization of protein structure. The free energy was evaluated using molecular force fields that describe the physical interactions between atoms, and two types of optimization methods, molecular dynamics and Brownian dynamics, were generally used.

Due to the huge search space for potential protein conformations, pure theoretical methods for protein structure prediction are not very successful in practice. Biochemists started to look for different approaches. A new type of protein structure prediction methods, referred to as *protein comparative modeling*, were developed based on the same concept that proteins with similar sequences often share structures. Browne and co-workers modeled the structure of $\alpha$-lactalbumin using the known lysozyme structure as a template, which is the first successful example of comparative modeling.[67] Since then, several generations of comparative modeling tools were developed and many protein structures were modeled.[68] The accuracy of comparative protein modeling programs depends on two factors: the identification of an appropriate (known) structure as template, and the alignment between the template and the protein to be modeled (target). When no close homolog exists for modeling, sophisticated methods to address these problems are needed to improve the quality of comparative modeling. *Threading* methods, which attempt to align the target protein sequence with template protein structure(s), can sometimes detect protein similarity beyond their sequence homology.[69] Other methods for achieving the same goal utilize multiple sequence alignment from the same protein families in template detection as well as template-target alignment to improve their sensitivity.[70]

A significant progress in this area is the recent development of *segment assembly* method for *ab initio* protein structure prediction. The ROSETTA program,[71] which pioneered this strategy to model protein structures by assembling predicted local structural segments, based on the assumption that short sequence segments in proteins almost determine their local structures, and the search space for the global protein structure can be narrowed to the arrangement of these structural segments. ROSETTA and several other programs using similar strategy have performed very well in a series of independent and blind tests, thus pushing forward the practical applications of protein structure prediction in molecular biology.[72]

## 3. Emerging Topics from New Genome Technologies

With the advancement of genome technologies, many new research topics in bioinformatics have emerged. Some of them relate to data analysis for specific experimental platforms, whereas others relate to integrating data generated using distinct techniques.

## 3.1. Comparative genomics: beyond genome comparison

In theory, the full sequence of a genome consists of the most heritable information of an organism. However, the sequence itself is not directly linked to the observable *phenotypes*, which are of the ultimate interests for life and medical scientists and will be the focus of analysis of the available genome sequences. Comparative genomics aims to discover the functional units by comparing multiple genomic sequences,[73] based on the principle that the functional units encoded in the genome, e.g. proteins, RNAs and regulatory elements, are conserved across species.[72]

The fundamental question that comparative genomics has to answer is how to discriminate conserved (and functional) sequence units from the rest of genomic sequences that are under neutral divergence. Depending on different phylogenetic distances between genomes, functional units within different biological systems can be discovered. The first few sequenced eukaryotic genomes, including the yeast, worm and fly genomes, are greater than 1 billion years apart. The comparison of these genomes can reveal a common set of proteins that are responsible for the basic biological functions.[74] Many genes involving in a large number of pathways can be commonly found in the worm and fly, but not in yeast, reflecting the higher cellular organization complexity of multi-cellular organisms. Despite the conservation of genes with essential function, other functional units, like non-coding RNA genes or the gene regulatory elements, are not anticipated to be conserved over such large evolution distances. In order to study those elements, multiple genomes at moderate evolution distances, e.g. 100 million years apart, should be used. Successful examples of such analysis include the comparison of human and mouse genomes,[21] two worm genomes,[75] and multiple yeast genomes.[76,77] Different biological questions can be addressed when comparing genomes that are very closely related. The comparison of human and chimpanzee genomes (5 million years apart) can reveal the key functional units that are responsible the phenotype difference between similar species.[78]

Genome alignment is the central computational technique used in comparative genomics. The recently developed methods for these problems are scalable to genome scale analysis. With the help of the power of supercomputers, the whole genome alignments now can be built soon after the availability of genome sequences, and made accessible through several

Table 1. Some integrated comparative genomics platforms.

| Platforms | Genomes covered | URL | Reference |
|---|---|---|---|
| EnteriX | Prokaryotes | http://globin.cse.psu.edu/enterix/ | 79 |
| PLATCOM | Prokaryotes | http://platcom.informatics.indiana.edu/ | 80 |
| Ensembl | Eukaryotes | http://www.ensembl.org/ | 81 |
| UCSC | Eukaryotes | http://genome.ucsc.edu | 82 |

integrated comparative genomics platforms, for prokaryotic and eukaryotic genomes (Table 1).

The interplay between evolutionary analysis and gene functional prediction is one of the themes in comparative genomics. On one hand, the functions of genes can be predicted through the comparative analysis of their occurrences across multiple genomes. Rigorous evolutionary analysis can distinguish *orthologous genes* from *paralogous genes* in large duplicated gene families. Orthologous genes often carry out the same biological function, thus can be used to improve the straightforward homolog-based gene function annotation.[83] Other information derived from comparative genomics, such as gene context[84] gene fusion,[84] and phylogenetic profile,[85] are useful in predicting functions of genes without function-known homologues. On the other hand, the genome-scale annotation of gene functions will provide a complete evolutionary scenario of the transfer and innovation of functions.[86]

Genomes evolve through not only point mutations in individual genes, but also the chromosomal *rearrangement* of gene contents and orders.[87] Genome duplications, including whole genome duplications and segmental duplications, are known to be important for evolution, in particular innovations of gene function. Comparative genomics can provide solid evidence to trace back those hypothetical events in history.[88] Chromosomal inversion, fusion/fission are frequently observed rearrangement events. Bioinformatics methods have been developed to elucidate them based on different mathematical models in the context of comparative genomics.[89]

### 3.2. *Pathway reconstruction*

Although comparative genomics approaches succeed in predicting the functions of many genes, they fail to annotate 20%–60% genes' function in most genomes, creating the well known *hypothetical proteins problem*.[84] Some of the hypothetical proteins are parts of key pathways, and hence, identification of the *missing* genes becomes an important problem for reconstructing the whole pathways in particular genomes. Combining evidences from multiple comparative genomics techniques, it is possible to infer the connection between a function unknown gene and certain cellular processes, thus to suggest putative missing genes for incomplete pathways. Application of this approach has produced valuable pathway reconstructions for newly sequence genomes.[90] Although these predictions are upon further experimental validation, they provide useful information for metabolic analysis and engineering of bacteria, and development of new medicine.[91] The pathways reconstructed from genomic sequences have been integrated into pathway databases, e.g. Kyoto Encyclopedia of Genes and Genomes (KEGG),[92] which are accessible through web-based searching.

### 3.3. Microarray analysis

The development of DNA *microarray* technique is a key technology that facilitates the genome wide analysis of gene expression levels.[93] Experimentally, a microarray is a tiny square array, on which thousands of *probes*, each corresponding to a specific gene of interest, are synthesized or placed at a high density. The mRNAs extracted from a sample are labeled with a fluorescent dye and hybridized to the microarray. The expression level of corresponding genes can be measured using the amount of mRNAs that stick to spot of probes.

According to the design of probes, DNA microarray can be classified into two broad categories: the oligonucleotide arrays and cDNA arrays. The oligonucleotide array technology (*GeneChip*) developed by Affymetrix (http:www.affimetrix.com) uses *situ* synthesized oligonucleotides as probes, whereas cDNA array technique places cDNA clones on the array as probes.[94] There are several levels of bioinformatics analysis for microarray experiments. On the bottom level, statistical methods are needed to analyze the scanned image from a microarray experiment to extract fluorescent intensities.[94] The resulting data needs to be further normalized within a single array to remove the background noise, and across multiple arrays to remove array specific biases.[96] After these steps, the expression level of each analyzed gene can be obtained and used for the next level analysis. The medium level data analysis involves hypothesis tests (for two sample comparison) or multi-variable variation analysis (for multiple sample comparison) in an attempt to detect differential gene expression. Finally the high level analysis aims at studying gene functions using gene expression levels from multiple samples and experiments, and also integrating other data resources.

Gene expression profiling is a straightforward application of microarray technique.[95] The result of a gene expression profiling experiment can be represented by a high dimensional matrix, in which each row represents an analyzed gene, and each column represents an individual microarray experiment, e.g. an environmental condition or a tissue sample. Since genes that are similarly expressed may be functionally related, various clustering methods have been used to recognize groups of genes sharing similar gene expression patterns, which can then be used ultimately to build a global gene regulatory network.[97] Gene expression profiling can also be used for biomarker discovery and disease diagnosis.[96] Conceptually, some genes may be differentially expressed in disease tissues and normal tissues, and can be used as biomarkers for early disease diagnosis. The biomarkers can also be used for a detailed classification of diseases that show no clear distinct phenotypes.

In addition to gene expression analysis, microarray techniques are also applied to other problems. Genome tiling arrays[98] utilize probes spanning the entire genome, thus can be used to detect genome variations, such as *single nucleotide polymorphisms (SNPs)*[99] and *copy number polymorphisms (CNPs)*,[100] by hybridizing to chromosomal DNAs, and to discover the transcription of new genes and alternative splicing[101] by hybridizing to mRNAs. Novel bioinformatics methods are required to analyze the data generated from these experiments.

### 3.4. Proteomics

While the genome encodes the entire genetic information of a living organism, it is proteins that carry out biological processes. Proteins are *synthesized* using amino acids molecules following the direction encoded in DNA or RNA. Proteomics aims to identify the whole set of proteins inside a cell (*proteome*) and to study their dynamic changes across different physiological conditions. In recent years, because of its high sensitivity, mass spectroscopy (MS) has become an essential analytical technology in proteomics. In a typical proteomics project, proteins are first separated by liquid chromatography (LC) or electrophoresis, then digested into peptides by proteases (e.g. trypsin) and finally analyzed by tandem mass spectroscopy (MS/MS).[102] In MS/MS instruments, the covalent bonds of peptides are broken at different energy levels and the masses of the resulting fragment ions are measured by MS, which provide valuable information for determining the covalent structures of peptides.

Many bioinformatics methods have been developed to interpret the peptide MS/MS spectra automatically. These methods are often classified into two types according to the methodology they adopt: database searching methods and *de novo* sequencing methods.[103] For examples, Sequest[104] and Mascot[105] are two most frequently used peptide database searching tools; algorithms have also been designed for *de novo* peptide sequencing.

Quantifying proteins in a complex proteome sample (or comparing protein abundances across different samples), is another focus in the field of proteomics, sometimes referred to as quantitative proteomics. Several labeling techniques applied to various MS instruments including isotopic coded affinity tag (ICAT), mass-coded abundance tagging (MCAT) stable isotopic labeling, and global internal standard technology (GIST).[106] On the other hand, label-free protein quantification approaches attempt to quantify protein abundances directly from high-throughput proteomics analysis. Different measures that can be derived from proteomics experiments and presumably correlated to protein abundance were proposed for different MS instruments. For instance, the integration of extracted ion chromatogram (XIC) peaks is thought to be a good measure for LC/MS experiments[107] and sophisticated data analysis tools have been proposed to improve its accuracy.[108]

Proteins undergo different types of modifications after they are translated from mRNA. Many of these post-translational modifications (PTMs) have important biological functions, e.g. phosphorylations in signal transduction. MS-based proteomics approaches have been applied to large scale analysis of site-specific modifications.[109] Although several algorithms have been developed to analyze these data, it remains a challenge in bioinformatics to automatically identify these sites from proteomics data.[110]

Table 2. Online resources for curated protein-protein interactions.

| Database | URL | Reference |
|---|---|---|
| Database of Interacting Proteins (DIP) | http://dip.doe-mbi.ucla.edu/ | 115 |
| The Biomolecular Interaction Network Database (BIND) | http://bind.ca | 116 |
| Munich Information Center for Protein Sequences (MIPS) | http://mips.gsf.de/ | 117 |

### 3.5. *Protein-protein interaction*

Proteins carry out their functions by cooperating with each other as well as other types of biomolecules. Recently, high throughput technologies have been developed to determine the interaction partners of proteins at genome scale.[111] *In vitro* methods like two hybrid technique[112] can determine a pair of proteins that can putatively interact with each other. MS-based methods can identify components of an *in vivo* trapped protein complex.[113] These data are being maintained as protein interaction databases (see Table 2). The availability of the interaction map on the whole proteome has inspired new computational methods to study protein functions and biological processes on a system level.[114]

## 4. Conclusion

Bioinformatics is still a young discipline and forming its core research topics. Nevertheless, its data-centric nature and the challenge of analyzing massive high dimensional data have drawn a lot of attention from computer scientists. Bioinformatics has been one of the major resources of new problems for computer science and it will remain in this way in the future.

### Acknowledgement

Sun Kim was partially supported by CAREER Award DBI-0237901 from National Science Foundations (USA).

### References

1. International Human Genome Sequencing Consortium, *Nature* **409**, 860 (2001).
2. G. McVean, C. C. Spencer and R. Chaix, *PLoS Genetics* **1**, e54 (2005).

3. P. O. Brown and D. Botstein, *Nature Genetics* **21**, 33 (1999).
4. C. L. de Hoog and M. Mann, *Annu. Rev. Genomics Hum. Genet.* **5**, 267 (2004).
5. G. Gibson and S. Muse, *A Primer of Genome Science*, Sinauer Associates, USA, (2004).
6. P. J. McLaughlin, L. T. Hunt and M. O. Dayhoff, *Journal of Human Evolution* **1**, 565 (1972).
7. K. D. Gibson and H. A. Scheraga *Proc. Nat. Acad. Sci., USA* **58**, 420 (1967).
8. M. Levitt, *Nature* **224**, 759 (1969).
9. D. Gusfield, *Algorithms on Strings, Trees, and Sequences: Computer Science and Computational Biology*, Cambridge University Press, England (1997).
10. S. B. Needleman and C. D. Wunsch, *Journal of Molecular Biology* **48**, 443 (1970).
11. T. F. Smith and M. S. Waterman, *Journal of Molecular Biology* **48**, 443 (1970).
12. S. Henikoff, *Curr. Opin. Struct. Biol.* **6**, 353 (1996).
13. D. J. Lipman and W. R. Pearson, *Science* **227**, 1435 (1985).
14. S. Altschul, W. Gish, W. Miller, E. Myers and J. Lipman, *Journal of Molecular Biology* **215**, 403 (1990).
15. J. Dumas and J. Ninio, *Nucleic Acids Research* **10**, 197 (1982).
16. W. Miller, *Bioinformatics* **17**, 391 (2001).
17. S. Batzoglou, *Brief in Bioinformics* **6**, 6 (2005).
18. J. H. Choi, H. G. Cho and S. Kim, *Comput. Biol. Chem.* **29**, 244 (2005).
19. A. L. Delcher, A. Phillippy, J. Carlton and S. L. Salzberg, *Nucleic Acids Research* **30**, 2478 (2002).
20. B. Ma, J. Tromp and M. Li, *Bioinformatics* **18**, 440 (2002).
21. L. Wang and T. Jiang, *Journal of Computational Biology* **1**, 337 (1994).
22. D. Feng and R. Doolittle, *Journal of Molecular Evolution* **25**, 351 (1987).
23. J. D. Thompson, D. G. Higgins and T. J. Gibson, *Nucleic Acids Research* **22**, 4673 (1994).
24. C. N. Dewey and L. Pachter, *Hum. Mol. Genet.* **15**, R51 (2006).
25. J. Stoye, *Gene* **211**, GC45 (1998).
26. C. Notredame, D. G. Higgins and J. Heringa, *Journal of Molecular Biology* **302**, 205 (2000).
27. F. Sanger, S. Nilken and A. R. Coulson, *Proc. Nat. Acad. Sci., USA* **74**, 5463 (1977).
28. R. Staden, *Nucleic Acids Research* **4**, 4037 (1977).
29. H. Peltola, H. Soderlund and E. Ukkonen, *Nucleic Acids Research* **12**, 307 (1984).
30. P. Green, *Documentation for Phrap* (1994).
31. J. C. Roach, C. Boysen, K. Wang and L. Hood, *Genomics* **26**, 345 (1995).
32. G. Myers, *IEEE Computing in Science and Engineering* **1**, 33 (1999).
33. P. A. Pevzner, H. Tang and M. S. Waterman, *Proc. Nat. Acad. Sci., USA* **98**, 9748 (2001).
34. M. Pop, S. L. Salzberg and M. Shumway, *IEEE Computer* **35**, 47 (2002).
35. M. Ronaghi, *Genome Research* **11**, 3 (2001).
36. M. Chaisson, P. A. Pevzner and P. H. Tang, *Bioinformatics* **20**, 2067 (2004).
37. M. Margulies, et al., *Nature* **437**, 376 (2005).
38. R. Staden and A. D. McLachlan, *Nucleic Acids Research* **10**, 141 (1982).
39. J. W. Fickett, *Nucleic Acids Research* **10**, 5318 (1982).
40. M. Borodovsky and J. McIninch, *Computers and Chemistry* **17**, 123 (1993).

41. C. Burge and S. Karlin, *Journal of Molecular Biology* **268**, 78 (1993).
42. M. R. Brent and R. Guigo, *Curr. Opin. Struct. Biol.* **14**, 264 (2004).
43. M. S. Gelfand, A. A. Mironov and P. A. Pevzner, *Proc. Nat. Acad. Sci., USA* **93**, 9061 (1996).
44. A. Lindlof, *Appl. Bioinformatics* **2**, 123 (2003).
45. M. R. Brent, *Genome Research* **15**, 1777 (2005).
46. I. Korf, P. Flicek, D. Duan and M. R. Brent, *Bioinformatics* **17**, S140 (2001).
47. S. S. Gross and M. R. Brent, *J. Comput. Biol.* **13**, 379 (2006).
48. M. Kellis, N. Patterson, M. Endrizzi, B. Birren and E. S. Lander, *Nature* **423**, 241 (2003).
49. R. Nussinov and A. B. Jacobson, *Proc. Nat. Acad. Sci., USA* **77**, 6309 (1980).
50. M. S. Waterman and T. F. Smith, *Math. Biosci.* **42**, 257 (1978).
51. M. Zuker, *Nucleic Acids Research* **31**, 3406 (2003).
52. I. L. Hofacker, W. Fontana, P. F. Stadler, L. S. Bonhoeffer, M. Tacker and P. Schuster, *Monatshefte ftir Chemie* **125**, 167 (1994).
53. S. R. Eddy, *Nature Rev. Genet.* **2**, 919 (2001).
54. E. Rivas and S. R. Eddy, *Bioinformatics* **16**, 583 (2000).
55. S. R. Eddy, *Cell* **109**, 137 (2002).
56. E. Rivas and S. R. Eddy, *BMC Bioinformatics* **2**, 8 (2001).
57. S. Washietl, I. L. Hofacker and P. F. Stadler, *Proc. Nat. Acad. Sci., USA* **102**, 2454 (2005).
58. M. S. Gelfand, A. A. Mironov, J. Jomantas, Y. I. Kozlov and D. A. Perumov, *Trends Genet.* **15**, 439 (1999).
59. B. J. Tucker and R. R. Breaker, *Curr. Opin. Struct. Biol.* **15**, 342 (2005).
60. R. F. Doolittle, *Science* **214**, 149 (1981).
61. G. D. Stormo, *Annu. Rev. Biophys. Biophys. Chem.* **17**, 241 (1988).
62. C. E. Lawrence, S. F. Altschul, M. S. Boguski, J. S. Liu, A. F. Neuwald and J. C. Wootton, *Science* **262**, 208 (1993).
63. T. L. Bailey and C. Elkan, *Proc. Int. Conf. Intell. Syst. Mol. Biol.* **2**, 28 (1994).
64. S. H. Sze and P. A. Pevzner, *Proc. Int. Conf. Intell. Syst. Mol. Biol.* **8**, 269 (2000).
65. J. Buhler and M. Tompa, *Journal of Computational Biololgy* **9**, 225 (2002).
66. E. Eskin and P. A. Pevzner, *Bioinformatics* **18**, S354 (2002).
67. M. A. Marti-Renom, A. C. Stuart, A. Fiser, R. Sanchez, F. Melo and A. Sali, *Annu. Rev. Biophys. Biomol. Struct.* **29**, 291 (2000).
68. W. J. Browne, A. C. North, D. C. Phillips, K. Brew, T. C. Vanaman and R. L. Hill, *Journal of Molecular Biology* **42**, 65 (1969).
69. S. H. Bryant and C. E. Lawrence, *Proteins* **16**, 92 (1993).
70. I. Friedberg, L. Jaroszewski, Y. Ye and A. Godzik, *Curr. Opin. Struct. Biol.* **14**, 307 (2004).
71. K. T. Simons, I. Ruczinski, C. Kooperberg, B. A. Fox, C. Bystroff and D. Baker, *Proteins* **34**, 82 (1999).
72. K. T. Simons, C. Strauss and D. Baker, *Journal of Molecular Biology* **306**, 1191 (2001).
73. R. C. Hardison, *PLoS Biology* **1**, e58 (2003).
74. G. M. Rubin, M. D. Yandell, J. R. Wortman, G. L. Miklos and C. R. Nelson, *Science* **287**, 2204 (2000).

75. L. D. Stein, Z. Bao, D. Blasiar, T. Blumenthal, M. R. Brent et al., *PLoS Biology* **1**, e44 (2003).
76. P. F. Cliften, L. W. Hillier, L. Fulton, T. Graves, T. Miner et al., *Genome Research* **11**, 1175 (2001).
77. M. Kellis, N. Patterson, M. Endrizzi, B. Birren and E. S. Lander, *Nature* **423**, 241 (2003).
78. The chimpanzee sequencing and analysis consortium, *Nature* **437**, 69 (2005).
79. L. Florea, C. Riemer, S. Schwartz, Z. Zhang, N. Stojanovic et al., *Nucleic Acids Research* **28**, 3486 (2000).
80. K. Choi, Y. Ma, J. H. Choi and S. Kim, *Bioinformatics* **21**, 2514 (2005).
81. W. J. Kent, C. W. Sugnet, T. S. Furey, K. M. Roskin, T. H. Pringle et al., *Genome Research* **12**, 996 (2002).
82. E. V. Koonin, *Annu. Rev Genet.* **39**, 309 (2005).
83. A. Osterman and R. Overbeek, Missing genes in metabolic pathways: a comparative genomics approach, *Curr. Opin. Chem. Biol.* **7**, 238 (2003).
84. E. M. Marcotte, M. Pellegrini, H. L. Ng, D. W. Rice, T. O. Yeates and D. Eisenberg, *Science* **285**, 5428 (1999).
85. M. Pellegrini, E. M. Marcotte, M. J. Thompson, D. Eisenberg, and T. O. Yeates, *Proc. Nat. Acad. Sci., USA* **96**, 4285 (1999).
86. L. Aravind, L. M. Iyer and E. V. Koonin, *Curr. Opin. Struct. Biol.* **16** (2006).
87. E. E. Eichler and D. Sankoff, *Science* **301**, 5634 (2006).
88. M. Kellis, B. W. Birren and E. S. Lander, *Nature* **428**, 617 (2004).
89. G. Bourque, G. Tesler and P. A. Pevzner, *Genome Research* **16**, 311 (2006).
90. T. Dandekar and R. Sauerborn, *Pharmacogenomics* **3**, 245 (2002).
91. M. W. Covert, C. H. Schilling, I. Famili, J. S. Edwards, I. I. Goryanin, E. Selkov and B. O. Palsson, *Trends Biochem. Sci.* **26**, 179 (2001).
92. M. Kanehisa, S. Goto, S. Kawashima and A. Nakaya, *Nucleic Acids Res.* **30**, 42 (2002).
93. V. G. Cheung, M. Morley, F. Aguilar, A. Massimi, R. Kucherlapati and G. Childs, *Nature Genetics* **21**, 15 (1999).
94. D. D. Bowtell, *Nature Genetics* **21**, 25 (1999).
95. P. C. Boutros and A. B. Okey, Unsupervised pattern recognition: an introduction to the whys and wherefores of clustering microarray data, *Brief Bioinformics* **6**, 331 (2005).
96. J. S. Verducci, V. F. Melfi, S. Lin, Z. Wang, S. Roy and C. K. Sen, *Physiol. Genomics* **25**, 355 (2006).
97. X. Wu and T. G. Dewey, *Methods Mol. Biol.* **316**, 35 (2006).
98. P. Kapranov, V. I. Sementchenko and T. R. Gingeras, *Brief Funct. Genomic Proteomic* **2**, 47 (2003).
99. A. E. Oostlander, G. A. Meijer and B. Ylstra, *Clin. Genet.* **66**, 488 (2004).
100. B. Ylstra, P. van den Ijssel, B. Carvalho, R. H. Brakenhoff and G. A. Meijer, *Nucleic Acids Res.* **34**, 445 (2006).
101. T. E. Royce, J. S. Rozowsky, P. Bertone, M. Samanta, V. Stolc, S. Weissman, M. Snyder and M. Gerstein, *Trends Genet.* **21**, 466 (2005).
102. J. R. Yates III, *Annu. Rev. Biophys. Biomol. Struct.* **33**, 297 (2004).
103. R. S. Johnson, M. T. Davis, J. A. Taylor and S. D. Patterson, *Methods* **35**, 223 (2005).
104. J. R. Yates, J. K. Eng, A. L. McCormack and D. Schieltz, *Anal. Chem.* **67**, 1426 (1995).

105. D. N. Perkins, D. J. Pappin, D. M. Creasy and J. S. Cottrell, *Electrophoresis* **20**, 3551 (1999).
106. X. Zhang, W. Hines, J. Adamec, J. M. Asara, S. Naylor and F. E. Regnier, *J. Am. Soc. Mass Spectrom.* **16**, 1181 (2005).
107. R. E. Higgs, M. D. Knierman, V. Gelfanova, J. P. Butler and J. E. Hale, *J. Proteome Res.* **4**, 1442 (2005).
108. K. C. Leptos, D. A. Sarracino, J. D. Jaffe, B. Krastins and G. M. Church, *Proteomics*, **157**, 1770 (2006).
109. S. A. Carr, R. S. Annan and J. Huddleston, *Methods Enzymol.* **405**, 82 (2005).
110. D. Tsur, S. Tanner, E. Zandi, V. Bafna and P. A. Pevzner, *Nat. Biotechnol.* **23**, 1562 (2005).
111. J. Piehler, *Curr. Opin. Struct. Biol.* **15**, 4 (2005).
112. J. Miller and I. Stagljar, *Methods Mol. Biol.* **261**, 247 (2004).
113. S. Kaveti and J. R. Engen, *Methods Mol. Biol.* **316**, 179 (2006).
114. M. E. Cusick, N. Klitgord, M. Vidal and D. E. Hill, *Hum. Mol. Genet.* **14**, R171 (2005).
115. L. Salwinski, C. S. Miller, A. J. Smith, F. K. Pettit, J. U. Bowie and D. Eisenberg, *Nucleic Acids Research* **32**, D449 (2004).
116. C. Alfarano, C. E. Andrade, K. Anthony *et al.*, *Nucleic Acids Research* **33**, D418 (2005).
117. H. W. Mewes, D. Frishman, K. F. X. Mayer, M. Munsterkotter, O. Noubibou, P. Pagel and T. Rattei, *Nucleic Acids Research* **34**, D169 (2006).

# CHAPTER 2

# AN INTRODUCTION TO SOFT COMPUTING

Amit Konar* and Swagatam Das

*Department of Electronics and Telecommunication Engineering*
*Jadavpur University, Kolkata 700032, India*
*\*konaramit@yahoo.co.in*

The chapter provides an introduction to soft computing. It defines soft computing in a comprehensive manner and examines the scope of soft computing to overcome the limitations of the traditional artificial intelligence (AI). The chapter briefly introduces various tools of soft computing, such as fuzzy logic, neural network, evolutionary algorithms and probabilistic reasoning. The synergistic behavior of the above tools on many occasions far exceeds their individual performance. A discussion on the synergistic behavior of neuro-fuzzy, neuro-GA, neuro-belief and fuzzy-belief network models is also included in the chapter. The chapter finally focuses on a few topics of current interest like the artificial life, swarm intelligence, rough sets and granular computing, and artificial immune system (AIS), which are gaining increasing importance in the field of soft computing.

## 1. Classical AI and its Pitfalls

Artificial Intelligence (AI) aims at designing machines capable of thinking and acting like the human beings. AI includes a vast discipline of knowledge,[1–8] such as automated reasoning,[9–10] machine learning, planning, intelligent search, language and image understanding.[11] In this section, we briefly review AI to examine the scope of soft computing models. Interested readers may consult any standard textbook[14] for a comprehensive discussion on AI.

Classical problem solvers in AI employ a set of rules to cause transition in problem states. The IF-parts of the rules are instantiated with the current problem states and on successful matching the selected rule is fired causing a transition in problem states. To continue firing of the rules until the goal is found, the knowledge base is usually enriched with a large number of rules. A large knowledge base, however, calls for more search time and thus is responsible for degradation in efficiency of a reasoning system. One approach to circumvent this problem is to organize the knowledge base with fewer rules but to allow **partial matching** of the problem states with the IF-part of the rules. The *logic of fuzzy sets*[15–19] that we shall introduce shortly is capable of such partial matching.

Traditional AI is very good in inductive[20] and analogy-based learning,[21] but it is inefficient to realize supervised learning.[22–27]

In supervised learning, the trainer provides a number of input/output training instances for the learning system. The learning system has to adapt its internal parameters to generate the correct output instance in response to a given input instance. *Neural network models*[28–37] that we shall introduce here can perform supervised learning very well. It may be added here that approximately 2/3-rd of the commercial applications of machine learning falls within the category of supervised learning.

Except the heuristic search algorithm, traditional AI is not competent enough to handle real world optimization problems. Classical derivative-based optimization techniques fail for most of the engineering problems because of severe roughness of the search landscape, framed by the system constraints. *Evolutionary algorithms*[38–44] employ derivative-free stochastic optimization techniques, and are currently gaining importance in handling complex optimization problems. Genetic Algorithm is one such evolutionary algorithm, which has proved itself successful in machine learning, optimization and intelligent search problems. Typical application of this algorithm includes optimization of rules in an expert system,[14] tuning of parameters[45] in a learning system and determining the optimal order of events in a scheduling problem.[14]

It is thus apparent that the traditional AI was incompetent in serving the increasing demand of search, optimization and machine learning especially in (i) large biological and commercial databases and (ii) factory automation for steal, aerospace, power and pharmaceutical industries. The shortcomings of AI became more and more pronounced with successive failures of the decade long Japanese project on *Fifth Generation Computing Machines*. Almost at the same time, the contemporary models of non-traditional machine intelligence such as rough sets, fuzzy logic, artificial neural networks, genetic algorithms, belief networks, computational learning theory and chaos theory could prove their theoretical basis. The failure of classical AI opened up new avenues for the non-conventional models in various engineering applications. These computational tools gave rise to a new discipline called **Soft Computing**.

## 2. What is Soft Computing?

Soft Computing refers to a collection of new computational techniques in computer science, artificial intelligence, machine learning, and many applied and engineering areas where one tries to study, model, and analyze very complex phenomena, those for which the classical and precise scientific tools were incapable of giving low cost, analytic, and complete solution. Scientific methods of previous centuries could model, and precisely analyze, merely, relatively simple systems of physics, classical Newtonian mechanics, and engineering. However, many complex cases, e.g. systems related to biology and medicine, humanities, management sciences, and similar fields remained outside of the main territory of successful applications of precise mathematical and analytical methods. The term soft computing was actually coined by Prof. L. A. Zadeh, the father of fuzzy set theory and fuzzy logic, in 1992. In Zadeh's own words:

*"Soft computing is an emerging approach to computing which parallels the remarkable ability of the human mind to reason and learn in an environment of uncertainty and imprecision"*.[19]

*"Soft computing is not a homogeneous body of concepts and techniques. Rather it is a collection of methodologies, which in one form or another reflect the guiding principle of soft computing: exploit the tolerance for imprecision, uncertainty, and partial truth to achieve tractability, robustness, and low solution cost. Viewed in a slightly different perspective, soft computing is a consortium of methodologies which, either singly or in combination, serve to provide effective tools for the development of intelligent systems"*.[46]

*"...a recent trend to view fuzzy logic (FL), neurocomputing (NC), genetic computing (GC) and probabilistic computing (PC) as an association of computing methodologies falling under the rubric of so-called soft computing. The essence of soft computing is that its constituent methodologies are for the most part complementary and synergistic rather than competitive. A concomitant of the concept of soft computing is that in many situations it is*

*advantageous to employ FL, NC, GC and PC in combination rather than isolation*".[47]

It is thus clear from the above definitions that soft computing techniques resemble human reasoning more closely than traditional techniques, which are largely based on conventional logical systems, such as sentential logic and predicate logic, or rely heavily on the mathematical capabilities of a computer. Now, we see that, the principal constituents of soft computing (SC) are fuzzy logic (FL), neural network theory (NN) and probabilistic reasoning (PR), with the latter subsuming belief networks, genetic algorithms, chaos theory and parts of learning theory. What is important to note is that SC is not a mélange of FL, NN and PR. Rather; it is a partnership in which, each of the partners contributes a distinct methodology for addressing problems in its domain. In this perspective, the principal contributions of FL, NN and PR are complementary rather than competitive.

The definitions of Prof. Zadeh also imply that, unlike hard computing schemes, which strive for exactness and for full truth, soft computing techniques exploit the given tolerance of imprecision, partial truth, and uncertainty for a particular problem. Another common contrast comes from the observation that inductive reasoning plays a larger role in soft computing than in hard computing. The novelty and strength of soft computing lie in its synergistic power through fusion of two or more computational models/techniques.

## 3. Fundamental Components of Soft Computing

In this section, we briefly outline the common soft-computing models based on their fundamental characteristics.

### 3.1. *Fuzzy sets and fuzzy logic*

In the real world, information is often ambiguous or imprecise. When we state that it is warm today, the context is necessary to approximate the temperature. A warm day in January may be $-3$ degrees Celsius, but a warm day in August may be 33 degrees. After a long spell of frigid days, we may call a milder but still chilly day relatively warm. Human reasoning filters and interprets information in order to arrive at conclusions or to dismiss it as inconclusive. Although machines cannot yet handle imprecise information in the same ways that humans do, computer programs with fuzzy logic are becoming quite useful when the sheer volume of tasks defies human analysis and action.

An organized method for dealing with imprecise data is called fuzzy logic. The data are considered as fuzzy sets. Traditional sets include or do not include an individual element; there is no other case than true or false. Fuzzy sets allow partial membership. Fuzzy Logic is basically a multi-valued logic that allows intermediate values to be defined between conventional evaluations like yes/no, true/false, black/white, etc. Notions like rather warm or pretty cold can be formulated mathematically and processed with the computer. In this way, an attempt is made to apply a more human-like way of thinking in the programming of computers. Fuzzy logic is an extension of the classical propositional and predicate logic that rests on the principles of the binary truth functionality. For instance, let us consider the well-known *modus ponens* rule in propositional logic, given by:

$$\frac{p, \quad p \to q}{q},$$

where p and q are binary propositions (i.e. facts with binary truth values) and → denotes a 'if-then' operator. The rule states that given p is true and 'if p then q' is true, it can be inferred that q is true. However, what happens when the truth or falsehood of p cannot be guaranteed? Three situations may arise in this context. First, we may take support or refutation of other statements in the system, if any, to determine the truth or falsehood of p. If majority support for p is obtained, then p may be accepted to be true, else, p is assumed false. This idea led to the foundation of a new class of logic, called **non-monotonic logic**.[47] However, what happens when the exact binary truth functional value of p cannot be ascertained. This called for the other two situations. Suppose p is partially true. This can be represented by a truth functional value in between 0 and 1, and this idea was later formalized as the basis of **multi-valued logic**. The second situation thus presumes a non-binary truth functional value of p and attempts to infer a truth functional value of q in the range [0, 1]. Now, consider the third situation where a fact approximately same as p, called p′, is available and the nearest rule whose antecedent part partially matches with p′ is p→q. Thus, formally

$$\frac{\begin{array}{c} p' \\ p \to q \end{array}}{q'},$$

where q′ is the inferred consequence. This partial matching of p′ with p, and thus generating the inference q′ comes under the scope of **fuzzy logic**. The truth functional value of the propositions here lies in the same interval of [0, 1] like that of multi-valued logic. Thus in one sense fuzzy logic is a multi-valued logic. However, the most pertinent feature of fuzzy logic for which it receives so much attention is its scope of **partial matching**, as illustrated with the last example. Another interesting situation occurs when a number of rules' antecedent parts match partially with the fact p′. The resulting inference, say $q_{final}$, under this circumstance is determined by the composite influence of all these rules. Therefore, in any real world system, the inferences guided by a number of rules follow a middle decision trajectory over time. This particular behavior of following a middle decision trajectory[48] is humanlike and is a unique feature of fuzzy logic, which made it so attractive!

Very recently, Prof. Zadeh highlighted another important characteristic[49] of fuzzy logic that can distinguish it from other multi-valued logics. He called it **f.g-generalization**. According to him any theory, method, technique or problem can be fuzzified (or **f-generalized**) by replacing the concept of a crisp set with a fuzzy set. Further, any theory, technique, method or problem can be granulated (or **g-generalized**) by partitioning its variables, functions and relations into granules (information cluster). Finally, we can combine f-generalization with g-generalization and call it f.g-generalization. Thus ungrouping an information system into components by some strategy and regrouping them into clusters by some other strategy can give rise to a new kind of information sub-systems. Determining the strategies for ungrouping and grouping, however, rests on the designer's choice. The philosophy of f.g-generalization undoubtedly will re-discover fuzzy logic in a new form.

## 3.2. Neural networks

Neurons are the fundamental building blocks of a biological nervous system. A neuron has four main structural components[50,51]: the dendrites, the cell body, the axon and the synapse. The dendrites act as receptors, thereby receiving signals from several neighboring neurons and passing these on to a little thick fiber, called dendron. In other words, dendrites are the free terminals of dendrons. The received signals collected by different dendrons are processed within the cell body and the resulting signal is transferred through a long fiber called axon. At the other end of the axon, there exists an inhibiting unit called synapse. This unit controls the flow of neuronal current from the originating neuron to receiving dendrites of neighborhood neurons. A schematic diagram, depicting the above concept is presented in Fig. 1.

An artificial neural net is an electrical analog of a biological neural net.[23] The cell body in an artificial neural net is modeled by a linear activation function. The *activation function* in general, attempts to enhance the signal contribution received through different dendrons. The axon behaves like a signal conductive (resistive) device. The synapse in the artificial neural net is modeled by a nonlinear *inhibiting function* for limiting the amplitude of the signal processed at cell body. The most common nonlinear functions

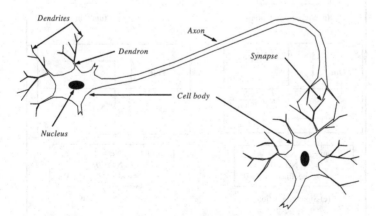

**Fig. 1.** A biological neuron comprising of a cell body, several dendrons, an axon and a synapse. The terminals of a dendron are called dendrites. The dendrites of a neuron transfer signal to dendrites of a second neuron through a synaptic gap.

used for synaptic inhibition are:

- sigmoid function
- tanh function
- signum function
- step function.

Sigmoid and tan hyperbolic (tanh) functions are grouped under soft nonlinearity, whereas signum and step functions under hard type nonlinearity. These functions are presented graphically in Fig. 2 for convenience. The schematic diagram of an artificial neuron, based on the above modeling concept is presented in Fig. 3.

Depending on the nature of problems, the artificial neural net is organized in different structural arrangements (topologies). Common topologies

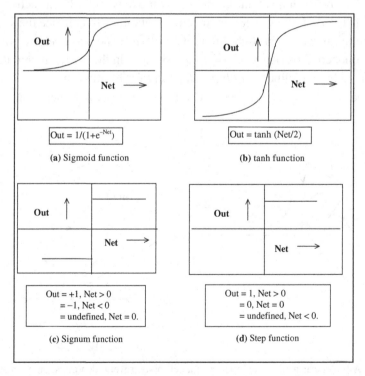

**Fig. 2.** Common nonlinear functions used for synaptic inhibition. Soft nonlinearity: (a) sigmoid and (b) tanh; hard nonlinearity: (c) signum and (d) step.

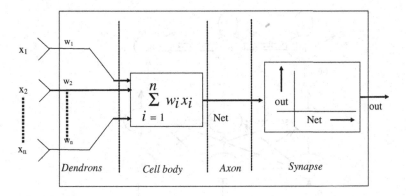

**Fig. 3.** An electrical equivalent of the biological neuron.

(vide Fig. 4) are:

- single layered recurrent net with lateral feedback structure,
- two layered feed-forward structure,
- two layered feedback structure,
- three layered feed-forward structure and
- single layered recurrent structure.

The single layered recurrent net with lateral feedback topology was proposed by Grossberg,[30] which has successfully been applied for classifying analog patterns. The feed-forward neural nets are most common structures for the well-known back-propagation algorithm.[52] Carpenter and Grossberg[30] have used two layered feedback structure, on the other hand for realization of adaptive resonance theory[31] and Kosko[53,54] used the same for realization of bi-directional associative memory. The last class of topology shown in Fig. 3(e) represents a recurrent net with feedback. Many cognitive nets[55] employ such topologies. Another interesting class of network topologies, where each node is connected to all other nodes bi-directionally and there is no direct self-loop from a node to itself, has been used by Hopfield in his studies.[56,57] We do not show the figure for this topology, as the readers by this time can draw it themselves for their satisfaction.

Fuzzy logic, which has proved itself successful especially in reasoning and unsupervised classification, however, is not directly amenable for automated machine learning.[6] Artificial neural nets (ANN) on the other hand can learn facts (represented by patterns) and determine the inter-relationship

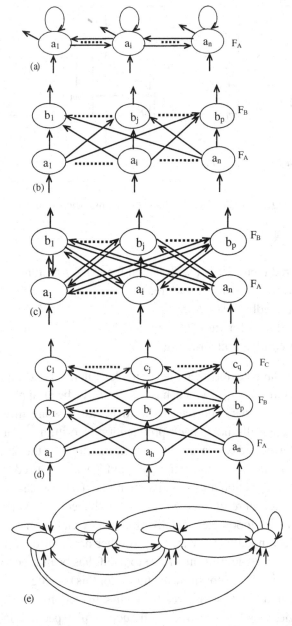

**Fig. 4.** Common topologies of artificial neural net: (a) single layered recurrent net with lateral feedback, (b) two layered feed-forward structure, (c) two layered structure with feedback, (d) three layered feed-forward structure, (e) a recurrent structure with self-loops.

among the patterns. The process of determining the interrelationships among patterns, in general, is informally called **encoding**. Some ANN parameters, usually the weights and the thresholds, are generally encoded to represent the interrelationship among the patterns. The process of determining the inter-related pattern(s) from one given pattern using the encoded network parameters is usually called **recall**. Depending on the type and characteristics of the encoding and recall process, machine-learning techniques are categorized into four major heads: (i) supervised learning, (ii) unsupervised learning, and (iii) reinforcement learning and competitive learning.

In a **supervised learning** system, an external teacher provides the input and the corresponding target (output) patterns. A learning algorithm is employed to determine a unique set of network parameters that jointly satisfy the input-output interrelationship of each two patterns. After the encoding process discussed above is over the network on excitation[58] with an unknown input pattern can generate its corresponding output pattern. The ANN when trained by a supervised learning algorithm, thus, behaves like a multi-input/multi-output function approximator.

A common question that naturally arises: how does a supervised learning system actually work? Generally, the network is initially assigned a random set of parameters such as weights and thresholds. Suppose that we want to train the network with a single input-output pattern. What should we do? We supply the network with our input pattern, and let the network generate its output pattern. The generated output pattern is compared with the target pattern. The difference of these two patterns results in an error vector. A supervised learning algorithm is then employed to adjust the network parameters using the error vector. For multiple input-output patterns, the error vector for each pattern set is determined and a function of these error vectors is used to adjust the network parameters. There exist quite a large number of supervised learning algorithms using neural nets. The most popular among them is the Back-propagation algorithm.

An **unsupervised learning system** employs no teacher, and thus the inter-relation among the patterns is not known. Generally, in an unsupervised learning process, one or more input patterns are automatically mapped to one pattern cluster. The encoding process in an unsupervised learning system varies greatly from one system to another. Most systems, however, employ a recursive learning rule that autonomously adjusts the network parameters for attaining some criteria like minimization of the network energy states. Among the unsupervised learning systems, Hopfield nets and associative memory are most popular.

The third category of the learning system that bridges the gap between the supervised and the unsupervised learning is popularly known as the **reinforcement learning**.[59] This learning scheme employs an **internal critic** that examines the response of the environment in turn of the action of the learning system on the environment. If the response is in favor of the goal, then the action is **rewarded** otherwise it is **penalized**. Determination of the status of the action: reward or penalty may, require quite a long time, until the goal is reached.

The last category of learning is well known as **competitive learning**. Neurons in this scheme compete with one another to satisfy a given goal. A competitive learning network usually consists of two layers: the input layer and the competitive/output layer. The input vector, submitted at the input layer, is passed on to the competitive layer through a set of forward connection of weights. The neurons in the competitive layer compete with each other by generating a positive signal to itself and a negative signal to the other neurons. Once the competition is complete, the connection weights of the winner neuron are update through some special weight adaptation dynamics.

### 3.3. *Genetic algorithms*

To tackle complex search problems (as well as many other complex computational tasks) computer-scientists have been looking into the nature for years — both as model and as metaphor-for inspiration. Optimization is at the heart of many natural processes like Darwinian Evolution itself. Through millions of years, every species had to optimize their physical structures to adapt to the environments they were in. This process of adaptation, this morphological optimization is so perfect that nowadays, the similarity between a shark, a dolphin or a submarine is striking.

A keen observation of the underlying relation between optimization and biological evolution has led to the development of a new paradigm of Computational Intelligence, marked as 'Evolutionary Algorithms (EA)' for performing very complex search and optimization. Below we illustrate the general principle of an EA with a simple pseudo-code. Here $P(t)$ denotes a population of chromosomes (trial solutions of the problem at hand) at time $t$.

The procedure initializes a population $P(t)$ randomly at iteration $t = 0$. The function: Evaluate $P(t)$ determines the fitness of the chromosomes by employing a specially constructed fitness measuring function. The 'while'-loop includes three main steps. First, it increases the iteration index by 1. Next it selects a population $P(t)$ from $P(t-1)$ based on the results of fitness evaluation. The function: Alter $P(t)$ evolves $P(t)$ by some complex non-linear operations. The while loop then re-evaluates $P(t)$ for the next iteration, and continues evolution until the terminating condition is reached.

**Procedure Evolutionary-Computation**
**Begin**
   $t \leftarrow 0$;
   Initialize $P(t)$;
   Evaluate $P(t)$;
   **While** (terminating condition not reached) **do**
    **Begin**
      $t \leftarrow t + 1$;
      Select $P(t)$ from $P(t-1)$;
      Alter $P(t)$;
      Evaluate $P(t)$;
   **End While;**
**End.**

The main functioning loop of an EA can also be illustrated by Fig. 5.

Depending on the choice of the function: alter $P(t)$, evolutionary computing algorithms can be of various types. Among the EAs one that has been attracting the attention of the largest number of researchers all over the globe till date is Genetic Algorithm (GA). GA is given the credit for being the algorithmic approach that most closely resembles real biological evolution. It encodes data and programs into DNA-like structures e.g. fixed length strings that may contain any alphabet. It then applies a wide variety of genetic operators on the population of such structures to create offspring for the next generation. These genetic operators, broadly classified into 'crossover' and 'mutation' operators, govern the speed of convergence of the GA, i.e. how fast it can find a considerably good solution. Selection of population members, which will survive to next generation and take part in reproduction, also plays a vital role in GA.

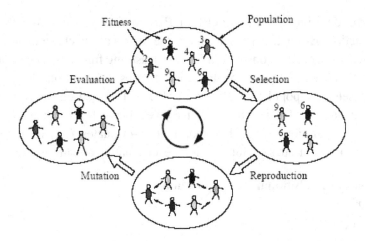

**Fig. 5.** The main loop of an EA.

Proposed by Holland[60] in early 1960s, this algorithm is gaining its importance for its wide acceptance in solving three classical problems, such as learning, search and optimization. A number of researchers throughout the world have developed their own ways to prove the convergence of the algorithm. Among these, the work by Goldberg,[61] De Jong,[62] Davis,[63] Muehlenbein,[41] Chakraborti,[64–66] Fogel,[67] and Vose[43,44] need special mention. A GA operates through a simple cycle of stages:[68]

(i) Creation of a "population" of strings,
(ii) Evaluation of each string,
(iii) Selection of best strings and
(iv) Genetic manipulation to create new population of strings.

Each cycle in GA produces a new generation of possible solutions for a given problem. In the first phase, an initial population, describing representatives of the potential solution, is created to initiate the search process. The elements of the population are encoded into bit-strings, called chromosomes. The performance of the strings, often called fitness, is then evaluated with the help of some functions, representing the constraints of the problem. Depending on the fitness of the chromosomes, they are selected for a subsequent genetic manipulation process. It should be noted that the **selection** process is mainly responsible for assuring survival of the best-fit individuals. After the selection of the population strings is over, the genetic

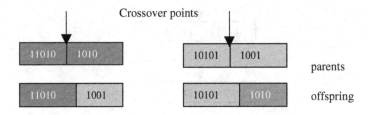

**Fig. 6.** A single point crossover after the 3rd bit position from the L.S.B.

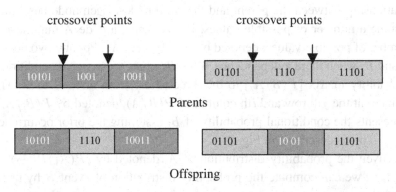

**Fig. 7.** Two point crossover: one after the 4th and the other after the 8th bit positions from the L.S.B.

manipulation process consisting of two steps is carried out. In the first step, **crossover** operation that recombines the bits (genes) of each two selected strings (chromosomes) is executed. Various types of crossover operators are found in the literature. The single point and two-points crossover operations are illustrated in Figs. 6 and 7 respectively. The **crossover points** of any two chromosomes are selected randomly. The second step in the genetic manipulation process is termed **mutation**, where the bits at one or more randomly selected positions of the chromosomes are altered (Fig. 8). The mutation process helps overcome trapping at local maxima. The offspring produced by the genetic manipulation process are the next population to be evaluated.

## 3.4. *Belief networks*

A Bayesian belief network[69,70] is represented by a directed acyclic graph or tree, where the nodes denote the events and the arcs denote the cause-effect

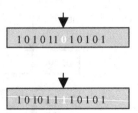

**Fig. 8.** Mutation of a chromosome at the 5th bit position.

relationship between the parent and the child nodes. Each node here, may assume a number of possible values. For instance, a node $A$ may have $n$ number of possible values, denoted by $A_1, A_2, \ldots, A_n$. For any two nodes, $A$ and $B$, when there a dependence $A \to B$, exists we assign a conditional probability matrix $[P(B/A)]$ to the directed arc from node $A$ to $B$. The element at the $j$th row and $i$th column of $P(B/A)$, denoted by $P(B_j/A_i)$, represents the conditional probability of $B_j$ assuming the prior occurrence of $A_i$. This is described in Fig. 9.

Given the probability distribution of $A$, denoted by $[P(A_1)P(A_2)\ldots P(A_n)]$, we can compute the probability distribution of event $B$ by using the following expression:

$$\begin{aligned}
\boldsymbol{P}(\boldsymbol{B}) &= [P(B_1)P(B_2)\cdots P(B_m)]_{1\times m} \\
&= [P(A_1)P(A_2)\cdots P(A_n)]_{1\times n} \cdot [\boldsymbol{P}(\boldsymbol{B}/\boldsymbol{A})]_{n\times m} \\
&= [\boldsymbol{P}(\boldsymbol{A})]_{1\times n} \cdot [\boldsymbol{P}(\boldsymbol{B}/\boldsymbol{A})]_{n\times m}.
\end{aligned} \quad (1)$$

We now illustrate the computation of $P(B)$ with an example.

**Example 1:** Consider a Bayesian belief tree describing the possible causes of a defective car.

Here, each event in the tree (Fig. 11) can have two possible values: true or false. Thus, the matrices associated with the arcs will have dimensions $(2 \times 2)$. Now, given $P(A) = [P(A = \text{true}) P(A = \text{false})]^T$, we can easily

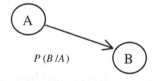

**Fig. 9.** Assigning a conditional probability matrix in the directed arc connected from $A$ to $B$.

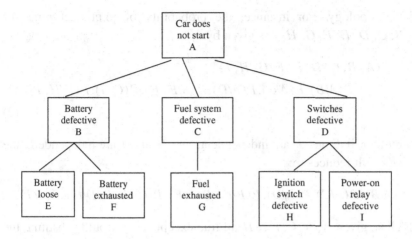

Fig. 10. A diagnostic tree for a car.

compute $P(B), P(C), P(D), P(E), \ldots, P(I)$ provided we know the transition probability matrices connected with the links.

Here, each event in the tree (Fig. 10) can have two possible values: true or false. Thus, the matrices associated with the arcs will have dimensions $(2 \times 2)$. Now, given $P(A) = [P(A = \text{true}) P(A = \text{false})]^T$, we can easily compute $P(B), P(C), P(D), P(E), \ldots, P(I)$ provided we know the transition probability matrices connected with the links. As an illustrative example, we compute $P(B)$ with $P(B/A)$ and $P(A)$.

$$\text{Let } P(A) = [P(A = \text{true}) P(A = [\text{false})]^T$$
$$= [0.7 \quad 0.3]$$

|  | $B_j$ | |
|---|---|---|
| $A_i$ | $B = \text{true}$ | $B = \text{false}$ |
| $P(B/A) = A = \text{true}$ | 0.8 | 0.2 |
| $A = \text{false}$ | 0.4 | 0.6 |

So,

$$P(B) = P(A) \cdot P(B/A) = [0.68 \quad 0.32].$$

One interesting property of the Bayesian network is that we can compute the probability of the joint occurrence easily with the help of the topology. For instance, the probability of joint occurrence of $A, B, C, D, E, F, G, H, I$ is given by:

$$P(A, B, C, D, E, F, G, H, I)$$
$$= P(A/B).P(A/C).P(A/D).P(B/E, F).P(C/G).P(D/H, I). \tag{2}$$

Further, if $E$ and $F$ are independent, and $H$ and $I$ are independent, the above result reduces to:

$$P(A/B).P(A/C).P(A/D).P(B/E).P(B/F).P(C/G).P(D/H).P(D/I).$$

Thus, given $A, B, C, \ldots, H$ all true except $I$, we would substitute the conditional probabilities for $P(A = \text{true}/B = \text{true})$, $P(A = \text{true}/C = \text{true})\ldots$ and finally $P(D = \text{true}/I = \text{false})$ in the last expression to compute $P(A = \text{true}, B = \text{true}, \ldots H = \text{true}, I = \text{false})$.

Judea Pearl,[69–71] proposed a scheme for propagating beliefs of evidence in a Bayesian network. We shall first demonstrate his scheme with a Bayesian tree like that in Fig. 10. It may, however, be noted that like the tree of Fig. 10 each variable, say $A, B \ldots$ need not have only two possible values. For example, if a node in a tree denotes German Measles (GM), it could have three possible values like severe-GM, little-GM, moderate-GM.

In Pearl's scheme for evidential reasoning, he considered both the causal effect and the diagnostic effect to compute the **belief function** at a given node in the Bayesian belief tree. For computing belief at a node, say $V$, he partitioned the tree into two parts: (i) the subtree rooted at $V$ and (ii) the rest of the tree. Let us denote the subset of the evidence, residing at the subtree of $V$ by $e_v^-$ and the subset of the evidence from the rest of the tree by $e_v^+$. We denote the belief function of the node $V$ by Bel($V$), where it is defined as

$$\begin{aligned} \text{Bel}(V) &= P(V/e_v^+, e_v^-) \\ &= P(e_v^-/V) \cdot P(V/e_v^+)/\alpha, \\ &= \lambda(V)\Pi(V)/\alpha \end{aligned} \tag{3}$$

where

$$\left. \begin{aligned} \lambda(V) &= P(e_v^-/V), \\ \Pi(V) &= P(V/e_v^+), \end{aligned} \right\}, \tag{4}$$

and $\alpha$ is a normalizing constant, determined by

$$\alpha = \sum_{v \in (\text{true,false})} P(e_v^-/V) \cdot P(V/e_v^+). \tag{5}$$

It seems from the last expression that $v$ could assume only two values: true and false. It is just an illustrative notation. In fact, $v$ can have a number of possible values.

Pearl designed an interesting algorithm for belief propagation in a causal tree. He assigned *a priori* probability of one leaf node, say node $H$ in Fig. 11, to be defective and then propagated the belief from this node to its parent and then from the parent to the grandparent, until the root is reached. Then he considered a downward propagation of belief from the root to its children, and from each child node to its children, and so on until the leaves are reached. The leaf having the highest belief is then assigned *a priori* probability and the whole process described above is repeated. Pearl has shown that after a finite number of up-down traversal on the tree, a *steady-state* condition is reached following which a particular leaf node in all subsequent up-down traversal yields a maximum belief with respect to all other leaves in the tree. The leaf thus selected is considered as the defective item.

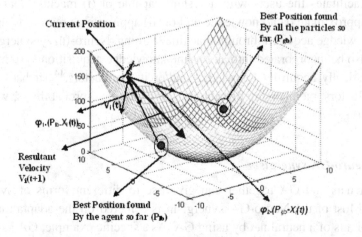

**Fig. 11.** A particle moving on a 3-dimensional fitness landscape.

## 4. Synergism in Soft Computing

The characteristics of the soft computing tools reveal that each tool has its own pros and cons. Fuzzy logic, for example, is good in approximate reasoning, but does not have much scope in machine learning or optimization applications. ANN on the other hand has a bigger role in machine learning. GA can be employed in intelligent search, optimization and some machine learning applications. Belief networks can be applied in diagnosis of systems from noisy sensory data. A co-operative synthesis of these tools thus may give rise to better computational models that can complement the limitation of one tool by the judicious use of the others.

Depending on the behavioral compatibility of the soft computing tools, we can classify their synergism into the following types. Each type again may be sub-divided into **strongly coupled** and **weakly coupled** synergism.[72] In a strongly coupled synergism, the individual tools are mixed in an inseparable manner, whereas in a weakly coupled synergism the individual tools have their structural/modular identity. More details on this issue will be discussed in a separate chapter.

### 4.1. *Neuro-fuzzy synergism*

Synergism of this type is needed to use the joint benefits of neural nets and fuzzy logic in a common platform.[73] A blending of neural nets and fuzzy logic facilitates the users with a system capable of (i) machine learning from approximate data/knowledge and/or (ii) approximate reasoning using the knowledge acquired through machine learning. Neuro-fuzzy synergism may also be incorporated into an on-line knowledge acquisition system for automatically acquiring knowledge and/or its parameters,[14] such as certainty factors, membership distributions or conditional probabilities, while reasoning.

### 4.2. *Neuro-GA synergism*

Neural nets and GA together can give rise to different forms of synergism. Most of the Neuro-GA synergism is centered on the adaptation of the weights of a neural net by using GA. As a specific example, GA can be employed to determine the weights of a Hopfield neural net.[74] The energy function of the Hopfield net under this circumstance may be regarded as the fitness function of the GA.[14] GA can also be used to determine the weights of a feed-forward neural net. A mean-square-error sum of the neurons at the output layer may, under this circumstance, be used as the fitness function. The training thus imparted to the neural net occasionally outperforms the typical back-propagation algorithm for the possibility of its being trapped at *local minima*.[52]

### 4.3. Fuzzy-GA synergism

Common forms of fuzzy-GA synergism include optimization of the parameters of a fuzzy system by using GA. A fuzzy system employs a fuzzifier to map the real world signal amplitudes into a range of [0, 1]. This is usually realized by a specialized curve called membership distributions. The $x$-axis of the membership distribution is the signal amplitude, while its $y$-axis denotes the fuzzy scale. The choice of the distributions has a significant impact in the response of a fuzzy system. The membership distributions in a fuzzy system may be adapted by a GA for optimization in the performance of a fuzzy system.

### 4.4. Neuro-belief network synergism

This is a relatively a new type of synergism and much discussion on this is not readily available in the current literature. A common-sense reasoning reveals 2 possible types of synergism under this category. Firstly, neural networks may be employed to determine the conditional probabilities in a belief network from the measurement data of known case histories. Secondly, a specialized type of belief computation, called the Dempster-Shfer theory, may by applied to fuse multi-sensory data, which consequently may be used as the training instances of a supervised neural net.[75]

### 4.5. GA-belief network synergism

Like Neuro-Belief network synergism, this too is new and unfortunately this has not been reported elsewhere. GA may be used to adapt the conditional probabilities of a belief network, and a performance criterion may be defined to test the success of the known case histories with the presumed conditional probabilities in each GA cycle until convergence of the GA occurs. In the coming future, GA-belief network synergism will be applied extensively in fault detection of complex engineering systems.

### 4.6. *Neuro-fuzzy-GA synergism*

This can be configured in different forms. One simple configuration of this includes a neural net as a pattern classifier, trained with fuzzy membership distributions, which has been pre-tuned by a GA. Among the other configurations, GA may be employed in a tightly coupled neuro-fuzzy system to optimize its parameters. GA can also be used to determine the best set of training instances of a tightly coupled neuro-fuzzy system.

## 5. Some Emerging Areas of Soft Computing

In this section, we briefly introduce some new members of the soft computing family that are currently gaining importance for their increasing applications in both science and engineering. The list of the new members includes Artificial Life, Particle Swarms, Artificial Immune Systems, Rough Set Theory and Granular Computing, and Chaos Theory.

### 5.1. *Artificial life*

Pioneered by Langton, Artificial Life ("ALife" or "ALife")[76] is the name given to a new discipline that studies "natural" life by attempting to recreate biological phenomena from scratch within computers and other "artificial" media. ALife complements the traditional analytic approach of traditional biology with a synthetic approach in which, rather than studying biological phenomena by taking apart living organisms to see how they work, one attempts to put together systems that behave like living organisms.

ALife embraces human-made systems that possess some of the fundamental properties of natural life.[77] While studying ALife, we are specifically interested in artificial systems that serve as models of living systems for the investigation of open questions in biology. Natural life on earth is organized into at least four fundamental levels of structure: the molecular level, the cellular level, the organism level and the population-ecosystem level. A living creature at any of these levels is a complex adaptive system exhibiting behavior that emerges from the interaction of a large number of elements from the levels below. Dealing with this multilevel complexity requires a broad methodological shift in the biological sciences today as a new collection of ALife models of natural biological systems. ALife thus amounts to the practice of "synthetic biology". By analogy with synthetic chemistry, we may note that the attempt to recreate biological phenomena in alternative media will result in not only better theoretical understanding of the phenomena under study, but also in practical applications of biological principles in the technology of computer hardware and software, mobile robots, spacecraft, medicine, nanotechnology, industrial fabrication and other vital engineering projects.

### 5.2. *Particle swarm optimization (PSO)*

James Kennedy and Russel C. Eberhart introduced the concept of function-optimization by means of a particle swarm in 1995.[78] PSO[79,80] is in principle such a multi-agent parallel search technique. Particles are conceptual entities, which fly through the multi-dimensional search space. In PSO a population of particles is initialized with random positions $X_i$ and velocities $V_i$, and a fitness function, $f$, is evaluated, using the particle's positional coordinates as input values. In an $n$-dimensional search space, $X_i = (x_{i1}, x_{i2}, x_{i3}, \ldots, x_{in})$ and $V_i = (v_{i1}, v_{i2}, v_{i3}, \ldots, v_{in})$. Positions and velocities are adjusted, and the function is evaluated with the new coordinates at each time-step. The velocity and position update equations for the $d$th dimension of $i$th particle in the swarm may be given as follows:

$$V_{id}(t+1) = \omega \cdot V_{id}(t) + C_1 \cdot \varphi_1 \cdot (P_{lid} - X_{id}(t))$$
$$+ C_2 \cdot \varphi_2 \cdot (P_{gd} - X_{id}(t)) \qquad (6)$$
$$X_{id}(t+1) = X_{id}(t) + V_{id}(t+1).$$

The variables $\varphi_1$ and $\varphi_2$ are random positive numbers, drawn from a uniform distribution, and with an upper limit $\varphi_{max}$, which is a parameter of the system. $C_1$ and $C_2$ are called acceleration constants, and $\omega$ is the inertia weight. $P_{li}$ is the best solution found so far by an individual particle, while $P_g$ represents the fittest particle found so far in the entire community.

The following figure illustrates how an individual particle (marked in the figure as a humanoid agent) moves over a three-dimensional fitness landscape.

### 5.3. Artificial immune system

In recent years, much attention has been focused on behavior-based AI for its proven robustness and flexibility in a dynamically changing environment. Artificial Immune Systems (AIS)[81,82] is one such behavior-based reactive system that aims at developing a decentralized consensus-making mechanism, following the behavioral characteristics of biological immune system.

The basic components of the biological immune system are macrophases, antibodies and lymphocytes, the last one being classified into two types: B-lymhocytes and T-lymhocytes are the cells stemming from the bone marrow. The human blood circulatory system contains roughly $10^7$ distinct types of B-lymphocytes, each of which has a distinct molecular structure and produces Y-shaped antibodies from its surface. Antibodies can recognize foreign substances, called antigens that invade living creature. Virus, cancer cells etc. are typical examples of antigens. To cope with continuously changing environment, living system possess enormous repertoire of antibodies in advance.

The AIS draws inspirations from theoretical immunology and observed immune functions, principles and models, and apply them to solving many complex engineering problems. The pioneering task of AIS is to detect and eliminate non-self materials or antigens such as virus or cancer cells. It also plays a great role to maintain its own system against dynamically changing environment thus providing a new methodology suitable for dynamic problems dealing with unknown/hostile environments. We present a very simple mathematical model of the AIS:

Let

$a_i(t)$ be the concentration of the $i$th antibody,
$m_{ji}$ be the affinity between antibody $j$ and antibody $i$,
$m_{ik}$ be the affinity between antibody $i$ and the detected antigen $k$,
$k_i$ be the natural decay rate of antibody $i$,
$N$ and $M$ respectively denote the number of antibodies that stimulate and suppress antibody $i$.

The growth rate of antibody $i$ is given below:

$$\frac{da_i}{dt} = \left\{ \sum_{j=1}^{N} m_{ji} a_j(t) - \sum_{k=1}^{N} m_{ik} a_k(t) - m_i - k_i \right\} a_i(t), \qquad (7)$$

and

$$a_i(t+1) = \frac{1}{1+\exp(0.5 - a_i(t))}. \tag{8}$$

The first and the second term in the right hand side of (7) respectively denote the stimulation and suppression by other antibodies respectively. The third term denotes the stimulation from the antigen, and the fourth term represents the natural decay of $i$th antibody. Equation (8) is a squashing function used to ensure the stability of the concentration. The selection of antibodies can be simply carried out by Roulette wheel bass according to the magnitude of concentration of the antibodies. It is to be noted that only one antibody is allowed to activate and act its corresponding behavior in its world.

### 5.4. *Rough sets and granular computing*

Founded by Pawlak,[83,84] the methodology of rough sets is concerned with the classificatory analysis of imprecise, uncertain or incomplete information or knowledge expressed in terms of data acquired from experience. The primary notions of the theory of rough sets are the approximation space and lower and upper approximations of a set. The approximation space is a classification of the domain of interest into disjoint categories. The classification formally represents our knowledge about the domain, i.e. the knowledge is understood here as an ability to characterize all classes of the classification, for example, in terms of features of objects belonging to the domain. Objects belonging to the same category are not distinguishable, which means that their membership status with respect to an arbitrary subset of the domain may not always be clearly definable. This fact leads to the definition of a set in terms of lower and upper approximations. The lower approximation is a description of the domain objects, which are known with certainty to belong to the subset of interest, whereas the upper approximation is a description of the objects that possibly belong to the subset.

We can approximately represent a rough set $A$ by its upper and lower approximations $R_1(A)$ and $R_2(A)$ respectively. The boundary $R_2(A)$ attempts to approximate $A$ from inside, while boundary $R_1(A)$ attempts to approximate $A$ from outside. The idea is illustrated in Fig. 12.

Granular computing[85] is another emerging conceptual and computational paradigm of information processing. It concerns processing of

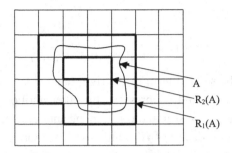

Fig. 12. The rough boundaries $R_1(A)$ and $R_2(A)$ of a given point set $A$.

complex information entities called "information granules", which arise in the process of abstraction of data and derivation of knowledge from information; this process is called information granulation. Information granules are collection of entities, usually originating at the numeric levels that are arranged together due to their similarity, functional adjacency, indistinguishability, coherency or the like. Granular computing can be conceived as a category of theories, methodologies, techniques and tools that make use of information granules in the process of problem solving. The theoretical foundations of granular computing involve set theory (interval analysis), fuzzy sets, rough sets and random sets.

### 5.5. *Chaos theory*

Chaos theory describes the behavior of certain nonlinear dynamical systems that under certain conditions exhibit a peculiar phenomenon known as chaos. One important characteristic of the chaotic systems is their sensitivity to initial conditions (popularly referred to as the butterfly effect). Because of this sensitivity, the behavior of these systems appears to be random, even though the dynamics is deterministic in the sense that it is well defined and contains no random parameters. Examples of such systems include the atmosphere, the solar system, plate tectonics, turbulent fluids, economics, and population growth.

Currently, fuzzy logic and chaos theory form two of the most intriguing and promising areas of mathematical research. Recently, fuzzy logic and chaos theory have merged to form a new discipline of knowledge, called fuzzy chaos theory.[86] The detailed implications of fuzzy chaotic models are out of the scope of the present chapter.

### 5.6. *Ant colony systems (ACS)*

It is a natural observation that a group of 'almost blind' ants can figure out the shortest route between a cube of sugar and their nest without any visual information. They are capable of adapting to the changes in the environment as well.[87] It is interesting to note that ants while crawling deposit trails of a chemical substance known as pheromone to help other members of their team to follow its trace. The resulting collective behavior can be described as a loop of positive feedback, where the probability of an ant's choosing a path increases as the count of ants that already passed by that path increases.[87,88]

The basic idea of a real ant system is illustrated in Fig. 13. In the left picture, the ants move in a straight line to the food. The middle picture illustrates the situation soon after an obstacle is inserted between the nest and the food. To avoid the obstacle, initially each ant chooses to turn left or right at random. Let us assume that ants move at the same speed depositing pheromone in the trail uniformly. However, the ants that, by chance, choose to turn left will reach the food sooner, whereas the ants that go around the obstacle turning right will follow a longer path, and so will take longer time to circumvent the obstacle. As a result, pheromone accumulates faster in the shorter path around the obstacle. Since ants prefer to follow trails with larger amounts of pheromone, eventually all the ants converge to the shorter path around the obstacle, as shown in the right picture.

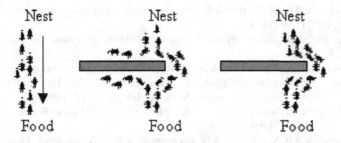

**Fig. 13.** Illustrating the behavior of real ant movements.

An artificial Ant Colony System (ACS) is an agent-based system, which simulates the natural behavior of ants and develops mechanisms of cooperation and learning. ACS was proposed by Dorigo et al.[89] as a new heuristic to solve combinatorial optimization problems. This new heuristic, called Ant Colony Optimization (ACO) has been found to be both robust and versatile in handling a wide range of combinatorial optimization problems.

## 6. Summary

The chapter introduced different fundamental components of soft computing and discussed their scope of possible synergism. It is clear from the discussions that fuzzy logic is a fundamental tool for reasoning with approximate data and knowledge. Neural network plays a significant role in machine learning and GA has an extensive application in intelligent search and optimization problems. Belief networks are capable of propagating beliefs of an event node based on the probabilistic support of its cause and effect nodes in the causal tree/graph. The chapter also provided a list of possible synergism of 2 or more computational models that fall under the rubric of soft computing. It ended with the brief exposure to some very recently developed methodologies, which are gaining importance in the realm of soft computing at a rapid pace.

## References

1. L. C. Jain and B. Lazzerini (eds.), *Knowledge-Based Intelligent Techniques in Character Recognition*, CRC Press (1999).
2. L. C. Jain and N. M. Martin (eds.), *Fusion of Neural Networks, Fuzzy Sets and Genetic Algorithms: Industry Applications*, CRC Press (1998).
3. L. C. Jain and C. W. De Silva (eds.), *Intelligent Adaptive Control: Industrial Applications*, CRC Press (1998).
4. L. C. Jain (ed.), *Intelligent Biometric Techniques in Fingerprint and Face Recognition*, CRC Press (1999).
5. M. Jamshidi, A. Titli, L. Zadeh and S. Boverie (eds.), *Applications of Fuzzy Logic: Towards High Machine Intelligence Quotient Systems*, Prentice-Hall (1997).
6. M. M. Mitchell, *Machine Learning*, McGraw-Hill, pp. 81–127 (1997).
7. D. W. Patterson, *Introduction to Artificial Intelligence and Expert Systems*, Prentice-Hall, pp. 107–119 (1990).
8. S. Russel and P. Norvig, *Artificial Intelligence: A Modern Approach*, Prentice-Hall, pp. 598–644 (1995).

9. E. H. Shortliffe and B. G. Buchanan, A model of inexact reasoning, *Mathematical Biosciences* **23**, 351–379 (1975).
10. E. H. Shortliffe, *Computer Based Medical Consultations: MYCIN*, American Elsevier (1976).
11. B. Biswas, A. Konar and A. K. Mukherjee, Image matching with fuzzy moment descriptors, *Engineering Applications of Artificial Intelligence* **14**, 43–49 (2001).
12. A. Konar and A. K. Mandal, Uncertainty management in expert systems using fuzzy petri nets, *IEEE Trans. on Knowledge and Data Engineering* **8**(1) (1996).
13. P. Saha and A. Konar, A heuristic approach for computing the max-min inverse fuzzy relation, *Int. J. of Approximate Reasoning* **30**, 131–147 (2002).
14. A. Konar, *Artificial Intelligence and Soft Computing: Behavioral and Cognitive Modeling of the Human Brain*, CRC Press (1999).
15. G. J. Klir and B. Yuan, *Fuzzy Sets and Fuzzy Logic: Theory and Applications*, Prentice-Hall (1995).
16. T. J. Ross, *Fuzzy Logic with Engineering Applications*, McGraw-Hill (1995).
17. L. A. Zadeh, Outline of a new approach to the analysis of complex systems and decision processes, *IEEE Trans. Systems, Man and Cybernetics* **3**, 28–45 (1973).
18. L. A. Zadeh, *Fuzzy Sets, Information and Control* **8**, 338–353 (1965).
19. L. A. Zadeh, *Fuzzy Logic, Neural Networks and Soft Computing*, 1 page course announcement of CS 294-4, Spring 1993, University of California at Berkeley (1992).
20. J. R. Quinlan, Induction of decision trees, *Machine Learning* **1**(1), 81–106.
21. P. Winston, *Learning Structural Descriptions from Examples*, PhD dissertation, MIT Technical Report AI-TR-231 (1970).
22. D. E. Rumelhart and J. L. McClelland, *Parallel Distributed Processing: Exploring in the Microstructure of Cognition*, MIT Press, Cambridge (1986).
23. D. E. Rumelhart, G. E. Hinton and R. J. Williams, Learning representations by back-propagation errors, *Nature* **323**, 533–536 (1986).
24. P. D. Wasserrman, *Neural Computing: Theory and Practice*, Van Nostrand Reinhold, New York, pp. 49–85 (1989).
25. B. Widrow and M, E. Hoff, Adaptive switching circuits, *1960 IRE WESCON Convention Record*, Part 4, pp. 96–104 (1960).
26. B. Widrow, Generalization and information storage in networks of ADALINE neurons, *Self-Organizing Systems*, M. C. Yovits, G. T. Jacobi and G. D. Goldstein, (eds), pp. 435–461 (1962).
27. R. J. Williams, On the use of back-propagation in associative reinforcement learning, *IEEE Int. Conf. on Neural Networks, NY* **1**, 263–270 (1988).
28. J. A. Anderson, A simple neural network generating an associative memory, *Mathematical Biosciences* **14**, 197–220 (1972).
29. E. A. Bender, *Mathematical Methods in Artificial Intelligence*, IEEE Computer Society Press, Los Alamitos, pp. 589–593 (1996).
30. G. A. Carpenter and S. Grossberg, A massively parallel architecture for a self-organizing neural pattern recognition machine, *Computer Vision, Graphics and Image Processing* **37**, 54–115 (1987).
31. G. A. Carpenter and S. Grossberg, ART2: self-organization of stable category recognition codes for analog input patterns, *Applied Optics* **23**, 4919–4930 (1987).
32. J. L. Castro, M. J. Carlos and J. M. Benitez, Interpretation of artificial neural networks by means of fuzzy rules, *IEEE Trans. on Neural Networks* **13**(1) (2002).

33. L. I. M. Fu, *Neural Networks in Computer Intelligence*, McGraw-Hill (1994).
34. S. Haykin, *Neural Networks: A Comprehensive Foundation*, Prentice- Hall (1999).
35. J. Hertz, A. Krogn and G. R. Palmer, *Introduction to the Theory of Neural Computation*, Addison-Wesley (1990).
36. T. Kohonen, G. Barna and R. Chrisley, Statistical pattern recognition using neural networks: benchmarking studies, *IEEE Conf. on Neural Networks*, San Diego **1**, 61–68.
37. T. Kohonen, *Self-Organization and Associative Memory*, Springer-Verlag (1989).
38. Z. Michalewicz, *Genetic Algorithms + Data Structures = Evolution Programs*, Springer-Verlag (1992).
39. M. M. Mitchell, *Machine Learning*, McGraw-Hill, New York, pp. 81–127 (1997).
40. M. M. Mitchell, *An Introduction to Genetic Algorithms*, MIT Press, Cambridge (1996).
41. H. Muehlenbein and U. K. Chakraborty, Gene pool recombination genetic algorithm and the onemax function, *Journal of Computing and Information Technology* **5**(3), 167–182 (1997).
42. M. Srinivas and L. M. Patnaik, Genetic search: analysis using fitness moments, *IEEE Trans. on Knowledge and Data Eng.* **8**(1), 120–133 (1996).
43. M. D. Vose and G. E. Liepins, Punctuated equilibrium in genetic search, *Complex Systems* **5**, 31–44 (1991).
44. M. D. Vose, *Genetic Algorithms*, MIT Press (1999).
45. J. R. Mc Donell, Control in *Handbook of Evolutionary Computation*, T. Back, D. B. Fogel and Z. Michalewicz (eds.), IOP and Oxford University Press (1998).
46. W. Pedrycz, *Fuzzy Sets Engineering*, CRC Press, pp. 73–106 (1996).
47. G. Antoniou, *Nonmonotonic Reasoning*, MIT Press (1997).
48. B. Kosko, *Neural Networks and Fuzzy Systems: A Dynamical Systems Approach to Machine Intelligence*, Prentice-Hall (1991).
49. W. Pedrycz and F. Gomide, *An Introduction to Fuzzy Sets: Analysis and Design*, MIT Press (1998).
50. S. K. Pal and S. Mitra, *Neuro-Fuzzy Pattern Recognition: Methods in Soft Computing*, John Wiley & Sons, Inc. (1999).
51. R. J. Schalkoff, *Artificial Neural Networks*, McGraw-Hill, New York, pp. 146–188 (1997).
52. D. E. Rumelhart and J. L. McClelland, *Parallel Distributed Processing: Exploring in the Microstructure of Cognition*, MIT Press, Cambridge, MA (1986).
53. B. Kosko, Adaptive bi-directional associative memories, *Applied Optics* **26**, 4947–4960 (1987).
54. B. Kosko, Bi-directional associative memories, *IEEE Trans. on Systems, Man and Cybernetics* **SMC-18**, 49–60 (1988).
55. A. Konar and S. Pal, Modeling cognition with fuzzy neural nets, *Neural Network Systems: Techniques and Applications*, Leondes, C. T. (ed.), Academic Press, New York, pp. 1341–1391 (1999).
56. J. J. Hopfield, Neural networks with graded response have collective computational properties like those of two state neurons, *Proc. Natl. Acad. Sci.* **81**, 3088–3092 (1984).
57. J. Hopfield, Neural nets and physical systems with emergent collective computational abilities, *Proc. Natl. Acad. Sci.* **79**, 2554–2558 (1982).
58. B. Paul, A. Konar and A. K. Mandal, Fuzzy ADALINEs for gray image recognition, *Neurocomputing* **24**, 207–223 (1999).

59. R. J. Williams and J. Peng, Reinforcement learning algorithm as function optimization, *Proc. Int. Joint Conf. on Neural Networks, NY* **II**, 89–95 (1989).
60. J. H. Holland, *Adaptation in Natural and Artificial Systems*, University of Michigan Press, Ann Arbor (1975).
61. D. E. Goldberg, *Genetic Algorithms in Search, Optimization and Machine Learning*, Addison-Wesley, Reading, MA (1989).
62. K. A. De Jong, *An Analysis of Behavior of a Class of Genetic Adaptive Systems*, Doctoral dissertation, University of Michigan (1975).
63. T. E. Davis and J. C. Principa, A Markov chain framework for the simple genetic algorithm, *Evolutionary Computation* **1**(3), 269–288 (1993).
64. U. K. Chakraborty, and D. G. Dastidar, Using reliability analysis to estimate the number of generations to convergence in genetic algorithm, *Information Processing Letters* **46**, 199–209 (1993).
65. U. K. Chakraborty and H. Muehlenbein, Linkage equilibrium and genetic algorithms, *Proc. 4th IEEE Int. Conf. on Evolutionary Computation*, Indianapolis, pp. 25–29 (1997).
66. U. K. Chakraborty, K. Deb and M. Chakraborty, Analysis of selection algorithms: a Markov chain approach, *Evolutionary Computation* **4**(2), 133–167 (1996).
67. D. B. Fogel, *Evolutionary Computation*, IEEE Press, Piscataway, NJ (1995).
68. D. W. Patterson, *Introduction to Artificial Intelligence and Expert Systems*, Prentice-Hall, Englewood Cliffs, NJ, pp. 107–119 (1990).
69. J. Pearl, Distributed revision of composite beliefs, *Artificial Intelligence* **33**, 173–213 (1987).
70. J. Pearl, Fusion, propagation and structuring in belief networks, *Artificial Intelligence* **29**, 241–288 (1986).
71. Y. Shoham, *Artificial Intelligence Techniques in PROLOG*, Morgan Kaufmann, San Mateo, CA, pp. 183–185 (1994).
72. A. Konar and L. C. Jain, An introduction to computational intelligence paradigms, *Practical Applications of Computational Intelligence Techniques*, L. C. Jain and P. D. Wilde (eds.), Kluwer (2001).
73. H. N. Teodorescu, A. Kandel and L. C. Jain (eds.), *Fuzzy and Neuro-Fuzzy Systems in Medicine*, CRC Press, London (1999).
74. D. W. Tank and J. J. Hopfield, Simple neural optimization networks: an A/D converter, signal decision circuit and a linear programming circuit, *IEEE Trans. on Circuits and Systems* **33**, 533–541 (1986).
75. J. Sil and A. Konar, Reasoning using a probabilistic predicate transition net model, *Int. J. of Modeling and Simulations* **21**(2), 155–168 (2001).
76. C. G. Langton (ed.), *Artificial Life: An Overview*, MIT Press, Cambridge, MA (1995).
77. C. Taylor and D. Jefferson, Artificial Life as a tool for biological inquiry, *Artificial Life: An Overview*, C. G. Langton (ed.), MIT Press, MA (1995).
78. J. Kennedy and R. C. Eberhart, Particle swarm optimization, *Proceedings of the 1995 IEEE International Conference on Neural Networks*, IEEE Press **4**, 1942–1948 (1995).
79. J. Kennedy and R. C. Eberhart, *Swarm Intelligence*, ISBN 1-55860-595-9, Academic Press (2001).

80. R. C. Eberhart and Y. Shi, Particle swarm optimization: developments, applications, and resources, *Proceedings of the 2001 Congress on Evolutionary Computation CEC2001*, IEEE, pp. 81–86 (2001).
81. N. K. Jerne, The immune systems, *Scientific American* **229**(1), 52–60 (1973).
82. N. K. Jerne, The generative grammar of the immune system, *EMBO Journal* **4**(4) (1983).
83. Z. Pawlak, Rough sets, *Int. J. of Computer and Information Science* **11**, 341–356 (1982).
84. Z. Pawlak, *Rough Sets: Theoretical Aspects of Reasoning About Data*, Kluwer Academic Press, Dordrecht (1991).
85. W. Pedrycz, *Granular Computing: An Introduction*, Kluwer Academic Press, Dordrecht (2002).
86. J. J. Buckley, Fuzzy dynamical systems, *Proc. IFSA'91*, Brussels, Belgium, pp. 16–20 (1991).
87. M. Dorigo and L. M. Gambardella, A study of some properties of ant Q, *Proc. PPSN IV — 4th Int. Conf. Parallel Problem Solving From Nature*, Berlin, Germany: Springer-Verlag, pp. 656–665 (1996).
88. M. Dorigo, G. Di Caro and L. M. Gambardella, Ant algorithms for discrete optimization, *Artif. Life* **5**(2), 137–172 (1999).
89. M. Dorigo and L. M. Gambardella, Ant colony system: a cooperative learning approach to the traveling salesman problem, *IEEE Trans. Evol. Comput.* **1**, 53–66 (1997).

# II.
# BIOLOGICAL SEQUENCE AND STRUCTURE ANALYSIS

# CHAPTER 3

# RECONSTRUCTING PHYLOGENIES WITH MEMETIC ALGORITHMS AND BRANCH-AND-BOUND

José E. Gallardo*, Carlos Cotta† and Antonio J. Fernández‡

*Dept. Lenguajes y Ciencias de la Computación, ETSI Informática*
*University of Málaga, Campus de Teatinos*
*29071 — Málaga, Spain*
*\*pepeg@lcc.uma.es*
*†ccottap@lcc.uma.es*
*‡afdez@lcc.uma.es*

A phylogenetic tree represents the evolutionary history for a collection of organisms. We consider the problem of inferring such a tree given a certain set of data (genomic, proteomic, or even morphological). Given the computational hardness of this problem, exact approaches are inherently limited. However, exact techniques can still be useful to endow heuristic approaches with problem-awareness. We analyze this hybridization in the context of memetic algorithms and branch-and-bound techniques. Focusing in the ultrametric model for phylogenetic inference, we show that this combination can be synergetic. We analyze the parameters involved in this hybrid model, and determine a robust setting for these. A summary of related work is also provided.

## 1. Introduction

Phylogenetic trees provide a hierarchical representation of the degree of closeness among a set of organisms. The reconstruction of phylogenetic trees from data is undoubtedly a task of paramount importance in molecular biology. It has direct implications in areas such as multiple sequence alignment,[20] protein structure prediction[53] or molecular of viruses,[48] just to cite a few. Unfortunately, this task turns out to be very difficult from the computational point of view. First of all, the phylogeny problem is intrinsically complex: *NP*-hardness has been shown for phylogenetic inference under several models.[11–13,17,62] Secondly, while the utilization of a quality measure for evaluating hierarchies implies the definition of an optimization problem, its global optimum does not have the same significance as in other classical problems: the existence of some uncertainty in the underlying empirical data may make high-quality suboptimal solutions equally valid.

Due to the reasons just mentioned, the use of classical exact techniques can be considered generally inappropriate in this context. Indeed, the use of heuristic techniques in this domain seems much more adequate. These can range from simple constructive heuristics (e.g. greedy agglomerative techniques such as UPGMA[58]) to complex metaheuristics (e.g. evolutionary algorithms[9]). At any rate, it is well-known that any heuristic method is going to perform in strict accordance with the amount of problem-knowledge it incorporates.[10,61] In this sense, while classical exact techniques are not adequate as stand-alone solvers for this problem, they can still be very useful in creating powerful problem-aware metaheuristics. We will precisely explore this possibility here, presenting a model for the integration of branch-and-bound techniques (BnB)[35] and memetic algorithms (MAs).[45–47] As it will be shown, this model can result in a synergistic combination yielding better results than those of the intervening techniques alone.

## 2. A Crash Introduction to Phylogenetic Inference

The inference of phylogenetic trees is one of the most important and challenging tasks in Systematic Biology. Such trees are used to represent the evolutionary history of a collection of $n$ organisms (or taxa) from their molecular sequence data, or from other form of dissimilarity information. The Phylogeny Problem can then be formulated as finding the phylogenetic tree that best — under a certain optimality criterion — represents the evolutionary history of a collection of taxa. For this purpose, it is clearly necessary to define an optimization criterion. Essentially, optimization criteria for assessing the goodness of a phylogenetic tree $T$ can fall within two major categories, *sequence*-based and *distance*-based.[29]

In sequence-based approaches, each node of $T$ is assigned a sequence. Such a sequence is known for the leaves (i.e. the taxa being classified) and can be inferred via pairwise alignments for internal nodes. Subsequently, the tree is evaluated using a criterion that in most situations is either *maximum likelihood* (ML) or *maximum parsimony* (MP). ML criteria are based on the assumption of a stochastic model of evolution,[14] e.g. the Jukes-Cantor model,[27] the Kimura 2-parameter model,[31] etc. Such a model is used in order to assess the likelihood that the current tree generated the observed data. The optimal tree would be the one that maximizes this likelihood. On the other hand, MP is grounded on Occam's razor, whereby given two equally predictive theories, the simpler should be chosen. Thus, the MP criterion specifies that the tree requiring the fewest number of evolutionary changes to explain the data is preferred.

As to distance-based approaches, they are founded on transforming the available sequence data into an $n \times n$ matrix $M$. This matrix is the unique information subsequently used. More precisely, edges in $T$ are assigned a weight. We denote $w(T, e)$ as the weight of edge $e$ in $T$. The basic idea here is that $M_{ij}$ represents the *evolutionary distance* or *dissimilarity* between taxa $i$ and $j$. We thus have an *observed* distance matrix $M$ (the input data) and an *inferred* distance matrix $\hat{M}$ obtained by making $\hat{M}_{ij}$ = distance from $i$ to $j$ in $T$. Let us represent trees using a LISP-like notation, i.e. trees are represented as $(r, L, R)$, where $r$ is the root, $L$ is the left subtree, and $R$ is the right subtree. A leaf $l$ is represented as $(l)$. Also, let $\mathcal{V}(T)$ and $\mathcal{E}(T)$ respectively denote the set of vertices and edges of tree $T$, and let $\mathcal{L}(T)$ denote the set of all leaves of tree $T$. We further use $T(v)$, where $v \in \mathcal{V}(T)$, to denote the subtree of $T$ rooted at node $v$, and $\pi(T, v)$ to denote the parent of $v$ in $T$. Then, $\hat{M}_{ij} = \Delta(T, i, j) + \Delta(T, j, i)$, where

$$\Delta(T, a, b) = \begin{cases} w(T, (a', a)) + \Delta(T, a', b) \\ \quad \text{if } b \notin \mathcal{L}(T(a)), \quad a' = \pi(T, a). \\ 0 \quad \text{if } b \in \mathcal{L}(T(a)) \end{cases} \quad (1)$$

The quality of the tree can now be quantified in a variety of ways. On one hand, it is possible to consider some meta-distance measure between observed and inferred distances, e.g. the $L_2$ metric:

$$L_2(M, \hat{M}) = \sum_{i=1}^{n} \sum_{j=1}^{n} (M_{ij} - \hat{M}_{ij})^2, \quad (2)$$

thus seeking a least-squares approximation of the observed distance. To do so, a quadratic program must be solved in order to find the best edge weights for a given topology.[40] This may be computationally demanding when used within a search-and-score method.

On the other hand, quality can be directly measured from $T$. This is typically the case when edge-weighting has been constrained so as to have $\hat{M}_{ij} \geqslant M_{ij}$, i.e. to have inferred distances greater than observed distances. This constraint is based on the fact that the observed distance between two taxa will always be a lower bound of the real evolutionary distance between them.[a] In this situation, minimizing the total weight of $T$ (i.e. the sum of all edge-weights) is usually the criterion. Notice that by taking $M_{ij}$ as the minimum number of evolutionary events needed to transform $i$ in $j$, this approach resembles MP. Actually, distance-based methods can be generally considered as an intermediate strategy between ML and MP, exhibiting good performance in practice as well.[24] For these reasons, we have focused on distance-based approaches in this work. To be precise, we have forced $\hat{M}$ to be *ultrametric*.[2] In this case, it holds that

$$\hat{M}_{ij} \leqslant \max\{\hat{M}_{ik}, \hat{M}_{jk}\}, \quad 1 \leqslant i, j, k \leqslant n. \tag{3}$$

If $\hat{M}$ is ultrametric, the distance in $T$ between any internal node $h$ and any leaf $l$ descendant of $h$ is the same — e.g. see Fig. 1 where the optimum ultrametric tree for an arthropod dataset[26] is depicted. This distance represents in each case the elapsed time since the corresponding evolutionary divergence event. Hence, a constant evolution rate is implicitly assumed in the line of the molecular-clock hypothesis.[39] Although this hypothesis is not in vogue nowadays, this condition still provides a very good approximation to the optimal solution under more relaxed assumptions such as mere additivity.[b] Much more important, it is also easy to compute: for a given tree $T$, and observed matrix $M$, edge weights can be determined in $O(n^2)$ time.[62] This can be solved as follows: let the *height* of a subtree be computed as:

$$height(T) = \begin{cases} 0 & \text{if } T = (l) \\ \max\left(height(L), height(R), \frac{D(L,R)}{2}\right) & \text{if } T = (h, L, R) \end{cases}, \tag{4}$$

where $D(L, R) = \max\{M_{ij} \mid i \in \mathcal{L}(L), j \in \mathcal{L}(R)\}$. Now, the weight of an edge connecting two nodes $i$ and $j$ is easily computed as

$$w(T, (i, j))|height(T(i)) - height(T(j))|. \tag{5}$$

---

[a] In essence, this is due to the existence of a set of phenomena — reversal, parallelism, and convergence — that make taxa appear more related than they really are. These phenomena are collectively termed *homoplasy*.[39]

[b] $\hat{M}$ is additive if for any $i, j, k, l$, the maximum of $\hat{M}_{ij} + \hat{M}_{kl}, \hat{M}_{ik} + \hat{M}_{jl}, \hat{M}_{il} + \hat{M}_{jk}$ is not unique.

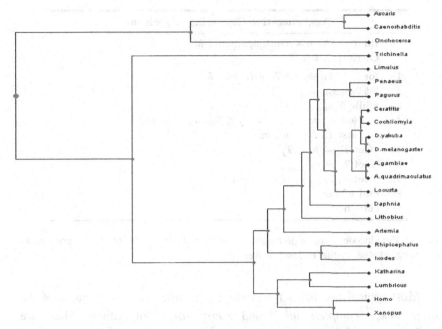

**Fig. 1.** An example of ultrametric tree. Branch lengths represent the elapsed evolutionary time between the corresponding species. Notice that the horizontal distance from an interior node to any of its leaves is the same.

Unlike this simple computation, finding optimal edge weights for an additive tree requires solving a linear program with $2n - 2$ variables (the number of internal edges), and $n(n - 1)/2$ binary constraints, one for each pair of taxa $(i,j)$, corresponding to $\hat{M}_{ij} \geq M_{ij}$.

Notice that the number of possible phylogenetic trees for a given set of $n$ taxa is huge: there are $(2n - 3)!!$ rooted trees,[24] where $k!!$ is the double factorial of $k$ (i.e. the difference between successive factors is 2 rather than 1 as in the standard factorial). For example, there exist $8.2 \times 10^{21}$ possible trees for 20 taxa. This clearly illustrates the impossibility of applying exhaustive search to this problem. Furthermore, finding provably good solutions constitutes a very hard combinatorial optimization problem for most optimality criteria, as anticipated in the previous section. Exact techniques such as branch-and-bound can be used, but they are computationally unaffordable for even moderate-size (say, 30–40 taxa) problem instances. Hence, the use of heuristic techniques seems appropriate.

|  |
| :--- |
| **Agglomerative Clustering Algorithm** |
| INPUT:   An $n \times n$ distance matrix $M$.<br>OUTPUT: A tree $T$.<br>1:   **for** $i = 1 : n$ **do let** $T_i \leftarrow (i)$ **end for**<br>2:   **let** $N_{clusters} \leftarrow n$<br>3:   **while** $N_{clusters} > 1$ **do**<br>4:        Select $T_i$ and $T_j$ $(1 \leqslant i < j \leqslant N_{clusters})$ for which $dist(T_i, T_j)$ is minimal.<br>5:        **let** $T_i \leftarrow (h, T_i, T_j)$<br>6:        **let** $T_j \leftarrow T_{N_{clusters}}$<br>7:        **let** $N_{clusters} \leftarrow N_{clusters} - 1$<br>8:   **end while**<br>9:   **return** $T_1$ |

**Fig. 2.** Pseudocode of an agglomerative clustering algorithm. In line 5, $h$ represents an undistinguishable internal node of the tree.

Most typical heuristics for phylogenetic inference are variants of the *single-link*,[57] *complete-link*,[32] and *average-link*[58] algorithms. These are agglomerative clustering algorithms whose functioning matches the generic template shown in Fig. 2. As it can be seen, these algorithms proceed by iteratively joining in a tree the two closest clusters, until just one group remains. They differ in the way intercluster distance is defined. To be precise, they consider the distance measures shown in Table 1. Notice that complete-link can be regarded as a greedy approach for the quality criterion we have chosen (the height of an internal node is always one half of the largest distance between any of its leaves — recall Eq. (4) — and complete-link makes local decisions trying to minimize this quantity).

Other popular heuristics for phylogenetic inference are based in metaheuristic approaches, such as for example evolutionary algorithms (EAs). In the next section will briefly overview previous work done in the application of EAs in this domain.

**Table 1.** Distance measures for determining the two *closest* clusters during agglomerative clustering.

| | | |
| ---: | :---: | :--- |
| Single-Link | : | $dist(T, T') = \min\{M_{ij} \mid i \in \mathcal{L}(T), j \in \mathcal{L}(T')\}$ |
| Complete-Link | : | $dist(T, T') = \max\{M_{ij} \mid i \in \mathcal{L}(T), j \in \mathcal{L}(T')\}$ |
| Average-Link | : | $dist(T, T') = \frac{1}{|\mathcal{L}(T)| \cdot |\mathcal{L}(T')|} \sum_{i \in \mathcal{L}(T), j \in \mathcal{L}(T')} M_{ij}$ |

## 3. Evolutionary Algorithms for the Phylogeny Problem

To the best of our knowledge, the first application of a genetic algorithm (GA) to phylogenetic inference was developed by Matsuda. The GA was used to construct phylogenetic trees from amino acid sequences using a ML approach.[41,42] It was shown that the performance of the method was comparable to those of other tree-construction methods (i.e. neighbor-joining,[54] MP, ML and UPGMA combined with different search algorithms). Later, an improved genetic algorithm was applied to *rbc*L sequence data of green plants.[38] The results were really promising as the GA required only 6% of the computational effort required by a conventional heuristic search using tree bisection/reconnection branch swapping to obtain the same maximum-likelihood topology.

Further work by Skourikhine[56] reported a self-adaptive genetic algorithm (GA) for the ML reconstruction of phylogenetic trees using nucleotide sequence data. The algorithm produced faster reconstructions of the trees with less computing power and automatic self-adjustment of settings of the optimization algorithm parameters. Other evolutionary proposals for searching the ML trees were developed by Meade *et al.*[43] and Katoh *et al.*[28] (considering simple GAs), and Lemmon and Milinkovitch[37] (using a multi-population GA). Parallel GAs under the ML optimality criterion were investigated by Brauer *et al.*[5] as well, with encouraging results. Also in the ML context, Shen and Heckendorn[55] showed that discretizing edge lengths changed the fundamental character of the search and could produce higher quality trees.

The representation issue has attracted a lot of interest in the application of EAs to the Phylogeny problem. Indeed, the choice of representation (and subsequently, the choice of operators) has a clear influence on the performance of the algorithm as shown by Reijmers *et al.*[51] A number of different EAs — based on the use of alternative representations (direct and indirect) and/or reproductive operators — were developed, compared and evaluated using a distance-based measure by Cotta and Moscato.[9] The conclusion was that directly evolving phylogenetic trees yields better results than indirect approaches using decoders. The exception to this rule was shown by a greedy permutational-decoder based EA that provided the overall best results, achieving near 100%-success at a lower computational cost than the remaining approaches. However this latter approach is less scalable than its direct counterpart. Poladian[49] also presented some results in this line of research and proposed different representations for the phenotype and the genotype in two existing algorithms for phylogenetic inference (i.e. neighbor-joining and ML) co-utilized within a GA. One conclusion reached is that working directly with the trees presents some disadvantages as for instance to establish the concept of distance between tree topologies. Also, separating the genotype from the phenotype requires the use of a good mapping from one space to the other (although this case allows the use of a different set of genetic operators). The direct representation is used in the Gaphyl package by Congdon,[6] whereas a permutational decoder is used by Kim et al.[30]

A potential difficulty for the inference problem may arise when different data sets provide conflicting information about the inferred 'best' tree(s). Poladian and Jermiin[50] have proposed the application of evolutionary multi-objective optimization algorithms (EMOOA) in this case. They showed that EMOOA can resolve many of the issues concerned with analyzing multiple data sets that give conflicting signals about evolutionary relationships.

There have been also evolutionary proposals based on the parsimony criterion. For example, Ribeiro and Vianna[52] described a genetic algorithm that makes use of an innovative optimized crossover strategy which is an extension of path relinking.[19] This proposal was shown to be computationally promising. Cotta[8] described an approach based on scatter search[34] that uses path relinking too (although not based on the parsimony criterion but on distance methods). Hill et al.[23] presented a GA-based approach for weighted maximum parsimony phylogenetic inference, especially for data sets involving a large number of taxa.

This summary of evolutionary approaches to phylogenetic inference is not exhaustive. The reader may check the survey by Fogel[16] for further information on applications of evolutionary computation to the inference of phylogenies, and highlights of new and current challenges for facing this problem.

## 4. A BnB Algorithm for Phylogenetic Inference

As it was stated in Sec. 2, the *min ultrametric tree with a given topology* problem ($MUTT(T)$) can be determined in time $O(n^2)$, for a topology

$T$ containing $n$ taxa. Hence, it is sufficient to solve this problem for each possible topology in order to find the best ultrametric tree (i.e. the one with minimum total weight — the MUT problem). However, this approach is not suitable in practice as the number of topologies increases exponentially with $n$. BnB algorithms can be used with the aim of reducing the size of the explored space. For this purpose, the BnB algorithm[62] (see Fig. 3) will explore a BnB tree (BBT) that contains in its leaves all possible topologies of ultrametric trees with leaf set $\{1, \ldots, n\}$. Let zero be the depth of the root of the BBT and $i + 1$ the depth of a child of a node with depth $i$. Therefore, internal nodes in the BBT at depth $i$ correspond to topologies with leaf set $\{1, \ldots, i + 2\}$. Observe that any topology corresponding to a node in the BBT at depth $i$ has $2 + 2i$ edges (see Fig. 4). Accordingly, a *branching rule* that inserts the next taxon into each edge of the current topology can be used to generate the BBT. In this way, a node at depth $i$ in the BBT has $2 + 2i$ children.

---

**Branch and Bound Algorithm for the MUT Problem**

INPUT: An $n \times n$ distance matrix $M$.
OUTPUT: The minimum ultrametric tree for $M$.

1: Relabel taxa such that $(1, 2, \ldots, n)$ is a maxmim permutation
2: Create the root node $R = (h, (1), (2))$ of the BBT representing the unique topology with leaves 1 and 2
3: let $U \leftarrow$ Complete Link$(M)$; $UB \leftarrow w(U)$
4: let *open* $\leftarrow \{R\}$
5: **while** *open* $\neq \emptyset$ **do**
6:    *open* $\leftarrow \{T \mid T \in \text{open}, LB(T) < UB\}$
7:    Extract some $T$ from *open*
8:    Compute $MUTT(T)$, $w(MUTT(T))$ and $LB(T)$
9:    **if** $\mathcal{L}(T) = \{1, 2, \ldots, n\}$ **then**
10:      **if** $w(MUTT(T)) < UB$ **then**
11:        let $U \leftarrow MUTT(T)$; $UB \leftarrow w(MUTT(T))$
12:      **end if**
13:    **else if** $LB(T) < UB$ **then**
14:      let *open* $\leftarrow$ *open* $\cup$ children of $T$
15:    **end if**
16: **end while**
17: **return** $U$

---

Fig. 3. Pseudocode of Branch and Bound Algorithm for the MUT Problem.

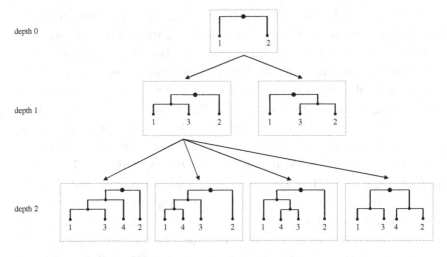

**Fig. 4.** Partial Branch and Bound Tree.

With the aim of pruning the BBT, the BnB maintains an upper bound corresponding to the weight of the best ultrametric tree found so far. For each internal node $T_i$ in the BBT corresponding to a minimum ultrametric tree with leaf set $\{1, \ldots, i\}$, a lower bound $LB(T_i)$ on the weight of the minimum ultrametric tree $T_n$ with leaf set $\{1, \ldots, n\}$ whose topology contains $T_i$ can be calculated using the following inequality:

$$w(T_n) \geqslant w(T_i) + \sum_{i<j\leqslant n} min\{M_{kj} \mid k < j\}/2. \tag{6}$$

If this bound exceeds the value of the current upper bound, the node $T_i$ and all its children can be safely pruned, as they will not lead to a better solution.

Initially, the complete-link algorithm described in Sec. 2 is used in line 3 to find a feasible solution, and the weight of this solution is set as the upper bound. Clearly, finding large lower bounds early in the exploration of the BBT is helpful as the number of pruned nodes increases during the execution of the BnB algorithm. With the aim of maintaining a lower bound as large as possible, a heuristic order for the $n$ taxa will be used. This way, as a preprocessing step, taxa are sorted to constitute a *maxmin* permutation, i.e. a permutation $(a_1, a_2, \ldots, a_n)$ of $\{1, \ldots, n\}$ such that $M_{a_1 a_2} = max(M)$ and $min_{k<i}\{M_{a_i a_k}\} \geqslant min_{k<i}\{M_{a_j a_k}\}$ for all $1 < i < j$. Note that a *maxmim* permutation for an $n \times n$ matrix can be found in $O(n^2)$ time.

As to the root node of the BBT, an interesting property regarding optimal ultrametric trees is that if $M_{uv} = max(M)$, then there exists a minimum ultrametric tree $T = (r, L, R)$, such that $u \in \mathcal{L}(L)$ and $v \in \mathcal{L}(R)$. Since taxa are organized in a *maxmim* permutation, $M_{12} = max(M)$, and hence the root of the BBT can be set as done in line 2.

The so generated BBT can be traversed in several ways. The most efficient (in terms of the number of iterations required to find the optimum and prove its optimality) is best-first, i.e. expanding firstly the most promising — according to the lower bound — nodes. However, the memory requirements can make this strategy unrealistic for large problem instances. The alternative is to use a depth-first traversal. This strategy does not require large amounts of memory, but can expand much more nodes than best-first. A third option is to use a breadth-first traversal (i.e. every node in a level is explored before moving to the next). In principle, this option would have the drawbacks of the previous two strategies, unless a heuristic choice is made: keep at each level just the best (according to some *quality* measure) $k$ nodes. This implies sacrificing exactness, but provides a very effective heuristic search approach. The name *beam search* (BS) has been coined to denote this strategy.[3,60]

Unfortunately, BnB algorithms alone need too much time to be practical, except for very small sets of taxa. As we will show, a hybrid algorithm based on this latter strategy (BS) and a MA can provide better performance.

## 5. A Memetic Algorithm for Phylogenetic Inference

This algorithm constitute a family of metaheuristics that blend together concepts from population-based techniques (e.g. evolutionary algorithms — EAs) and trajectory-based techniques (e.g. simulated annealing). Their central philosophy thus relies on two major pillars: individual improvement plus populational cooperation. More precisely, a MA can be characterized as a population of *agents* that alternate periods of self-improvement with periods of cooperation, and competition. The term 'agent' is purposefully used to indicate that the population constituents are not mere 'individuals', that is passive entities; on the contrary, they are active, and try to exploit all available knowledge about the target problem. This concern for adapting to the problem is backed up by ground-breaking theoretical results such as the *No Free Lunch Theorem*,[61] and it is ultimately responsible for the impressive success record of MAs.

Adaptation to the target problem is achieved via the use of constructive heuristics, local-search methods, exact techniques, etc. For this reason, it is often the case that MAs are used under different names, such as 'Lamarckian EAs', or more generally 'hybrid EAs'. For the particular application we consider, namely the inference of phylogenetic trees from distance matrices, our MA uses problem-specific operators for manipulating solutions, and an *ad hoc* local-improvement scheme. The overall process is shown in Fig. 5.

---

**Memetic Algorithm**

INPUT: An $n \times n$ distance matrix $M$.
OUTPUT: A tree $T$, inducing a near optimal matrix $\hat{M}$.

1: **for** $i = 1 : popsize$ **do**
2:   let $pop[i] \leftarrow \text{RANDOMTREE}(n)$
3:   EVALUATE($pop[i], M$)
4: **end for**
5: let $N_{eval} \leftarrow 0$.
6: **while** $N_{eval} < maxevals$ **do**
7:   **for** $i = 1 : offsize$ **do**
8:     **if** recombination is performed **then**
9:       let $parent_1 \leftarrow \text{SELECT}(pop); parent_2 \leftarrow \text{SELECT}(pop)$
10:       let $offspring[i] \leftarrow \text{RECOMBINATION}(parent_1, parent_2)$
11:     **else let** $offspring[i] \leftarrow \text{SELECT}(pop)$
12:     **end if**
13:     **if** mutation is performed **then**
14:       let $offspring[i] \leftarrow \text{MUTATE}(offspring[i])$
15:     **end if**
16:     Evaluate($offspring[i], M$)
17:     let $N_{eval} \leftarrow N_{eval} + 1$
18:     **if** local improvement is performed **then**
19:       let $offspring[i] \leftarrow \text{IMPROVE}(offspring[i])$
20:     **end if**
21:   **end for**
22:   $pop \leftarrow \text{REPLACE}(pop, offspring)$
23: **end while**
24: let $T \leftarrow \text{BEST}(pop)$.
25: **return** $T$

---

**Fig. 5.** Pseudocode of the memetic algorithm.

First of all, the population for this MA must be initialized. Each agent in the MA population represents a feasible tree, and hence the search is directly performed in the space of all possible phylogenetic trees.[9] These trees are initially created at random. Subsequently, the MA enters the main reproductive loop which comprises selection, recombination, mutation, local improvement, and replacement. Among these, the first and fifth components are rather problem independent, whereas the second to the fourth components are specifically crafted to the problem. Let us focus on these problem-aware operators.

The recombination operator is intended to produce new solutions by mixing the information comprised in several parental solutions. In this context, this mixing process amounts to the transference of tree-topology information from parents to offspring. To do so, a branch-exchange procedure can be used, i.e. a subtree can be pruned from one of the parents, and grafted onto the second one, much like it is done in the field of Genetic Programming.[33] Note however that phylogenetic trees are constrained to have exactly $n$ leaves, each one representing a different species. Therefore, the recombination operator has to be careful of removing duplicated elements. Taking into account these considerations, the Prune-Delete-Graft (PDG) tree crossover operator (see the pseudocode in Fig. 6) has been used.[9,44] PDG is a three-step procedure for recombing two trees $T_1$ and $T_2$ by pruning a subtree $T$ from $T2$, deleting from $T_1$ all leaves occurring in $T$, and grafting $T$ at a randomly selected point in $T1$. This is illustrated in Fig. 7.

---

**Prune-Delete-Graft Recombination**

---

INPUT:   Two trees, $T_1$ and $T_2$ to be recombined.
OUTPUT: A tree, corresponding to the offspring.

1:   Select a subtree $T$ from $T_2$.
2:   **for each** $l \in \mathcal{L}(T)$ **do**
3:     Find subtree U from $T_1$ such that $U = (h, (l), U')$
      or $U = (h, U', (l))$.
4:     Replace $U$ by $U'$ in $T_1$.
5:   **end for**
6:   Select a random subtree $V$ from $T_1$.
7:   Replace $V$ by $V' = (h', T, V)$ in $T_1$, where $h'$ is a new vertex.
8:   **return** $T_1$

---

Fig. 6. Pseudocode of the PDG recombination.

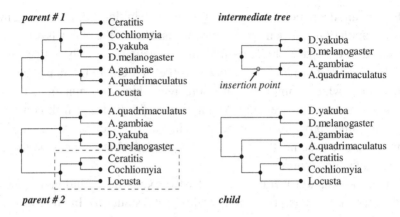

**Fig. 7.** An example of PDG recombination.

As to mutation, its purpose is to introduce new information (i.e. new topological relationships) in a certain solution. This can be accomplished in several ways. In this case, the MA considered uses the so-called Scramble operator that first selects a subtree $Q$ from the tree being mutated, and then rearranges its topology in a random way — see Fig. 8. Note that this mutation operator fulfills the previously mentioned constraint regarding the presence of exactly $n$ leaves in the tree.

Finally, local search has been applied to individuals in the population with the aim of improving solutions produced by the recombination and mutation methods. In general, this is achieved by applying small changes to a solution, keeping them if they produce a quality increase, or discarding them otherwise. Different strategies would have been possible given the tree

---

**Scramble Mutation**

INPUT:   A tree $T$ to be mutated.
OUTPUT: The same tree $T$ after mutation.

1: Select a random subtree $Q$ from $T$.
2: let $k \leftarrow |\mathcal{L}(Q)|$
3: let $R \leftarrow \text{RandomTree}(k)$
4: Relabel leaves in $\mathcal{L}(R)$ as leaves in $\mathcal{L}(Q)$.
5: Replace $Q$ by $R$ in $T$.
6: **return** $T$

---

**Fig. 8.** Pseudocode of the SCRAMBLE mutation.

representation used in this problem[1]. In this work, the improvement method chosen is based on performing rotations within the tree.

Four symmetric operations are used within the local search improvement method. The first one, $ROT_R^1$, is defined as

$$ROT_R^1[\,(h,(h',T_{LL},T_{LR}),T_R)\,](h,T_{LL},(h',T_{LR},T_R)), \qquad (7)$$

where $h$ and $h'$ are internal nodes. This operation moves $T_{LR}$, the right subtree of the left subtree of $h$, to the right so it becomes the left subtree of the right subtree of $h$. A $ROT_R^2$ operation would have performed the same movement on $T_{LL}$ rather than on $T_{LR}$.

Analogously, $ROT_L^1$ and $ROT_L^2$ are mirror-inverted versions of the previous operations. For each interior node of the tree, it is first checked whether a $ROT_R$ movement is possible, and if so, whether $ROT_R^1$ or $ROT_R^2$ produce an improvement. If this were the case, the change would be retained, and the improvement method would stop. Otherwise, $ROT_L$ movements would be analogously attempted. If no improvement is possible either, the procedure is recursively applied to the left and right subtrees of $h$.

The application of this local improvement strategy may be costly if applied to every new solution created by the MA. This circumstance has been also recognized in other domains,[7,25] where the use of partial Lamarckism has been proposed. We have opted for a variant of this strategy, performing local search only when the quality of a solution improves the current incumbent.

## 6. A Hybrid Algorithm

Branch-and-Bound and memetic algorithms represent two very different approaches for tackling the MUT problem. These approaches are not incompatible though. In this section, we describe a hybrid model that combines these two techniques. To be precise, it executes the BnB and MA in an interleaved way. The goal is combining synergistically these two different solving approaches, exploiting the capability of BnB for identifying provably good regions of the search space, and the power of the MA for exploring these. This hybrid algorithm is described in Fig. 9.

The algorithm starts by executing BS (i.e. *Beam Search*, see Sec. 4) for $l_0$ levels of the search tree. Afterwards, the MA and BS are interleaved until a termination condition is reached. Every time the MA is run, its population

| Hybrid Algorithm for the MUT Problem |
|---|
| 1:   **for** $l_0$ levels **do** run BS |
| 2:   **do** |
| 3:     **select** randomly *popsize* nodes from problem queue |
| 4:     **initialize** MA population with selected nodes |
| 5:     **run** MA |
| 6:     **if** MA solution < BS solution **then** |
| 7:       **let** BS solution $\leftarrow$ MA solution |
| 8:     **for** $l$ levels **do** run BS |
| 9:   **until** timeout **or** tree exhausted |
| 10: **return** BS solution |

Fig. 9. Pseudocode of the hybrid algorithm.

is initialized using random nodes in the BS queue. Let us note that nodes in the BS queue represent partial trees in which some taxa are included but another ones are not, so they must first be converted in full trees. For this purpose, the greedy completion algorithm described in Fig. 10 is applied to every partial tree.

The intended goal of the initialization explained above is to lead the MA to these regions of the search space (recall that the nodes in the queue represent subsets of the search space considered promising by the BnB; hence, the MA is used for finding probably good solutions in this region).

| Greedy Completion Algorithm |
|---|
| INPUT:   A tree $T$ and a list of nodes $ns$. |
| OUTPUT: The tree obtained by inserting all nodes in $ns$ into $T$. |
| 1:   **while** $ns \neq \varnothing$ **do** |
| 2:     Extract $n$ from $ns$ |
| 3:     **let** $best \leftarrow \infty$ |
| 4:     **for each** insertion point $p$ in $T$ **do** |
| 5:       **let** $T_{trial} \leftarrow$ insert $n$ at position $p$ of $T$ |
| 6:       **if** $w(T_{trial}) < best$ **then** |
| 7:         **let** $best \leftarrow w(T_{trial})$; $T_{best} \leftarrow T_{trial}$ |
| 8:       **end if** |
| 9:     **end for** |
| 10:    **let** $T \leftarrow T_{best}$ |
| 11:  **end while** |
| 12:  **return** $T$ |

Fig. 10. Pseudocode of the Greedy Completion Algorithm.

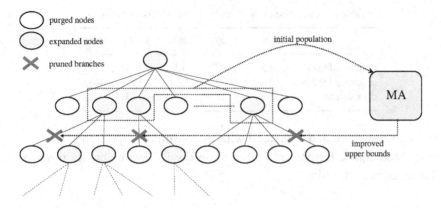

**Fig. 11.** Sketch of the hybrid algorithm.

Upon stabilization of the MA, control is returned to the BnB algorithm. The lower bound for the optimal solution obtained by the MA is then compared to the current incumbent in the BnB, updating the latter if necessary. This may lead to new pruned branches in the BS tree; see Fig. 11 for an outline of the general process. This is repeated until the search tree is exhausted or a time limit is reached.

Several parameters can be tuned to control the algorithm. The aim of parameters $l_0$ and $l$ is to control the balance between MA and BS. Parameter $l_0$ indicates how many levels the BS descends before starting running the MA. A lower value for $l_0$ would give more computation time to the MA component of the algorithm. On the other hand, parameter $l$ indicates how often, after this initial descent, should the MA be run. In this case, a low value for this parameter would execute more often the MA component of the algorithm, thus increasing its influence in the final outcome. Furthermore, parameter $k$ controls the maximum number of partial solutions maintained on each level of the BS. Increasing this parameter improves the quality of the results obtained by the BS component, but a very high value may delay the descent along the search tree.

## 7. Experimental Results

The evaluation of the hybrid model has been approached in two stages. Firstly, a sensitivity analysis was performed to determine an appropriate parameterization. Subsequently, the model was deployed on a full set of

Table 2. Description of the data sets used in the experimentation.

|  | M420 | M1097 | M877 | M971 | M808 |
|---|---|---|---|---|---|
| number of taxa | 85 | 107 | 134 | 158 | 178 |
| sequence length | 1016 | 2084 | 2684 | 1193 | 3453 |
| data source | 59 | 18 | 21 | 4 | 22 |

problem instances, and compared with its constituent algorithms as standalone techniques. Before presenting these results, lets us briefly describe the experimental setting.

### 7.1. Experimental setting

The experiments have been performed in a Pentium IV PC (2400 MHz and 512 MB of main memory). All algorithms were coded in C and compiled using gcc 3.2. For the MA and hybrid algorithm, the experiments were done using a steady-state evolutionary algorithm (*popsize* $= 100$, $p_m = 1/(2n-1)$, $p_X = 0.9$, binary tournament selection). With the aim of maintaining diversity, duplicated individuals were not allowed in the population.

The different algorithms have been applied to five data sets comprising real biological data. These have been downloaded from TreeBASE,[c] an online repository of publicly available data, and comprise DNA sequences for a number of taxa ranging from 85 up to 178 (see Table 2 for details). The selection of this test suite is intended to cover uniformly a reasonable range of instance sizes. These are well beyond the tenable limit for exact techniques such as branch and bound. In all cases, distance matrices have been computed using the DNADIST program of the Phylip package.[d] To be precise, the Kimura 2-parameter model (with default parameterization) has been used.

### 7.2. Sensitivity analysis on the hybrid algorithm

As noted in Sec. 6, the functioning of the hybrid algorithm can be tuned using several parameters. With the aim to determine good values for these parameters, a sensitivity analysis was done for problems in the data

---

[c]http://www.treebase.org
[d]http://evolution.genetics.washington.edu/phylip.html

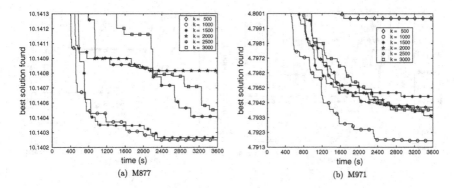

**Fig. 12.** Evolution of best fitness for different values of $k$ for different instances.

set. First of all, different values for $k$ (the search breadth in BS) were tested. Figure 12 shows the evolution of the hybrid algorithm for values of $k \in \{500, 1000, 1500, 2000, 2500, 3000\}$. As seen, the best quality of the solution is obtained for $k = 1000$. For $k = 500$, the performance of the hybrid algorithm degrades because too few nodes are kept by the BS. For greater values, effectiveness also decreases since most of the computation is dedicated to the BS part of the hybrid algorithm.

Next, the $l_0$ parameter was analyzed. Figure 13 shows the evolution of the hybrid algorithm for different values of $l_0$ (percentages are with respect to the number of taxa). The best performance of the algorithm is obtained when the MA starts running after the BS has descended 50% of the number of taxa. For $l_0 \in \{30\%, 40\%\}$ performance is slightly worse and it decreases considerably for greater values of $l_0$. Lastly, for the $l$ parameter, the best value is $l = 3$ (see Fig. 14). Efficiency clearly improves with lower values for $l$.

## 7.3. Analysis of results

After the preliminary study in Sec. 7.2, the parameters for the hybrid algorithm were set to: $k = 1000$, $l_0 = 0.5n$ and $l = 3$. For the BS algorithm alone, the greedy algorithm described in Fig. 10 was applied to the best partial solution in each level of the BBT, to improve the incumbent solution. A single execution for each instance was performed for the BS algorithm (it is a deterministic method) whilst 25 independent runs per instance were

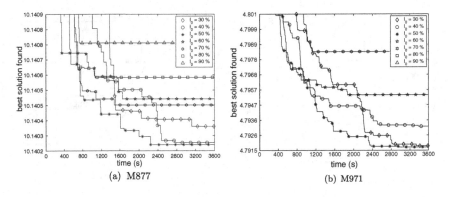

**Fig. 13.** Evolution of best fitness for different values of $l_0$ for different instances.

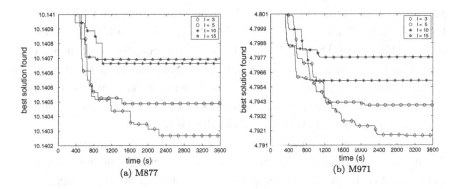

**Fig. 14.** Evolution of best fitness for different values of $l$ for different instances.

carried out for the MA and hybrid algorithm. All algorithms were run for 3600 seconds.

First of all, results for the three basic agglomerative algorithms, as well as for the neighbor-joining[54] and Fitch-Margoliash[15] algorithms are shown in Table 3. As expected, the complete-link algorithm provides much better performance for the quality criterion selected than the other approaches. According to the experience with smaller instances (for which the optimal solution can be calculated), this value usually lies near the optimum for this evaluation model.

The results of the BS, MA and hybrid algorithms are shown in Table 4. Figure 15 compares the temporal evolution of the best solution found for the BS, MA and hybrid algorithm for two different instances in the data set.

Table 3. Results of the classical heuristics on the data sets used.

|                  | M420    | M1097   | M877    | M971    | M808     |
|------------------|---------|---------|---------|---------|----------|
| single-link      | 2.93600 | 1.26385 | 53.82255| 6.47470 | 66.51205 |
| complete-link    | 2.35685 | 1.01205 | 10.15965| 4.82025 | 11.58550 |
| average-link     | 2.50110 | 1.04855 | 10.89800| 5.15390 | 12.45225 |
| neighbor-joining | 3.44775 | 1.24985 | 18.21510| 5.46825 | 38.20275 |
| Fitch-Margoliash | 2.63675 | 1.09675 | 29.09415| 5.91770 | 31.86065 |

Table 4. Results of the hybrid algorithm, MA and BS on the data sets used.

| Problem | best | mean ± std. dev. | worst | median |
|---------|------|------------------|-------|--------|
| Hybrid Algorithm | | | | |
| M420  | 2.34905  | 2.34918 ± 0.00028 | 2.34995  | 2.34910  |
| M1097 | 1.01160  | 1.01177 ± 0.00010 | 1.01190  | 1.01170  |
| M877  | 10.14010 | 10.14025 ± 0.00025 | 10.14080 | 10.14010 |
| M971  | 4.78780  | 4.79303 ± 0.00253 | 4.79920  | 4.79235  |
| M808  | 11.51415 | 11.51685 ± 0.00237 | 11.52525 | 11.51610 |
| Memetic Algorithm | | | | |
| M420  | 2.35555  | 2.37937 ± 0.01516 | 2.41785  | 2.37475  |
| M1097 | 1.01315  | 1.02676 ± 0.01024 | 1.04570  | 1.02660  |
| M877  | 10.19010 | 10.25138 ± 0.03557 | 10.36410 | 10.24875 |
| M971  | 4.83480  | 4.88207 ± 0.03494 | 4.94110  | 4.88340  |
| M808  | 11.70265 | 11.81517 ± 0.07383 | 12.03925 | 11.80730 |
| Beam Search Algorithm | | | | |
| M420  | 2.355350  | n.a. | n.a. | n.a. |
| M1097 | 1.012050  | n.a. | n.a. | n.a. |
| M877  | 10.159650 | n.a. | n.a. | n.a. |
| M971  | 4.820250  | n.a. | n.a. | n.a. |
| M808  | 11.585500 | n.a. | n.a. | n.a. |

As seen, the BS algorithm does not improve the initial bound provided by the complete link algorithm. Results for the MA alone are worse than the ones provided by BS. The hybrid algorithm provides better results than the constituent algorithms all over the run. This indicates the synergy of this combination, thus supporting the idea that this is a profitable approach for tackling phylogenetic inference.

To test the significance of these results, a statistical analysis has been conducted. A Wilcoxon rank sum test — also known as Mann-Whitney U test[36] — has been used for this purpose. Unlike t-test, this test does

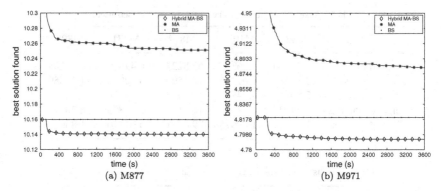

**Fig. 15.** Evolution of best fitness for the hybrid algorithm, MA and BS for different instances. Curves for the Hybrid and MA are averaged over 25 runs.

not assume normality of the samples. In this case, the test indicates that the difference of the hybrid algorithm with respect to the MA is always significant (at the standard 5% significance level).

## 8. Conclusions

In this paper we have investigated the effects of hybridizing an exact method (i.e. Branch and Bound) and a heuristic technique (i.e. a memetic algorithm) for the reconstruction of phylogenetic trees, a computationally hard task. Our hybrid model tries to combine the best advantages of both techniques as well as minimize the drawbacks of both methods as stand-alone techniques. The proposal is based on an interleaved collaboration of both approaches where, basically, the MA provides improved bounds that the BnB algorithm can use to purge the problem queue, whereas the BnB guides the MA to look into promising regions of the search space.

A performance analysis conducted on data sets comprising real biological data shows that the hybrid model outperformed each of its constituent techniques, as well as classical agglomerative algorithms and other efficient tree-construction methods. At any rate, it must be noted that the hybrid model represents a scheme whose performance depends on several parameters. To gain a better understanding of the model, we have done a dynamics and sensitivity analysis of the model and provided some guidelines about the rationale behind the setting of the parameters.

There is still room for improvements. For instance, more results could be obtained and analyzed by taking into account another different ways to traverse the search tree in the BnB part, or by considering the possibility of transforming the hybrid algorithm in a complete anytime algorithm (via replacing BS by alternatives models of BnB). Future work will also focus on the parallelization of the MA and BnB, leading to better performance. In this case, a parallelized model would demand a new analysis about parameter selection in order to achieve better understanding of the model.

## Acknowledgment

This work was partially supported by Spanish MCyT under contracts TIN2004-7943-C04-01 and TIN2005-08818-C04-01.

## References

1. A. A. Andreatta and C. C. Ribeiro, Heuristics for the phylogeny problem, *Journal of Heuristics* **8**, 429–447 (2002).
2. H. J. Bandelt. Recognition of tree metrics, *SIAM Journal on Discrete Mathematics*, **3**(1), 1–6 (1990).
3. A. Barr and E. A. Feigenbaum, *Handbook of Artificial Intelligence*. Morgan Kaufmann, New York NY (1981).
4. M. Binder, D. S. Hibbett and H. P. Molitoris, Phylogenetic relationships of the marine gasteromycete Nia vibrissa, *Mycologia* **93**, 679–688 (2001).
5. M. J. Brauer, M. T. Holder, L. A. Dries, D. J. Zwickle, P. O. Lewis and D. M. Hillis, Genetic algorithms and parallel processing in maximum-likelihood phylogeny inference, *Molecular Biology and Evolution* **19**(10), 1717–1726 (2002).
6. C. B. Congdon, Gaphyl: a genetic algorithms approach to cladistics, L. De Raedt and A. Siebes (eds), *5th European Conference on Principles of Data Mining and Knowledge Discovery — PKDD 2001. Lecture Notes in Computer Science*. Freiburg, Germany, September, Springer-Verlag **2168**, 67–78 (2001).
7. C. Cotta, Memetic algorithms with partial lamarckism for the shortest common supersequence problem, J. Mira and J. R. Álvarez (eds), *Artificial Intelligence and Knowledge Engineering Applications: A Bioinspired Approach, Lecture Notes in Computer Science*, Berlin, Heidelberg, Springer-Verlag **3562**, 84–91 (2005).
8. C. Cotta, Scatter search with path relinking for phylogenetic inference, *European Journal of Operational Research* **169**(2), 520–532 (2006).
9. C. Cotta and P. Moscato, Inferring phylogenetic trees using evolutionary algorithms, J. J. Merelo *et al.* (eds), *Parallel Problem Solving From Nature VII, Lecture Notes in Computer Science*, Springer-Verlag, Berlin **2439**, 720–729 (2002).
10. J. Culberson, On the futility of blind search: an algorithmic view of "No Free Lunch", *Evolutionary Computation* **6**(2), 109–128 (1998).

11. W. H. E. Day, Computationally difficult problems in phylogeny systematics, *Journal of Theoretic Biology* **103**, 429–438 (1983).
12. W. H. E. Day, Computational complexity of inferring phylogenies from dissimilarity matrices, *Bulletin of Mathematical Biology* **49**(4), 461–467 (1987).
13. W. H. E. Day, D. S. Johnson and D. Sankoff, The computational complexity of inferring rooted phylogenies by parsimony, *Mathematical Biosciences* **81**, 33–42 (1986).
14. J. Felsenstein, Statistical inference of phylogenies (with discussion), *Journal of the Royal Statistical Society A* **146**, 246–272 (1983).
15. W. M. Fitch and E. Margoliash, Construction of phylogenetic trees, *Science* **155**, 279–284 (1967).
16. G. B. Fogel, Evolutionary computation for the inference of natural evolutionary histories, *IEEE Connections* **3**(1), 11–14 (2005).
17. L. R. Foulds and R. L. Graham, The Steiner problem in phylogeny is NP-complete, *Advances in Applied Mathematics* **3**, 439–449 (1982).
18. L. M. Giussani, J. H. Cota-Sanchez, F. O. Zuloaga and E.A. Kellogg, A molecular phylogeny of the grass subfamily Panicoideae (Poaceae) shows multiple origins of C4 photosynthesis, *American Journal of Botany* **88**, 1993–2012 (2001).
19. F. Glover, Scatter search and path relinking, D. Corne, M. Dorigo and F. Glover (eds), *New Methods in Optimization*, McGraw-Hill, London, pp. 291–316 (1999).
20. J. Hein, A new method that simultaneously aligns and reconstructs ancestral sequences for any number of homologous sequences, when the phylogeny is given, *Molecular Biology and Evolution* **6**, 649–668 (1999).
21. D. S. Hibbett and M. J. Donoghue, Analysis of character correlations among wood decay mechanisms, mating systems, and substrate ranges in homobasidiomycetes, *Systematic Biology* **50**, 1–27 (2001).
22. D. S. Hibbett, L.-B. Gilbert and M. J. Donoghue, Evolutionary instability of ectomycorrhizal symbioses in basidiomycetes, *Nature* **407**, 506–508 (2000).
23. T. Hill, A. Lundgren, R. Fredriksson and H. B. Schiöth, Genetic algorithm for large-scale maximum parsimony phylogenetic analysis of proteins, *Biochim Biophys Acta* **1725**(1), 19–29 (2005).
24. S. Holmes, Phylogenies: an overview, M. E. Halloran and S. Geisser (eds), *Statistics and Genetics*, Springer-Verlag, New York NY, pp. 81–119 (1999).
25. C. Houck, J. A. Joines, M. G. Kay, and J. R. Wilson, Empirical investigation of the benefits of partial lamarckianism. *Evolutionary Computation*, **5**(1), 31–60, 1998.
26. U. W. Hwang, M. Friedrich, D. Tautz, C. J. Park and W. Kim, Mitochondrial protein phylogeny joins myriapods with chelicerates, *Nature* **413**, 154–157 (2001).
27. T. H. Jukes and C. R. Cantor, Evolution of protein molecules, H. N. Munro (ed), *Mammalian Protein Metabolism*, Academic Press, New York NY **3**, 21–132 (1969).
28. K. Katoh, K.-I. Kuma and T. Miyata, Genetic algorithm-based maximum-likelihood analysis for molecular phylogeny, *Journal of Molecular Evolution* **53**, 477–484 (2001).
29. J. Kim and T. Warnow, Tutorial on phylogenetic tree estimation, T. Lengauer *et al.* (eds), *Proceedings of the 7th International Conference on Intelligent Systems for Molecular Biology*, Heidelberg, The American Association for Artificial Intelligence Press (1999).
30. Y.-H. Kim, S.-K. Lee, and B.-R. Moon, Optimizing the order of taxon addition in phylogenetic tree construction using genetic algorithm, E. Cantú-Paz *et al.*, *Genetic and Evolutionary Computation — GECCO 2003, LNCS*, Chicago, 12–16 July, Springer-Verlag **2724**, 2168–2178 (2003).

31. M. Kimura, Estimation of evolutionary distances between homologous nucleotide sequences, *Proceedings of the Natural Academy of Sciences of the United States of America* **78**, 454–458 (1981).
32. B. King, Step-wise clustering procedures, *Journal of the American Statistical Association* **69**, 86–101 (1967).
33. J. R. Koza, *Genetic Programming*, MIT Press, Cambridge, MA (1992).
34. M. Laguna and R. Martí, *Scatter Search. Methodology and Implementations*, C. Kluwer Academic Publishers, Boston, MA (2003).
35. E. L. Lawler and D. E. Wood, Branch and bounds methods: a survey, *Operations Research* **4**(4), 669–719 (1966).
36. E. L. Lehmann, *Nonparametric Statistical Methods Based on Ranks*, McGraw-Hill, New York NY (1975).
37. A. R. Lemmon and M. C. Milinkovitch, The metapopulation genetic algorithm: an efficient solution for the problem of large phylogeny estimation, *Proceedings of the Natural Academy of Sciences of the United States of America*, **99**(16), 10516–10521 (2002).
38. P. O. Lewis, A genetic algorithm for maximum-likelihood phylogeny inference using nucleotide sequence data, *Molecular Biology and Evolution*, **15**(3), 277–283 (1998).
39. W. H. Li, *Molecular Evolution*, Sinauer, Boston (1997).
40. J. A. Lozano and P. Larrañaga, Applying genetic algorithms to search for the best hierarchical clustering of a dataset, *Pattern Recognition Letters*, **20**(9), 911–918 (1999).
41. H. Matsuda, Construction of phylogenetic trees from amino acid sequences using a genetic algorithm, *Genome Informatics Workshop*, Universal Academy Press, pp. 19–28 (1995).
42. H. Matsuda, Protein phylogenetic inference using maximum likelihood with a genetic algorithm, L. Hunter and T. E. Klein (eds), *1st Pacific Symposium on Biocomputing'96*, London, January, World Scientific, pp. 512–523 (1996).
43. A. Meade, D. Corne, M. Pagel and R. Sibly, Using evolutionary algorithms to estimate transition rates of discrete characteristics in phylogenetic trees, *2001 Congress on Evolutionary computation (CEC 2001)*, Seoul, Korea, **2**, pp. 1170–1176 (2001).
44. A. Moilanen, Searching for the most parsimonious trees with simulated evolution, *Cladistics*, **15**, 39–50 (1999).
45. P. Moscato, Memetic algorithms: a short introduction, D. Corne, M. Dorigo and F. Glover (eds), *New Ideas in Optimization*, McGraw-Hill, Maidenhead, Berkshire, England, UK, pp. 219–234 (1999).
46. P. Moscato and C. Cotta, A gentle introduction to memetic algorithms, F. Glover and G. Kochenberger (eds), *Handbook of Metaheuristics*, Kluwer Academic Publishers, Boston MA, pp. 105–144 (2003).
47. P. Moscato, A. Mendes and C. Cotta, Memetic algorithms, G.C. Onwubolu and B. V. Babu (eds), *New Optimization Techniques in Engineering*, Springer-Verlag, Berlin, Heidelberg, pp. 53–85 (2004).
48. C.-K. Ong, S. Nee, A. Rambaut, H.-U. Bernard and P. H. Harvey, Elucidating the population histories and transmission dynamics of papillomaviruses using phylogenetic trees. *Journal of Molecular Evolution*, **44**, 199–206 (1997).

49. L. Poladian, A GA for maximum likelihood phylogenetic inference using neighbour-joining as a genotype to phenotype mapping, H.-G. Beyer and U.-M. O'Reilly (eds), *Genetic and Evolutionary Computation Conference GECCO 2005*, Washington DC, USA, June, ACM, pp. 415–422 (2005).
50. L. Poladian and L. S. Jermiin, Multi-objective evolutionary algorithms and phylogenetic inference with multiple data sets, *Soft Computing* **10**(4), 359–368 (2005).
51. T. H. Reijmers, R. Wehrens, and L. M. C. Buydens, Quality criteria of genetic algorithms for construction of phylogenetic trees. *Journal of Computational Chemistry* **20**(8), 867–876 (1999).
52. C. C. Ribeiro and D. S. Vianna, A genetic algorithm for the phylogeny problem using an optimized crossover strategy based on path-relinking, S. Lifschitz *et al.* (eds), *II Brazilian Workshop on Bioinformatics (WOB 2004)*, Macaé, RJ, Brazil, pp. 97–102 (2003).
53. B. Rost and C. Sander, Prediction of protein secondary structure at better than 70% accuracy. *Journal of Molecular Biology* **232**, 584–599 (1993).
54. N. Saitou and M. Nei, The neighbor-joining method: a new method for reconstructing phylogenetic trees. *Molecular Biology and Evolution* **4**, 406–425 (1987).
55. J. Shen and R. B. Heckendorn, Discrete branch length representations for genetic algorithms in phylogenetic search, G.R. Raidl *et al.* (eds), *Applications of Evolutionary Computing 2004, Lecture Notes in Computer Science*, Coimbra, Portugal, Springer, **3005**, 94–103 (2004).
56. A. N. Skourikhine, Phylogenetic tree reconstruction using self-adaptive genetic algorithm, N. G. Bourbakis (ed), *1st IEEE International Symposium on Bioinformatics and Biomedical Engineering BIBE 2000*, Arlington, Virginia, USA, November, IEEE Computer Society, pp. 129–134 (2000).
57. P. H. Sneath and R. R. Sokal, *Numerical Taxonomy*. Freeman, London, UK (1973).
58. R. R. Sokal and C. D. Michener, A statistical method for evaluating systematic relationships, *University Kansas Science Bulletin* **38**, 1409–1438 (1958).
59. M. F. Whiting, J. C. Carpenter, Q. D. Wheeler, and W. C. Wheeler, The strepsiptera problem: phylogeny of the holometabolous insect orders inferred from 18S and 28S ribosomal DNA sequences and morphology, *Systematic Biology* **46**(1), 1–68 (1997).
60. P. Wiston, *Artificial Intelligence*, Addison-Wesley, Reading MA (1984).
61. D. H. Wolpert and W. G. Macready, No free lunch theorems for optimization, *IEEE Transactions on Evolutionary Computation* **1**(1), 67–82 (1997).
62. B. Y. Wu, K.-M. Chao, and C. Y. Tang, Approximation and exact algorithms for constructing minimum ultrametric trees from distance matrices, *Journal of Combinatorial Optimization* **3**(2), 199–211 (1999).

# CHAPTER 4

# CLASSIFICATION OF RNA SEQUENCES WITH SUPPORT VECTOR MACHINES

Jason T. L. Wang

*Bioinformatics Center and
Department of Computer Science, New Jersey
Institute of Technology, Newark, NJ 07102, USA*

Xiaoming Wu

*Department of Computer Science
Math & Engineering, Shepherd University
Shepherdstown, WV 25443, USA*

Support vector machines (SVMs) are a state-of-the-art machine learning tool widely used in speech recognition, image processing and biological sequence analysis. An essential step in SVMs is to devise a kernel function to compute the similarity between two data points. In this chapter we review recent advances of using SVMs for RNA classification. In particular we present a new kernel that takes advantage of both global and local structural information in RNAs and uses the information together to classify RNAs. Experimental results demonstrate the good performance of the new kernel and show that it outperforms existing kernels when applied to classifying non-coding RNA sequences.

## 1. Introduction

Support vector machines (SVMs), originally proposed by Vapnik,[22] have been highly successful in many application areas such as text categorization, handwriting recognition and face detection. SVMs are based on a simple yet intuitive idea, namely, the best hyperplane to separate two groups of points in a Euclidean space $R^n$ is the hyperplane with the maximum margin (Fig. 1).

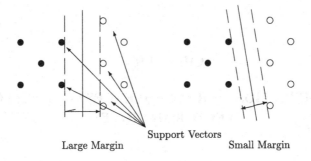

**Fig. 1.** Maximization of the margin of the hyperplane separating two groups of points.

For data points that cannot be separated by a linear hyperplane, we can map the data points to a feature space so that the images of the data points in the feature space can be linearly separated. In general, the dimension of the feature space can be very high and the mapping would be highly complicated. Fortunately, we do not need to construct the mapping explicitly; as in SVMs, the mapping is implicit in the kernel function used in a quadratic programming problem. It is only necessary to solve the quadratic programming problem to accomplish the learning task of SVMs. In fact, the kernel function represents the inner product of images of two data points in the feature space. Some commonly used kernel functions include

Gaussian RBF $\quad K(x, y) = e^{(-|x-y|^2/c)}$
Polynomial $\quad K(x, y) = ((x \cdot y) + \theta)^p$
Sigmoidal $\quad K(x, y) = \tanh(\kappa(x \cdot y) + \theta)$

The theoretical foundation of SVMs is the Vapnik-Chervonenkis theory and structural risk minimization principle.[22] For more information on SVMs, see Ref. 5. For information on implementations of SVMs, see Refs. 3 and 18.

Recently SVMs have been applied to many pattern recognition and classification problems in biological sequence analysis, ranging from protein homology detection, microarray gene expression analysis, to recognition of translation start sites.[17] Noble[17] stated the motivations behind the application of SVMs to computational biology and bioinformatics. First, many biological problems involve high-dimensional, noisy data, for which SVMs are known to behave well compared to other statistical or machine learning methods. Second, in contrast to most machine learning methods, kernel methods like SVMs can easily handle non-vector inputs, such as variable-length sequences or graphs. These types of data are common in computational biology and bioinformatics.

This chapter discusses the application of SVMs to automatically classifying RNA sequences. An SVM for this purpose is based on a kernel function that computes the similarity between two RNA sequences. Existing methods for computing the similarity (or distance) of two strings such as edit distance[20] or alignment score[6] are not sufficient for RNA sequences because they do not take into account the secondary structures of RNA. An RNA molecule is not only a string of letters taken from the alphabet $\Sigma = \{a, c, g, u\}$; it also folds itself into a-u, c-g and g-u pairs. These base pairs form the secondary structure of RNA that consists of geometric patterns such as bulges, hairpins and stacked pairs (Fig. 2). The secondary structure of an RNA molecule determines its 3D shape and hence its functional role. Therefore an important step of using SVMs for RNA classification is to devise a kernel function that takes into account the secondary structure information of RNA.

There are different ways to represent RNA secondary structures, for example, using dual graphs[7] or trees.[16] Kernels that deal with structured data such as graphs and trees are proposed in Refs. 7 and 14. Only a handful of kernels are specifically devised for RNA secondary structures. For example, Tsuda et al.[21] associated each base in an RNA sequence with a state $L$, $R$ or $P$ (for the meaning of these states, see Sec. 2 or Ref. 21), which reflects the secondary structure information of RNA. For RNA sequences with known secondary structures, Tsuda's method considers the occurrence

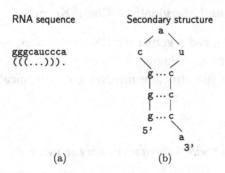

Fig. 2. (a) A hypothetical RNA sequence where parentheses represent base-pairings. (b) The secondary structure of the RNA sequence in (a).

frequencies of bi-grams such as ((a, $L$), (c, $P$)) in the sequences. Here (a, $L$) and (c, $P$) are base-state pairs. For RNA sequences with unknown secondary structures, Tsuda's method uses a probabilistic model, called stochastic context free grammar, to estimate the probability that each state $L$, $R$ or $P$ occurs at each base and uses this probability to calculate the kernel between two RNA sequences.

Another kernel for RNA was proposed in Ref. 13. In this approach, an RNA sequence is first converted to a graphical representation, called labeled dual graph, based on its secondary structure. The similarity of two labeled dual graphs, and thus the kernel function, can then be computed by taking random walks in the two graphs and measuring similarities between sequences of labels resulted from these random walks.

In this chapter we present a new kernel-based approach to RNA classification using support vector machines. We extract recurring substrings from RNA molecules; part of the kernel is based on these recurring substrings. The other part of the kernel is based on counting bi-grams in RNA molecules, as done in Tsuda's method.[21] The idea behind the proposed kernel is to utilize both global and local structural information of RNA for classification.

The rest of the chapter is organized as follows. Section 2 reviews Tsuda's method.[21] Section 3 describes the kernel based on labeled dual graphs.[13] Section 4 presents the new kernel based on bi-gram occurrence frequencies and recurring substrings. Section 5 reports experimental results and Sec. 6 concludes the chapter.

## 2. Count Kernels and Marginalized Count Kernels

Tsuda *et al.*[21] developed a kernel for RNA sequences with either known secondary structures or unknown secondary structures. It was assumed that an RNA secondary structure involves only canonical base pairs a-u and g-c.

### 2.1. *RNA sequences with known secondary structures*

An RNA secondary structure can be represented by a context free grammar (CFG). For example, the RNA secondary structure in Fig. 2 can be represented as generative rules of the following CFG:

$$S \to R_1, R_1 \to P_1 \text{a}, P_1 \to \text{g} P_2 \text{c}, P_2 \to \text{g} P_3 \text{c},$$
$$P_3 \to \text{g} L_1 \text{c}, L_1 \to \text{c} L_2, L_2 \to \text{a} L_3, L_3 \to \text{u} E.$$

Here $P$ is the state that emits a canonical base pair. $L$ and $R$ are states that emit only one base to the left or right. $S$ and $E$ are special states that represent the beginning and end, respectively, of the CFG rules. The generative rules can be represented as a state-path in a matrix, called *CFG matrix*, which demonstrates how the states in the CFG are associated with the bases in the RNA secondary structure. For example, Fig. 3 shows the CFG matrix of the RNA sequence in Fig. 2. In Fig. 3, $P$ at $(1, 9)$ corresponds to the first base g and the ninth base c; $L$ at $(4, 6)$ corresponds to the fourth base c; $R$ at $(1, 10)$ corresponds to the tenth base a.

We can define two "count kernels" with respect to the CFG representation: the first order count kernel and the second order count kernel. In these kernels we convert an RNA sequence to a vector and the kernel of two RNA sequences is defined as the inner product of their corresponding vectors. The CFG matrix plays a central role in constructing the count kernels, as explained below.

In the first order count kernel, we count the occurrence numbers of base-state combinations and use them as the coordinates of the *count*

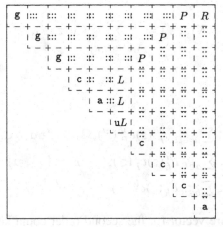

CFG Matrix

**Fig. 3.** The CFG matrix of the RNA sequence in Fig. 2.

*feature vector* of an RNA sequence. For example, for the RNA sequence in Fig. 2, the occurrence numbers of base-state combinations are 1 for (a, R), 3 for (gc, P), 1 for (c, L), 1 for (a, L), and 1 for (u, L) respectively; the occurrence number of other combinations such as (au, P) and (g, L) is 0. More formally, let $x = x_1, x_2, \ldots, x_n$ be an RNA sequence and let $W$ be the CFG matrix of the sequence. The count feature vector of $z = \{x, W\}$ is defined as follows:

$$c_P^{ab}(z) = \frac{1}{n} \sum_{i=1}^{n} \sum_{j=i+1}^{n} \delta[W(i,j) = P, x_i = a, x_j = b], \quad (1)$$

$$c_L^a(z) = \frac{1}{n} \sum_{i=1}^{n} \sum_{j=i}^{n} \delta[W(i,j) = L, x_i = a], \quad (2)$$

$$c_R^b(z) = \frac{1}{n} \sum_{i=1}^{n} \sum_{j=i}^{n} \delta[W(i,j) = R, x_j = b]. \quad (3)$$

Here $z = \{x, W\}$ is the sequence $x$ combined with the CFG matrix $W$; $a, b$ are nucleotides; $\delta[W(i,j) = P, x_i = a, x_j = b] = 1$ if $W(i,j) = P, x_i = a, x_j = b$ and 0 otherwise; $1/n$ is a normalization factor with respect to the sequence length $n$.

The *first order count kernel* of two RNA sequences is defined as the inner product of their count feature vectors:

$$K(z, z') = \sum_{V \in \{P, L, R\}} C_V(z, z'),$$

where

$$C_V(z, z') = \begin{cases} V = P : \sum_{ab \in \Omega} c_P^{ab}(z) c_P^{ab}(z'), & \Omega = \{\text{au, ua, gc, cg}\} \\ V = L : \sum_{a \in B} c_L^a(z) c_L^a(z'), & B = \{\text{a, u, c, g}\} \\ V = R : \sum_{b \in B} c_R^b(z) c_R^b(z') \end{cases}.$$

The count feature vector for the second order count kernel is obtained by counting the occurrence numbers of combinations of two consecutive base-state pairs in $z$. The following shows the formal definition of the count feature vector for the second order count kernel (only three out of nine equations are listed here):

$$c_{PP}^{abcd}(z) = \frac{2}{n} \sum_{i=1}^{n-1-\Delta_l^P-\Delta_r^P} \sum_{j=i+1+\Delta_l^P+\Delta_r^P}^{n} d_{PP}^{abcd}(i,j), \quad (4)$$

$$d_{PP}^{abcd}(i,j) \equiv \delta\big[W(i,j) = P, W(i+\Delta_l^P, j-\Delta_r^P) = P,$$
$$x_i = a, x_j = b, x_{i+\Delta_l^P} = c, x_{j-\Delta_r^P} = d\big]$$

$$c_{PL}^{abc}(z) = \frac{2}{n} \sum_{i=1}^{n-1-\Delta_l^P} \sum_{j=i+1+\Delta_r^P}^{n} d_{PL}^{abc}(i,j), \quad (5)$$

$$d_{PL}^{abc}(i,j) \equiv \delta\big[W(i,j) = P, W(i+\Delta_l^P, j-\Delta_r^P) = L,$$
$$x_i = a, x_j = b, x_{i+\Delta_l^P} = c\big]$$

$$c_{PR}^{abd}(z) = \frac{2}{n} \sum_{i=1}^{n-1-\Delta_l^P} \sum_{j=i+1+\Delta_r^P}^{n} d_{PR}^{abd}(i,j), \quad (6)$$

$$d_{PR}^{abd}(i,j) \equiv \delta\big[W(i,j) = P, W(i+\Delta_l^P, j-\Delta_r^P) = R,$$
$$x_i = a, x_j = b, x_{j-\Delta_r^P} = d\big].$$

Here $\Delta_l^V$ and $\Delta_r^V$ are flags of a binary value indicating whether state V emits a symbol to the left or to the right (see the table in Fig. 4).

The *second order count kernel* of two RNA sequences is defined as the inner product of their count feature vectors:

$$K(z, z') = \sum_{VY \in \Psi} C_{VY}(z, z'), \quad (7)$$

| | V | P | L | R |
|---|---|---|---|---|
| Values of $\Delta_{l|r}^V$ represent whether state V emits a symbol to the left or to the right; "1" indicates emission and "0" indicates no emission. | $\Delta_l^V$ | 1 | 1 | 0 |
| | $\Delta_r^V$ | 1 | 0 | 1 |

**Fig. 4.** Definition of $\Delta_{l|r}^V$.

where

$$\Psi = \{PP, PL, PR, LP, LL, LR, RP, RL, RR\}$$

$$C_{VY}(z, z') = \begin{cases} VY = PP : \sum_{abcd \in \Omega \times \Omega} c_{PP}^{abcd}(z) c_{PP}^{abcd}(z') \\ VY = PL : \sum_{abc \in \Omega \times B} c_{PL}^{abc}(z) c_{PL}^{abc}(z') \\ VY = PR : \sum_{abd \in \Omega \times B} c_{PR}^{abd}(z) c_{PR}^{abd}(z') \\ \vdots \end{cases}$$

### 2.2. RNA sequences with unknown secondary structures

If we do not know the secondary structure of an RNA sequence *a priori*, we can use a stochastic context free grammar (SCFG) to estimate the probability that each state appears at entry $(i, j)$ of the CFG matrix of the sequence. We then use these state probabilities instead of explicit states to construct count feature vectors. These vectors are called *marginalized count feature vectors* and kernels based on these vectors are called *marginalized count kernels* (MCKs). Let $x = x_1, x_2, \ldots, x_n$ be an RNA sequence and let $W$ be the CFG matrix of the sequence. The *first order marginalized count feature vector* of $z = \{x, W\}$ is defined as follows:

$$g_P^{ab}(z) = \frac{1}{n} \sum_{i=1}^{n} \sum_{j=i+1}^{n} p(W(i, j) = P|x)\delta[x_i = a, x_j = b], \quad (8)$$

$$g_L^{a}(z) = \frac{1}{n} \sum_{i=1}^{n} \sum_{j=i}^{n} p(W(i, j) = L|x)\delta[x_i = a], \quad (9)$$

$$g_R^{b}(z) = \frac{1}{n} \sum_{i=1}^{n} \sum_{j=i}^{n} p(W(i, j) = R|x)\delta[x_j = b]. \quad (10)$$

Here $p(W(i, j) = V)$ is the probability that state $V$ occurs at $W(i, j)$ and $0 \leq p(W(i, j) = V) \leq 1$. The *first order marginalized count kernel* of two RNA sequences is defined as

$$K(z, z') = \sum_{V \in \{P, L, R\}} G_V(z, z'), \quad (11)$$

where

$$G_V(z, z') = \begin{cases} V = P : \sum_{ab \in \Omega} g_P^{ab}(z) g_P^{ab}(z'), \Omega = \{\text{au, ua, gc, cg}\} \\ V = L : \sum_{a \in B} g_L^{a}(z) g_L^{a}(z'), \quad B = \{\text{a, u, c, g}\} \\ V = R : \sum_{b \in B} g_R^{b}(z) g_R^{b}(z') \end{cases}.$$

Let $\xi_{VY}(i, j)$ be the joint probability of having state V at $W(i, j)$ followed by state Y at $W(i + \Delta_l^V, j - \Delta_r^V)$. We have

$$\xi_{VY}(i, j) \equiv p\big(W(i, j) = V, W(i + \Delta_l^V, j - \Delta_r^V) = Y | x\big).$$

We define the *second order marginalized count feature vector* of an RNA sequence as follows (only three out of nine equations are listed here):

$$g_{PP}^{abcd}(z) = \frac{2}{n} \sum_{i=1}^{n-1-\Delta_l^P-\Delta_r^P} \sum_{j=i+1+\Delta_l^P+\Delta_r^P}^{n} h_{PP}^{abcd}(i, j), \tag{12}$$

$$h_{PP}^{abcd}(i, j) \equiv \xi_{PP}(i, j) \delta[x_i = a, x_j = b, x_{i+\Delta_l^P} = c, x_{j-\Delta_r^P} = d],$$

$$g_{PL}^{abc}(z) = \frac{2}{n} \sum_{i=1}^{n-1-\Delta_l^P} \sum_{j=i+1+\Delta_l^P}^{n} h_{PL}^{abc}(i, j), \tag{13}$$

$$h_{PL}^{abc}(i, j) \equiv \xi_{PL}(i, j) \delta[x_i = a, x_j = b, x_{i+\Delta_l^P} = c],$$

$$g_{PR}^{abd}(z) = \frac{2}{n} \sum_{i=1}^{n-1-\Delta_l^P} \sum_{j=i+1+\Delta_r^P}^{n} h_{PR}^{abd}(i, j), \tag{14}$$

$$h_{PR}^{abd}(i, j) \equiv \xi_{PR}(i, j) \delta[x_i = a, x_j = b, x_{j-\Delta_r^P} = d].$$

The *second order marginalized count kernel* of two RNA sequences is defined as:

$$K(z, z') = \sum_{VY \in \Psi} G_{VY}(z, z'), \tag{15}$$

where

$$\Psi = \{PP, PL, PR, LP, LL, LR, RP, RL, RR\}$$

$$G_{VY}(z, z') = \begin{cases} VY = PP : \sum_{abcd \in \Omega \times \Omega} g_{PP}^{abcd}(z) g_{PP}^{abcd}(z') \\ VY = PL : \sum_{abc \in \Omega \times B} g_{PL}^{abc}(z) g_{PL}^{abc}(z') \\ VY = PR : \sum_{abd \in \Omega \times B} g_{PR}^{abd}(z) g_{PR}^{abd}(z') \\ \vdots \end{cases}$$

For more detailed information regarding how to use an SCFG to obtain the probabilities $p(W(i, j) = V)$ and $p(W(i, j) = V, W(i + \Delta_l^V, j - \Delta_r^V) = Y|x)$, see Refs. 6 and 21.

In Ref. 21, three approaches for classifying RNA sequences were compared in a computational experiment: SVMs based on the first order MCK, SVMs based on the second order MCK, and a pure SCFG likelihood method. The dataset used in the experiment contained 74 sequences extracted from three human tRNA families. The experimental results showed that for all the three families, SVMs based on the second order MCK performed consistently better than the other two approaches.

## 3. Kernel Based on Labeled Dual Graphs

The idea behind this approach is to first convert an RNA secondary structure into a graph representation called labeled dual graph (LDG). The LDG is expected to capture some of the key topological features of the RNA secondary structure. Then a kernel is constructed to compute the similarity of the LDG representations of two RNA secondary structures. Karklin et al.[13] recently adopted this approach in which they used the marginalized kernel for labeled graphs[14] to classify non-coding RNA sequences.

### 3.1. Labeled dual graphs

The dual graph[7] representation of RNA captures important topological characteristics of RNA secondary structures such as the number and relative positions of helical regions. In this representation, a helical region is represented by a vertex in the graph. Single RNA strands are represented by edges; for example, an internal loop, multi-branched loop or bulge is represented by an edge connecting two vertices representing two helical regions and an external loop is represented by an edge connecting a vertex to itself. Thus a dual graph is a multi-graph in which two vertices are connected by at most two edges. We can attach biologically meaningful labels to the vertices and edges in a dual graph. For example, a vertex can be labeled with the number of base-pairs of the corresponding helical region and an edge can be labeled with the length of the corresponding single RNA strand (number of nucleotides) and the type of the corresponding loop (internal/external). The resulting graph is called a *labeled dual graph* (LDG). Figure 5 illustrates two RNA secondary structures and their LDG representations; (a) gives the secondary structure diagrams of the two RNA molecules; (b) shows the labeled dual graph representations of the RNA secondary structures in (a) where a vertex represents a helical region and an edge represents a single RNA strand; (c) gives a subset of label sequences produced by taking random walks in the two labeled dual graphs in (b) where $L$ is the length of a random walk; and (d) shows an example of the label sequence kernel applied to the highlighted pair of label sequences in (c).

### 3.2. *Marginalized kernel for labeled dual graphs*

With the marginalized kernel for labeled dual graphs, we take random walks in two graphs and calculate the similarity of the label sequences produced by these random walks.[14] A score based on the similarity of the label sequences reflects the similarity between the two graphs, which is defined as the marginalized kernel for the two graphs.

In order to generate random walks in an LDG, we assume a uniform starting probability over all vertices, a uniform probability to transit from a vertex to one of its neighbors, and a constant probability to terminate the walks at any step.[13] Suppose $z = v_1, e_{12}, v_2, e_{23}, v_3, \ldots$ and $z' = v'_1, e'_{12}, v'_2, e'_{23}, v'_3, \ldots$ are two sequences of vertex and edge labels produced by random walks $h$, $h'$ in graphs $G$ and $G'$, respectively. The similarity measure between $z$ and $z'$ is defined as:[13]

$$K_z(z, z') = K_v(v_1, v'_1) K_e(e_{12}, e'_{12}) K_v(v_2, v'_2) \ldots . \quad (16)$$

(a)

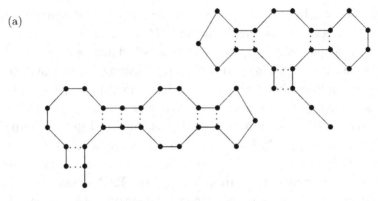

(b)

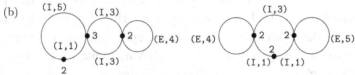

(c)

$L = 1$   {2}, {3}, {2}            {2}, {2}, {2}
$L = 2$   {2, (I, 1), 3}, {2, (E, 4), 2}, ...   {2, (I, 1), 2}, {2, (E, 4), 2}, ...
$L = 3$   {2, **(I, 1)**, **3**, **(I, 3)**, 2}, ...    {2, **(I, 1)**, 2, **(I, 3)**, 2}, ...
$L = 4$   {2, (I, 5), 3, (I, 3), 2, (E, 4), 2},   {2, (I, 1), 2, (E, 4), 2, (I, 3), 2},

⋮                                 ⋮

(d)
$$K_z = K_v(2,2)K_e(1,1)K_v(3,2)K_e(3,3)K_v(2,2)$$
$$= \exp(-(\log\frac{2}{2})^2 - (\log\frac{1}{1})^2 - (\log\frac{3}{2})^2 - (\log\frac{3}{3})^2 - (\log\frac{2}{2})^2)$$

**Fig. 5.** Two RNA secondary structures and their LDG representations.

Here $K_v$ and $K_e$ are kernels defined for vertices and edges respectively. For two vertices $v_i$ and $v_j$,

$$K_v(v_i, v_j) = \exp(-(\log(v_i/v_j))^2). \tag{17}$$

For two edges $e_{ij}$ and $e_{kl}$, representing two loops of the same type (internal or external), the kernel is defined as:[13]

$$K_e(e_{ij}, e_{kl}) = \exp(-(\log(e_{ij}/e_{kl}))^2). \tag{18}$$

If two loops are of different types, the kernel $K_e$ is defined as 0. Finally, we calculate the expected value of $K_z(z, z')$ over all possible random walks, weighted by the probability generating the random walks. The result is defined as the marginalized kernel of $G$ and $G'$, denoted $K(G, G')$:

$$K(G, G') = \langle K_z(z, z') \rangle_{h,h'}. \tag{19}$$

Obviously, $0 \le K(G, G') \le 1$. The more similar the two graphs $G$ and $G'$, the higher the score $K(G, G')$ will be.

Karklin et al.[13] conducted experiments on RNA sequences taken from families in the Rfam database[9] to evaluate the performance of this kernel. An SVM classifier based on this kernel was trained to find out if the kernel was able to pick up important topological information of RNA secondary structures so as to differentiate sequences in a family from random sequences with the same di-nucleotide statistics. In their experiments, sequences of a family were treated as positive samples. The nucleotides in those sequences were shuffled randomly to generate negative samples. For 22 of 25 families tested, $A_{ROC}$ was greater than 0.7. ($A_{ROC}$ is the area under the receiver operating characteristic (ROC) curve, which is a general measure of the discrimination ability of a classification algorithm.[2]) A multi-class classifier based on the kernel was trained and tested on nine families in the Rfam database. The experimental results showed good performance of this classifier for many of the tested families.

## 4. A New Kernel

We propose here a new kernel that takes into account both global and local structural information of RNA molecules. The global structural information is obtained by counting occurrence numbers of bi-grams in the molecules and the local structural information is obtained by considering recurring substrings in these molecules. The meaning of these terminologies will be made clearer in the following subsections. This kernel is an extension of our previously developed method[23] in which Bayesian neural networks (BNNs) were employed to classify protein sequences. In that method, the occurrence frequencies of all bi-grams in a protein sequence were calculated, which formed features reflecting the global structural information of the protein sequence. Then, the local motifs of a given protein family were extracted. These motifs were used to calculate another feature that reflected the local structural information of a protein sequence. The features of the training sequences used in protein classification were fed into a BNN to train its weight parameters.

In the case of proteins, the number of bi-grams is large. In fact, there are $20 \times 20 = 400$ possible bi-grams because there are 20 different amino acids. The number of local motifs may also be very large. If all these features are used to train the BNN, there would be too many weight parameters. This would make the training process hard, leading to a phenomenon called "curse of dimensionality." Therefore, a procedure was taken[23] to select only those features that, by intuition, had the greatest capability to differentiate sequences in a family from sequences outside the family. Here, the "curse of dimensionality" is not a serious problem because SVMs are known to perform well even with high-dimensional data. Furthermore, in applications like ours, SVMs usually yield better performance than BNNs.[17]

### 4.1. *Extracting features for global structural information*

First, we obtain the secondary structure of an RNA sequence using the Vienna RNA folding prediction package.[26] Second, for each RNA molecule, we transform it to a sequence that reflects both the primary structure and the secondary structure of the molecule. The transformation is performed as follows:

$$
\begin{array}{lll}
(a \to A & a) \to B & a. \to C \\
(c \to D & c) \to E & c. \to F \\
(g \to G & g) \to H & g. \to I \\
(u \to U & u) \to V & u. \to W
\end{array}
$$

For example, with this transformation, nucleotide a that is on the left hand side of a base pair is converted to $A$. If nucleotide a is on the right hand side of a base pair, it is converted to $B$. If nucleotide a is unpaired, it is converted to $C$. Therefore, each RNA sequence can be transformed to a sequence of letters from alphabet $\Sigma' = \{A, B, C, D, E, F, G, H, I, U, V, W\}$. As an example, the RNA sequence in Fig. 2 is transformed to

$$GGGFCW\,EEFC. \tag{20}$$

In the following, the new sequence obtained based on this transformation scheme will be referred to as an *adjusted RNA sequence*.

We map each adjusted RNA sequence to a vector, called the *feature vector* of the sequence, in a Euclidean space $R^k$. The dimension $k$ will be determined later. Each coordinate of the feature vector represents a feature of the sequence. After all adjusted RNA sequences are mapped to vectors in $R^k$, the kernel function of two adjusted RNA sequences can be as simple as the inner product of their corresponding feature vectors, or can be one of the commonly used kernel functions (Gaussian RBF, polynomial, or sigmoidal) defined in Sec. 1. In our approach, we use the Gaussian RBF kernel function.

Similar to Refs. 21 and 23, we adopt the bi-gram model to calculate the first set of features in a feature vector. The bi-gram model is known to be able to represent implicit but essential information for identifying biological sequences.[21] In our approach, a bi-gram is a combination of any two letters taken from the alphabet $\Sigma'$ for adjusted RNA sequences. There are 12 letters in $\Sigma'$, so the total number of bi-grams is $12 \times 12 = 144$. The occurrence frequency of a bi-gram in an adjusted RNA sequence is defined as the number of occurrences of the bi-gram in the sequence divided by the total number of bi-grams in the sequence, which is obviously the length of the sequence minus 1.

For example, the occurrence frequency of $GF$ in the adjusted RNA sequence shown in (20) is $1/9 = 0.111111$. The occurrence frequency of $EE$ in this adjusted RNA sequence is $2/9 = 0.222222$. The first 144 coordinates of the feature vector of an adjusted RNA sequence are the occurrence frequencies of all 144 bi-grams in the sequence. We can adopt the formal notation in Ref. 21; namely, letting $z = z_1, z_2, \ldots, z_n$ be an adjusted RNA sequence, the coordinate for a bi-gram e.g. $GF$, is defined as:

$$c_{GF} = 1/(n-1) \sum_{i=1}^{n-1} \delta[z_i = G, z_{i+1} = F], \qquad (21)$$

where $\delta[z_i = G, z_{i+1} = F] = 1$ if $z_i = G, z_{i+1} = F$ and 0 otherwise; $n$ is the length of the sequence. These coordinates can be thought of as features representing global RNA structural information. The second set of coordinates of the feature vector, discussed below, represents local RNA structural information.

## 4.2. Extracting features for local structural information

A recurring substring is a substring that occurs frequently in a set of biosequences. If significantly many sequences in a set share a similar substring, we can infer that the substring doesn't occur by chance.[25] Instead, the substring is related to some biological function. For example, the biological function of a shared substring might be a common protein binding site in the case of DNA sequences, or the active site of related enzymes in the case of protein sequences.[1] The problem of finding interesting recurring substrings in biological sequences is important and has been studied extensively. There are a number of algorithms that can be used, for example, MEME[1] and PRATT[12]; other algorithms can be found in Refs. 4 and 10.

In our approach, we use our previously developed tool **sdiscover**[24] that is suitable for the application at hand to find recurring substrings. This tool can be downloaded from http://datalab.njit.edu/biodata/ssi/software.shtml. It can find frequently occurring substrings with length greater than or equal to *Len* that occur in at least *Occur* sequences within *Mut* edit operations in a set of sequences where *Len*, *Occur* and *Mut* are user-specified parameters. The discovered substrings can be of several different forms. In the study presented here, a substring is simply a segment of letters taken from the alphabet $\Sigma'$. An edit operation can be an insertion, deletion or modification of a letter in our case.

Given a set of sequences, **sdiscover** works by selecting a small subset of the sequences and builds a generalized suffix tree (GST)[11] for the sequences in the subset. GST is a compact data structure that enables us to locate all the substrings of a set of sequences efficiently. Sequences in the small subset are regarded as random samples of the original set. Candidate substrings, which are recurring (or frequently occurring) substrings of these sample sequences, are selected based on a statistical heuristic. The purpose of this heuristic is to select only substrings that are most likely to meet the length (*Len*), occurrence (*Occur*) and similarity (*Mut*) requirements specified by the user. The selected candidate substrings are then compared with all sequences in the original set. Those substrings that satisfy the length (*Len*), occurrence (*Occur*) and similarity (*Mut*) constraints are returned as the final result of **sdiscover**.

**Training and Testing the Proposed SVM Classifier** Our classifier works in two phases. In the training phase, the classifier is trained by labeled adjusted RNA sequences where the label (class) of each sequence is specified. Let $S_p$ be the set of positive training sequences and let $S_n$ be the set of negative training sequences. We use sdiscover to extract recurring, or frequently occurring, substrings from the sequences in $S_p$ (we do not extract recurring substrings from the sequences in $S_n$). Let $N_p$ be the number of recurring substrings extracted from $S_p$. Each recurring substring $X$ in $S_p$ contributes to one coordinate in the feature vector of an adjusted RNA sequence, whether it is positive or negative. So, the dimension of the feature vector equals the total number of bi-grams, namely 144, plus $N_p$.

Let $z$ be an adjusted RNA sequence in $S_p \cup S_n$ and let $V$ be the feature vector of $z$. We calculate the coordinate in $V$ that corresponds to the recurring substring $X$ in $S_p$ by matching $X$ with $z$. If $X$ occurs in $z$ within $Mut$ edit operations, the coordinate is $1/N_p$; otherwise, the coordinate is 0. More formally, the coordinate in $V$ that corresponds to the recurring substring $X$ in $S_p$ is defined as

$$c_X = \begin{cases} 1/N_p & \text{if } dist(*X*, z) \leq Mut \\ 0 & \text{otherwise} \end{cases}, \qquad (22)$$

where $dist(*X*, z)$ is the edit distance between the substring $*X*$ and the sequence $z$ and $*$ is a variable length don't care (VLDC) symbol. In calculating $dist(*X*, z)$, the VLDCs may substitute for zero or more letters in $z$. The coordinates obtained from recurring substrings can be thought of as features that represent important local structural information shared by the majority of adjusted RNA sequences in the positive training data set.

In the testing phase, the proposed SVM classifier uses the following formula to calculate a score for an unlabeled test sequence:

$$u = \sum_{j=1}^{N} y_j \alpha_j K(x_j, x) - b. \qquad (23)$$

Here $K$ is the Gaussian RBF kernel function as defined in Sec. 1 and $x$ is the feature vector of the unlabeled test sequence. Like feature vectors for training sequences, $x$ has $144 + N_p$ dimensions; the coordinates of $x$ corresponding to the recurring substrings in $S_p$ are calculated based on the formula

(22). $N$ is the total number of support vectors (the support vectors are special training sequences on the boundary of a class or family); each $x_j$, $j = 1, 2, \ldots, N$, is a support vector obtained through training the SVM classifier; $y_j$ is the label of $x_j$ where $y_j = +1$ or $-1$ depending on whether $x_j$ is positive or negative. Each $\alpha_j$ is a Lagrange multiplier and $b$ is a threshold parameter also obtained through training the SVM classifier. These parameters come from the quadratic programming problem associated with the SVM classifier.[18] The score $u$ is the output value for the test sequence.

## 5. Experiments and Results

### 5.1. *Data and parameters*

To evaluate the performance of the proposed SVM classifier, we used nine families of non-coding RNA sequences taken from the Rfam database[9] (Table 1). For each family, we trained the SVM classifier using the standard *one versus all* methodology.[13] That is, the sequences in the family were considered as positive sequences while sequences in all other families were considered as negative sequences. For an unlabeled test sequence, we used the SVM classifier trained for each of the nine families to calculate an output value; cf. formula (23). The test sequence was assigned to the family that had the largest output value. We did our classification experiments on the seed sequences[9] of the nine families. The lengths of the sequences in the nine families ranged from 70 to 120 nucleotides. Although most of the nine families had less than 90 sequences, one family, namely Retroviral_psi, had

Table 1. RNA families used in classification.

| Family number | Family name |
| --- | --- |
| 1 | U6 |
| 2 | yybP-ykoY |
| 3 | S_box |
| 4 | SRP_bact |
| 5 | Retroviral_psi |
| 6 | sno_14q_I_II |
| 7 | ctRNA_pND324 |
| 8 | Entero_5_CRE |
| 9 | K_chan_RES |

168 sequences. Half of the sequences in each family were used for training; the other half were used for testing.

The *Len, Occur* and *Mut* parameters used in the sdiscover tool were tuned in such a way that the number of substrings extracted by sdiscover was in the range 30–60. In general, the sdiscover tool may find hundreds of recurring substrings from the positive training data set. Although we could use all the substrings to train our SVM classifier, the data is more than needed. In our experiments, we used a heuristic to select 30–60 most "informative" substrings to train the SVM classifier. This heuristic is based on our previously developed minimum description length principle[23] originated from information theory.[19] In a nutshell, assume all sequences are encoded in bits using some optimal coding scheme. If a recurring substring represents important information for the sequences in a set, we are able to save a significant amount of bits if we encode the sequences with the aid of that substring. The amount of bits that we save indicates the informativeness of the substring. We used this heuristic to select the most informative recurring substrings, i.e. substrings that contained the most information, to differentiate positive sequences from negative sequences. In general, this heuristic favors longer substrings that occur more frequently, with less frequent letters, in the positive sequences.[23]

After calculating the global and local features of an adjusted RNA sequence as discussed in Sec. 4, the sequence can be represented by a vector in a Euclidean space $R^k$ where $k$ equals the total number of bi-grams, namely 144, plus the number of recurring substrings found in the positive training data set. We use the Gaussian RBF kernel to train our SVM classifier. If $x$ and $y$ are two feature vectors, the kernel is

$$K(x, y) = e^{(-|x-y|^2/2\sigma)}. \tag{24}$$

Here $\sigma$ is a parameter that can be tuned. Basically, choosing a very small $\sigma$ results in "over-fitting". In other words, the SVM classifier can fit the training data very well, but its generalization ability suffers. It cannot correctly predict labels for untrained data in general. A good indication of whether "over-fitting" occurs is when there are a large number of support vectors. Choosing a very large $\sigma$ renders the SVM classifier incapable of fitting the training data accurately. Therefore, in choosing the $\sigma$ value there is a trade-off between accuracy and generalization. In our experiments, for the data on which we tested our SVM classifier, $\sigma$ was in the range $0.15 - 0.5$.

Table 2. Performance evaluation of the proposed classifier on nine RNA families.

| Family | 1 | 2 | 3 | 4 | 5 | 6 | 7 | 8 | 9 | $Q^D$ |
|---|---|---|---|---|---|---|---|---|---|---|
| 1 | 27 | 0 | 0 | 0 | 0 | 0 | 0 | 0 | 0 | 1.00 |
| 2 | 0 | 36 | 0 | 1 | 0 | 0 | 0 | 0 | 0 | 0.97 |
| 3 | 0 | 0 | 36 | 0 | 0 | 0 | 0 | 0 | 0 | 1.00 |
| 4 | 1 | 0 | 0 | 33 | 1 | 0 | 0 | 0 | 0 | 0.94 |
| 5 | 0 | 0 | 0 | 0 | 84 | 0 | 0 | 0 | 0 | 1.00 |
| 6 | 0 | 0 | 0 | 0 | 0 | 29 | 0 | 0 | 1 | 0.97 |
| 7 | 0 | 0 | 0 | 0 | 0 | 0 | 24 | 0 | 0 | 1.00 |
| 8 | 0 | 0 | 0 | 0 | 0 | 0 | 0 | 30 | 0 | 1.00 |
| 9 | 0 | 0 | 0 | 0 | 0 | 0 | 0 | 1 | 42 | 0.98 |
| $Q^M$ | 0.96 | 1.00 | 1.00 | 0.97 | 0.99 | 1.00 | 1.00 | 0.97 | 0.98 | |

We used Platt's sequential minimal optimization approach to implement the SVM classifier.[18]

### 5.2. Results

Table 2 shows the result of our experiments. In Table 2 we evaluate the performance of our SVM classifier using the sensitivity and specificity measures originally developed in Ref. 2. For each family, the sensitivity ($Q^D$) represents the percentage of test sequences correctly classified relative to the total number of test sequences in that family; the specificity ($Q^M$) represents the percentage of test sequences correctly predicted relative to the total number of test sequences predicted to be in that family. Formally, let $x_{ij}$ be the entry at row $i$ and column $j$ in Table 2; $x_{ij}$ is the number of sequences of family $i$ predicted to be in family $j$. The sensitivity for family $i$ is

$$Q^D = \frac{x_{ii}}{\sum_{k=1}^{9} x_{ik}}.$$

The specificity for family $i$ is

$$Q^M = \frac{x_{ii}}{\sum_{k=1}^{9} x_{ki}}.$$

Table 3. Comparison of the three studied kernels.

| Family | Sensitivity | Specificity |
|---|---|---|
| *Kernel of bi-gram count and recurring substring* | | |
| tmRNA | 61.3 | 66.4 |
| U6 | 90.0 | 98.3 |
| RRE | 100.0 | 100.0 |
| THI | 88.1 | 96.1 |
| S_box | 95.7 | 100.0 |
| SRP_bact | 91.4 | 98.8 |
| *Kernel of bi-gram count only* | | |
| tmRNA | 73.9 | 67.4 |
| U6 | 94.0 | 92.0 |
| RRE | 100.0 | 100.0 |
| THI | 78.9 | 83.0 |
| S_box | 90.7 | 92.8 |
| SRP_bact | 87.1 | 87.8 |
| *Kernel based on labeled dual graph* | | |
| tmRNA | 74.5 | 65.5 |
| U6 | 82.1 | 72.4 |
| RRE | 76.9 | 80.0 |
| THI | 84.5 | 78.5 |
| S_box | 85.9 | 88.2 |
| SRP_bact | 88.4 | 89.3 |

Table 3 compares the proposed approach that uses both bi-gram occurrence frequencies and recurring substrings with the approach using only bi-gram occurrence frequencies as suggested in Ref. 21, as well as the labeled dual graph (LDG) kernel described in Ref. 13. The comparison of these three approaches was done on six families, namely tmRNA, U6, RRE, THI, S_box and SRP_bact taken from the Rfam database. For each of these families, positive sequences were sequences in the family; negative sequences were obtained from randomly shuffling the nucleotides of the sequences while preserving the di-nucleotide frequencies in them, as done in Ref. 13. For each family, the performance of a classifier was evaluated using 10-fold cross-validation. The results for the LDG kernel in the table are taken from Ref. 13.

Table 3 shows that for all the tested families except tmRNA, our kernel outperforms the other two approaches. One possible reason that the performance of our kernel deteriorates for family tmRNA is that this family may have no recurring substrings that are statistically significant. Under this circumstance, the use of recurring substrings does not improve the performance of the kernel. Rather, it may even weaken the kernel's performance because of including statistically insignificant information.

## 6. Conclusion

In this chapter we presented several different approaches that use support vector machines for RNA classification. We proposed a new kernel that uses a bi-gram model to calculate global features, and uses recurring substrings to calculate local features of RNA molecules. This kernel considers both the primary structure and the secondary structure of an RNA molecule. Experimental results showed that the SVMs based on this new kernel perform well, and are better than existing methods in many cases.

The accuracy of our SVM classifier depends on the accuracy of the algorithm used to predict RNA secondary structures. We believe the accuracy of our classifier can be further improved if a more accurate folding algorithm is employed. One interesting direction of future research is to modify our approach so that it does not need to use any folding algorithm. This is reminiscent of the approach developed by Tsuda et al.,[21] which can handle the situation where the secondary structure of an RNA sequence is unknown. In this case, they used a probabilistic model, namely stochastic context free grammar, to compute the kernel. We may change our kernel so that it is based on a similar probabilistic model. The challenge is to figure out a way to exploit local structural information in the kernel when the secondary structure of RNA is only implied in the probabilistic model. Under this circumstance, the technique for extracting recurring substrings described in Sec. 4.2 needs to be modified or new techniques need to be developed.

## Acknowledgment

We would like to thank Taishin Kin, Koji Tsuda, Kiyoshi Asai, Yan Karklin, Richard Meraz and Stephen Holbrook for useful discussions and for permissions to present their work in Refs. 13 and 21.

# References

1. T. L. Bailey and C. Elkan, Fitting a mixture model by expectation maximization to discover motifs in biopolymers, *Proceedings of the Second International Conference on Intelligent Systems for Molecular Biology*, AAAI Press, pp. 28–36 (1994).
2. P. Baldi, S. Brunak, Y. Chauvin, C. A. Andersen and H. Nielsen, Assessing the accuracy of prediction algorithms for classification: an overview, *Bioinformatics* **16**(5), 412–424 (2000).
3. B. E. Boser, I. Guyon and V. N. Vapnik, A training algorithm for optimal margin classifiers, *Proceedings of the Fifth Annual Workshop on Computational Learning Theory*, pp. 144–152 (1992).
4. A. Brazma, I. Jonassen, E. Ukkonen and J. Vilo, Discovering patterns and subfamilies in biosequences, *Proceedings of the Fourth International Conference on Intelligent Systems for Molecular Biology*, pp. 34–43 (1996).
5. C. J. C. Burges, A tutorial on support vector machines for pattern recognition, *Data Mining and Knowledge Discovery* **2**(2), 955–974 (1998).
6. R. Durbin, S. Eddy, A. Krogh and G. Mitchison, *Biological Sequence Analysis*, Cambridge University Press, England (1998).
7. H. H. Gan, S. Pasquali and T. Schlick, Exploring the repertoire of RNA secondary motifs using graph theory; implications for RNA design, *Nucleic Acids Res.* **31**(11), 2926–2943 (2003).
8. T. Gartner, A survey of kernels for structured data, *ACM Special Interest Group on Knowledge Discovery and Data Mining Explorations* **5**(1), 49–58 (2003).
9. S. Griffiths-Jones, A. Bateman, M. Marshall, A. Khanna and S. R. Eddy, Rfam: an RNA family database, *Nucleic Acids Res.* **31**(1), 439–441 (2003).
10. R. Hart, A. K. Royyuru, G. Stolovitzky and A. Califano, Systematic and automated discovery of patterns in PROSITE families, *Proceedings of the Fourth Annual International Conference on Computational Molecular Biology*, pp. 147–154 (2000).
11. L. C. K. Hui, Color set size problem with application to string matching, A. Apostolico, M. Crochemore, Z. Galil and U. Manber, (eds), *Combinatorial Pattern Matching, Lecture Notes in Computer Science*, Springer-Verlag **644**, 230–243 (1992).
12. I. Jonassen, J. F. Collins and D. G. Higgins, Finding flexible patterns in unaligned protein sequences, *Protein Science* **4**, 1587–1595 (1995).
13. Y. Karklin, R. F. Meraz and S. R. Holbrook, Classification of non-coding RNA using graph representations of secondary structure, *Proceedings of the Pacific Symposium on Biocomputing*, pp. 5–16 (2005).
14. H. Kashima, K. Tsuda and A. Inokuchi, Marginalized kernels between labeled graphs, *Proceedings of the International Conference on Machine Learning*, AAAI Press, pp. 321–328 (2003).
15. J. Liu, J. T. L. Wang, J. Hu and B. Tian, A method for aligning RNA secondary structures and its application to RNA motif detection, *BMC Bioinformatics* **6**(89) (2005).
16. B. Ma, L. Wang and K. Zhang, Computing similarity between RNA structures. *Theoretical Computer Science* **276**(1–2), 111–132 (2002).
17. W. S. Noble, Support vector machine applications in computational biology, B. Schoelkopf, K. Tsuda and J.-P. Vert, (eds), *Kernel Methods in Computational Biology*, MIT Press, pp. 71–92 (2004).

18. J. Platt, Sequential minimal optimization: a fast algorithm for training support vector machines, Technical Report 98-14, Microsoft Research, Redmond, Washington (1998).
19. C. E. Shannon, A mathematical theory of communication, *Bell System Technical Journal* **27**, 379–423, 623–656 (1948).
20. T. F. Smith and M. S. Waterman, Comparison of biosequences, *Advances in Applied Mathematics* **2**, 482–489 (1981).
21. K. Tsuda, T. Kin and K. Asai, Marginalized kernels for biological sequences, *Proceedings of the Tenth International Conference on Intelligent Systems for Molecular Biology*, pp. 268–275 (2002).
22. V. Vapnik, *The Nature of Statistical Learning Theory*, Springer-Verlag, New York (1995).
23. J. T. L. Wang, Q. Ma, D. Shasha and C. Wu, New techniques for extracting features from protein sequences, *IBM Systems Journal*, Special Issue on Deep Computing for the Life Sciences **40**(2), 426–441 (2001).
24. J. T. L. Wang, T. Marr, D. Shasha, B. A. Shapiro and G. Chirn, Discovering active motifs in sets of related protein sequences and using them for classification, *Nucleic Acids Res.* **22**(14), 2769–2775 (1994).
25. J. T. L. Wang, M. J. Zaki, H. T. T. Toivonen and D. Shasha, (eds.), *Data Mining in Bioinformatics*, Springer, London, New York (2005).
26. S. Wuchty, W. Fontana, I. L. Hofacker and P. Schuster, Complete suboptimal folding of RNA and the stability of secondary structures, *Biopolymers* **49**(2), 145–165 (1999).

# CHAPTER 5

# BEYOND STRING ALGORITHMS: PROTEIN SEQUENCE ANALYSIS USING WAVELET TRANSFORMS

Arun Krishnan

*Institute for Advanced Biosciences*
*Keio University, Yamagata, Japan*
*krishnan@ttck.keio.ac.jp*

Kuo-Bin Li

*Bioinformatics Research Center*
*National Yang-Ming University, Taipei, Taiwan*
*kbli@ym.edu.tw*

Wavelet transform is an efficient and fast method for capturing the hidden information from many different types of data. Recently there has been a growing interest in using wavelet transforms in the analysis of biological sequences and biology related signals. This review goes through several aspects of bioinformatics, especially those relevant to the functional studies of protein sequences. We focus on applications that use wavelet transforms as the main technology. The areas covered include motif search, sequence comparison, protein secondary structure prediction, detection of transmembrane proteins and hydrophobic cores, mass spectrometry and disease related applications.

## 1. Introduction

There is no doubt that bioinformatics has become an active field in recent years. Biologists, biochemists, pharmacologists, pharmacists and medical doctors have benefited from using computers to solve their daily tasks. One of the recent developments in bioinformatics is to apply signal processing techniques to analyze various types of biological signals. This article provides an overview of the applications of signal processing techniques, in particular wavelet transforms, to perform protein sequence analysis.

## 1.1. *String algorithms*

Although it is relatively easy to obtain a DNA or a protein sequence in a laboratory, to determine its function or structure by direct experiment remains a tedious and costly procedure. Hence, the use of computers to infer biological information from DNA or protein sequences has become widely accepted. The computational methods for solving these problems fall under the sequence analysis.

A strong assumption about protein structure says that all the information needed for correct three-dimensional folding is contained in the protein sequence itself.[1] Since the structure of a protein is the most important factor for its biological function, along with other factors like the environment that the protein lives in, computer scientists has since 1980s considered that many protein function related problems can be defined primarily on sequences. The general way of accomplishing this is to consider proteins as made up of a string of characters, each character representing an amino acid. Techniques from natural language processing and algorithms from computer engineering can then be applied to analyze protein sequences.

## 1.2. *Sequence analysis*

Here we introduce popular computer programs in bioinformatics and the string algorithms used by these programs.

The type of information that can be extracted from a sequence alone is restricted compared to what can be obtained from analyzing many sequences together. As a result, significant effort has been placed in solving the problems of sequence comparisons, which includes database similarity search and sequence alignment.

Similarity searches on sequence databases are perhaps the most popular bioinformatics tasks. Given a protein or DNA sequence, similarity search is able to identify sequences from databases that are similar to the query sequence, or more precisely, the sequences that might derive from the same ancestor. The general assumption is that if two sequences are similar, they probably have the same structure and a similar biological function. This assumption even works when the sequences come from very different organisms. Note that in exceptional cases, proteins with similar sequences may possess quite different biological functions.

BLAST[2,3] is without a doubt the most widely used program to search a database. The BLAST program applies various heuristics to speedup the potentially lengthy search. Among the most important heuristics, BLAST aims at identifying core similarities for later extension. The core similarity is defined by a window with a certain number of matches on DNA (the default is 11 nucleotides) or with an amino acid score above some threshold for proteins. Washington University has an alternative implementation of BLAST,[4] which has many features that are not seen in the NCBI implementation of BLAST. Those two versions of BLAST also have different default parameters. PatternHunter,[5] BLASTZ[6] are two of the recent programs using discontinuous seeds, which were meant to improve the sensitivity of the database searching.

Instead of finding similar sequences in a database, pairwise sequence alignment is a scheme for writing one sequence on top of another so that portions sharing the same evolutionary origin may become identifiable. Due to insertions or deletions in sequences, the aligned sequences may have gaps introduced in them which are represented by dashes in one or both of the sequences. There are two kinds of alignments: global alignment, where the two sequences are aligned over their entire lengths, and local alignments, where only the similar portions of the two sequences are aligned. The general idea of pairwise alignment is to assign a score to an alignment and then to minimize or maximize the scores over all possible alignments. Needleman and Wunsch first proposed a global alignment algorithm using dynamic programming[7] in 1970. Still using dynamic programming, Smith and Waterman developed a local alignment algorithm[8] in 1984. Although dynamic programming algorithms are guaranteed to find the optimal scoring alignment, the long execution time makes them less attractive for large scale alignments. Commonly used heuristic methods include Lalign[9] (best with proteins), BLAST and MegaBLAST[10] (best with DNA), SSAHA[11] (aligning sequence reads to genome), and BLAT[12] (aligning cDNA to genome).

### 1.3. *Wavelet transform*

The concept of wavelets can be viewed as a synthesis of theories which originated from several different fields, such as engineering (subband coding, quadrature mirror filter, pyramid schemes), physics and mathematics. It has provided a resurgence of interest in the area of multi-resolution signal processing and analysis for several reasons, (1) the development of efficient computational algorithms to compute the wavelet transform[13] and (2) the introduction of a family of orthonormal wavelet functions.[14]

The main feature of wavelet transforms is their ability to represent signals so as to obtain simultaneous time and scale (analogous to frequency) localization of the signal. Excellent introductions to wavelet theory can be found in Refs. 15–17.

The wavelet transform involves representing general functions in terms of simple, fixed building blocks at different scales and locations. The term *wavelets* was first suggested by Meyer and Morlet.[16] Wavelets are generated from a single basic function called a *mother wavelet*, ($\Psi(t)$), by translation and dilation (scaling) operations. The scaling operation can be equated to performing *stretching* and *compression* operations on the mother wavelet, which in turn can be used to analyze the frequency information of the function. The translation information, involves *shifting* the mother wavelet along the time/space axis. Thus, translations are used to analyze the time/space information of the function.

Multi-resolution analysis provides a formal approach to the construction of the orthonormal basis. As such, multi-resolution analysis can explain the wavelet transformation in terms of representing a function as a series of successive approximations to the original function. The idea is to represent a function, $f(t)$, as a limit of successive approximations, each of which is a smoother version of $f(t)$. The successive approximations correspond to different resolutions, hence the name multi-resolution analysis.

The representation of $f(t)$ at the $m$th scale is given as:[18]

$$f^{(m)}(t) = \sum_{j=-\infty}^{j=+\infty} f(m, j)\phi(2^m t - j), \quad m, j \in \mathcal{Z}. \tag{1}$$

If the $(m + 1)$th approximation is a refinement of the $m$th approximation, the function $\phi(2^m t)$ should satisfy the following criterion:

$$\phi(2^m t) = \sum_j h(j)\phi(2^{m+1} t - j), \tag{2}$$

where, the right-hand side is a weighted sum of the translated (indexed by $j$) and dilated (indexed by $m$) basis functions spanning the space of the $(m + 1)$th approximation and $h(j)$ are the *scaling function coefficients*.

The concept of a wavelet transform can be explained by multi-resolution analysis by considering the incremental detail added in obtaining the $(m + 1)$th scale approximation given the $m$th one. Let $V^{(m+1)}$ represent the space of all functions spanned by the orthogonal set, $\{\phi(2^{m+1}t - k); k \in Z\}$ corresponding to a finer approximation of functions in $V^{(m)}$, the space of the all functions spanned by the orthogonal set, $\{\phi(2^m t - l); l \in Z\}$. Then, $V^{(m)} \subset V^{(m+1)}$ and this relation in terms of sets can be stated as

$$V^{(m+1)} = V^{(m)} \oplus W^{(m)}. \quad (3)$$

$W^{(m)}$ is orthogonal to $V^{(m)}$ and is the space that contains the information or detail in going from the coarser, $f^{(m)}(t)$, to the finer, $f^{(m+1)}(t)$, representation of the original function $f(t)$. The symbol $\oplus$ is the orthogonal sum.

The basis of the space $W^{(m)}$ is spanned by the orthogonal translates of a single *wavelet function*, $\psi(2^m t)$. Similar to Eq. (2), this is expressed as:[19]

$$f^{(m+1)}(t) = f^{(m)}(t) + \sum_{j=-\infty}^{j=+\infty} \delta f(m, j) \psi(2^m t - j), \quad m, j \in Z, \quad (4)$$

where $\delta f(m, j)$ are the wavelet coefficients. Not every wavelet function forms an orthogonal set.[16] The *wavelet* function $\psi(2^m t)$ is related to the *scaling* function $\phi(2^{m+1} t)$ through the relationship:[19]

$$\psi(2^m t) = \sum_k g(j) \phi(2^{m+1} t - j), \quad (5)$$

where $g(j)$ are the *wavelet coefficients*.

Combining Eqs. (1), (2) and (5) yields the synthesis form of the wavelet transform:

$$\sum_{j=-\infty}^{+\infty} f(m+1, j) \phi(2^{m+1} t - j)$$

$$= \sum_{j=-\infty}^{+\infty} f(m, j) \sum_k h(k) \phi(2^{m+1} t - j - k)$$

$$+ \sum_{j=-\infty}^{+\infty} \delta f(m, j) \sum_k g(k) \phi(2^{m+1} t - j - k) \quad m, j \in Z. \quad (6)$$

This defines a dynamic relationship between the scaling coefficients $f(m + 1, j)$ at one scale and the scaling and wavelet coefficients respectively ($f(m, j)$ and $\delta f(m, j)$) at the next. The coefficients $f(m, j)$

and $\delta f(m, j)$ are given by

$$f(m, j) = \sum_k h(2j - k) f(m + 1, k)$$
$$\delta f(m, j) = \sum_k g(2j - k) f(m + 1, k) \quad . \quad (7)$$

## 2. Motif Searching

### 2.1. Introduction

Motifs are small conserved regions within protein sequences. They usually carry specific structural or functional significance. Detection of common motifs among proteins with low sequence identities provides important clues to the function of the proteins or to classify unknown proteins into proper families, since similarity search is often incapable of identifying proteins with less than 30% identity.[20]

Most classical methods for motif detection can be divided into two major categories. The first one involves crafting a consensus sequence or pattern to reflect conserved amino acids in the motif. Many methods for obtaining consensus patterns are based on multiple alignments of known motif sequences.[21–23] The second category of motif detecting algorithm involves using a scoring or weight matrix.[24–28]

Since motifs are portions of protein sequences having structural significance, many recent motif detecting algorithms make use of the structural data of proteins. For example, multiple structure alignment of proteins[29] was shown to be able to detect common motifs. Repeating motifs could also be detected by continuous wavelet transforms.[30]

Alternatively, statistics or data mining techniques were also applied to locate protein motifs. The widely used MEME[31,32] discovers motifs from unaligned sequences by a statistical algorithm called expectation-maximization (EM).[33] Narasimhan et al.[34] described an automatic approach for motif detection using data mining and knowledge discovery techniques.

Krishnan et al.[35] adopted a new approach to motif detection. By representing a set of aligned protein sequences numerically, and then utilizing pattern recognition (similarity matching) techniques using wavelet transforms, they demonstrated that structural motifs do carry certain common components across structurally similar proteins.

## 2.2. Methods

With the application of signal processing techniques like Discrete Fourier Transforms (DFT) to protein sequences, it is hardly surprising that Discrete Wavelet Transforms (DWT) have also been utilized for analyzing the interaction of protein sequences. For example, DWTs has been used to study the similarity of two sequences across scales by taking the DWT of two protein sequences followed by cross correlation analysis at different scales.[36] The maximum absolute value of the correlation coefficient at each decomposition level is regarded as the similarity score for these two proteins at that level.

Signal processing techniques have been used extensively for pattern recognition in other fields. There is extensive literature dealing with the application of such techniques in the data mining field for determining the similarity of two sets of data, especially in the time-series domain.[37–40] A common feature of all these methods is the use of DFT to map sequences into frequency domain and to keep the first few coefficients in the index since this is thought, in general, to characterize the sequence. Two sequences are considered if their Euclidean distance is less than a user-defined threshold. However, there is a vast class of applications, where it is important to pose queries in terms of *similarity* of data sequences or objects rather than on equality or inequality.[41] The work done here, falls within this class of problems. There are often protein sequences that have very low sequence homology; yet have regions of mutually conserved motifs. Such regions are very difficult to predict from a simple sequence similarity analysis such as BLAST/ClustalW searches.

Our algorithm is shown in Algorithm 1. The algorithm first does a multiple sequence alignment in order to obtain equal sized sequences and then applies wavelet deconstruction and reconstruction to obtain the reconstructed details. At each level and between every pair of sequences and for a given window size $k$, a distance measure (MSVS metric[35]) is calculated. A threshold $\epsilon$ is then applied in order to identify regions that are similar across sequences.

**Algorithm 1** rapid motif searching using wavelet transforms

1: INPUTS ← $n$ sequences $\vec{S}_i, i = 1, \ldots, n, |S_i| = length(S_i)$
2: Get multiple sequence alignment (MSA) of $S$
3: $\mathcal{T} : S \to X$ where $S = \{strings\}, X = \{numeric\}$
4: Normalize values by $\vec{X}_i = \frac{\vec{S}_i - \mu_S}{\sigma_S}$
5: $k$ ← Window Size; $\epsilon$ ($\epsilon < 1$) ← Threshold
6: $\mathcal{W}$ ← 1D Wavelet Transform using Haar wavelet up to scale $\lceil \log_2 N \rceil$
7: $RD$ ← $Reconstructed\ Detail(\mathcal{W})$
8: **for all** Levels $\mathcal{M}$ **do**
9:   **for all** pairs of sequences $i, j$ **do**
10:     $C_{i,j}$ ← $CorrCoeff(RD_i, RD_j)$
11:   **end for**
12: **end for**
13: $M \subset \mathcal{M}$ with significant correlation
14: **for all** Levels $M$ **do**
15:   **for all** windows $K_i, i = 1, \ldots, N - k + 1; |K| = k\ N = $ size of sequences after MSA **do**
16:
$$D_{\vec{X},k} = \sum_{r,s=1; r \neq s}^{r,s=n} \left\{ \sum_{j=i}^{i+k-1} [(x_{r_j} - X_{rA}) - (x_{s_j} - X_{sA})]^2 \right\}^{1/2}$$
17:     Normalize $D_{\vec{X},k}$ to $[0, 1]$
18:     **if** $D_{\vec{X},k} > \epsilon$ **then**
19:       Dismiss sequences as dissimilar
20:     **end if**
21:   **end for**
22: **end for**
23: Plot of $1 - D_{\vec{X},k}$ gives regions of similarity

## 2.3. Results

We have applied the algorithm to data taken from BAliBASE,[42] a database collecting hundreds of manually aligned protein sequences mostly based on three-dimensional structural superimpositions. Figure 1 shows the results from the application of the wavelet-based rapid motif detection (RMD)

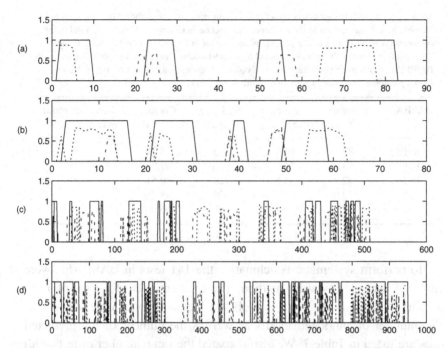

**Fig. 1.** The figure shows the results from the application of the technique on four different sets of data. The legends and threshold levels for the four figures (from top to bottom) are as follows: **(a) (1aboA)** :- Legend *solid line*: Reference, *dashed line*: $d_1$ Level, *dotted line*: $d_3$ Level; Thresholds $d_1$: 0.4, $d_2$: 0.2. **(b) (1aho)** :- Legend *solid line*: Reference, *dashed line*: $d_1$ Level, *dotted line*: $d_4$ Level; Thresholds $d_1$: 0.4, $d_4$: 0.4. **(c) (2myr)** :- Legend *solid line*: Reference, *dashed line*: $d_1$ Level, *dotted line*: $d_3$ Level; Thresholds $d_1$: 0.4, $d_3$: 0.2. **(d) (1taq)** :- Legend *solid line*: Reference, *dashed line*: $d_1$ Level, *dotted line*: $d_4$ Level; Thresholds $d_1$: 0.4, $d_4$: 0.2.

technique to four different data sets from BAliBASE. A window size of $k = 5$ was used for all the cases. A threshold $\epsilon = 0.4$ was used for the $d_1$ level. The threshold values for the other levels were smaller than that for $d_1$ and are indicated in the captions. As can be observed from the figures, the algorithm is able to correctly identify the different conserved regions. The figures also show that not all conserved motifs are seen in the same frequency band.

The results in Figs. 1(a) and 1(c) indicate that wavelet transforms can detect conserved motifs in proteins with low sequence similarities. In our tests, the similarity is below 25%. Most conserved motifs indicated by BAliBASE are clearly identifiable by our method.

**Table 1.** Column 1: Different sections of protein sequences taken from the BAliBASE. Column 2: Number of tests carried out in the corresponding section in Column 1. Column 3: Total number of annotated motif blocks of the corresponding section in Column 1 in BAliBASE. Column 4: Total number of times the motifs predicted by wavelet analysis matched the motifs annotated in BAliBASE. Column 5: Percentage ratio of column 4 against column 3. Column 6: Number of motifs predicted by wavelet analysis on shuffled sequences.

| BAliBASE | Number of tests | Annotated core blocks | Predicted motifs | Correct predictions | Predicted motifs over shuffled seq. |
|---|---|---|---|---|---|
| Ref 1 | 82 | 616 | 525 | 85.22 | 78 |
| Ref 2 | 23 | 241 | 236 | 97.92 | 0 |
| Ref 3 | 12 | 13 | 124 | 90.51 | 0 |
| Ref 4 | 12 | 32 | 30 | 93.75 | 0 |
| Ref 5 | 12 | 50 | 48 | 96.00 | 1 |

To perform systematic benchmarks, the 141 tests in BAliBASE were analyzed using the wavelet method. The tests consists of a total of 1293 protein sequences. On average each test includes about nine sequences. The number of annotated motifs as well as the number of the predicted ones are listed in Table 1. We also prepared the same number of tests with shuffled sequences as the control data.

## 2.4. *Allergenicity prediction*

Allergens are proteins that induce allergic responses. More specifically, they elicit IgE antibodies and cause the symptoms of allergy, which has been a major health problem in developed countries.[43]

With many transgenic proteins introduced into the food chain, the need to predict their potential allergenicity has become a crucial issue. Bioinformatics, more specifically, sequence analysis methods have an important role in the identification of allergenicity.[44,45] To identify allergenicity of a novel protein, World Health Organization (WHO) and the Food and Agriculture Organization (FAO) provide a guideline.[46] In addition to biological tests, this guideline says that a query protein is potentially allergenic if it either has an identity of at least six contiguous amino acids or minimum 35% sequence similarity over a window of 80 amino acids when compared with known allergens.[47] The six amino-acid identity rule is not practical because it produces a large number of false positive hits. The criterion of having a minimum 35% sequence similarity is on the other hand too stringent to find most true allergens. In any case, it has been shown[48] that the precision is low for methods solely relying on sequence similarities.

Analysis using protein motifs are an emerging method for the prediction of potential allergenicity.[48] The methods involve the following steps. First, a reference allergen database is to be created based on known allergens. Then a motif identification tool, such as MEME,[31] is used to identify candidate motifs among all proteins in the allergen database. The candidate motifs are then used to create a protein profile. Finally, a query protein is predicted to be allergenic if its sequence matched an allergen motif with a high score.

Our allergenicity prediction system is derived from the above mentioned motif approach but is based on the motif detection algorithm described above.

The flowchart in Fig. 2 describes the method that we used for allergenicity prediction. The distances between each pair of allergen protein sequences were calculated using ClustalW's global alignment functionality. These

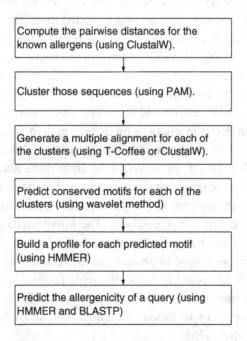

**Fig. 2.** Flowchart of the entire prediction system.

distances were then extracted and used to cluster the sequences into different groups so that sequences in a cluster have the same conserved regions. This was done through the partitioning around medoids method[49] using the R package.[50] The reference allergen sequences were thus divided into many smaller clusters. The sequences in each cluster were then multiple-aligned using either ClustalW[51] or T-Coffee.[52] This step was needed since the motif prediction in the next step required the input sequences to be aligned. Each of the cluster of sequences were then analyzed to identify allergen motifs using the wavelet analysis method and all the allergen motifs were thus extracted. We had found that there were about 20% of the sequences for which no conserved motif was detected. To reduce the number of such unique sequences, we implemented a two-step clustering procedure. In the first step, the initial unique sequences are matched against the motifs already obtained. The remaining unique sequences go through a re-clustering procedure to obtain all new motifs. The new motifs are added to the allergen motif list. Following this, an HMM profile for each motif was created using the HMMER package.[53] The allergen protein sequences, belonging to the clusters where no motifs were detected, were grouped together into another database. Any protein sequence can now be tested for allergenicity by scoring it against the database of allergen motif profiles first and then doing a similarity search using BLASTP against all the sequences in the second database. If either of these tests resulted in scores with E-values less than a threshold, the query sequence is predicted to be allergenic. Each of the main steps above will be elaborated in the sections below.

The performance of our allergen prediction system was assessed by a ten-fold cross validation experiment. The performance was measured as *recall* and *precision*. The recall measures the algorithm's ability to pick out true allergens, while the precision is a measurement of the ability to separate true allergens from the false ones. The performance of our methodology was tested on the entire Swiss-Prot database. The allergenicity was predicted for each of the 135 850 Swiss-Prot proteins (as of 11-Nov-2003). All 664 reference allergens were used to prepare the allergen motifs. We predicted 62 different motifs that could identify allergenic proteins.

The results are summarized in Table 2. A total of 2042 potential allergens are predicted using the motif-based prediction. Adding BLASTP searches, the number goes up to 4768.

**Table 2.** The motif-based allergenicity predictions for the 135 850 Swiss-Prot proteins. Known allergens are defined as the sequences that can be found both in our reference allergen database (664 sequences) and in Swiss-Prot, using 100% sequence identity as the criterion. We found 341 known allergens. The recalls are the ratios between the predicted known allergens and the 341 total known allergens. The "motif+BLASTP" method involves subsequent BLASTP searches, E-value of $10^{-3}$ was used.

| Prediction methods | Predicted allergen | % ratio in Swiss-Prot | Predicted known allergen | % recall |
|---|---|---|---|---|
| Motif+BLASTP | 4768 | 3.5 | 319 | 93.6 |
| Motif only | 2042 | 1.5 | 295 | 86.5 |

## 3. Transmembrane Helix Region (HTM) Prediction

HTM location predictions are important since a large portion (10–35%) of proteins are membrane proteins. However, the determination of the structure of TM proteins is difficult both by NMR and X-ray spectroscopy. Hence predictive methods have assumed enormous importance. Classical prediction methods use amino acid propensities or hydropathy scales[54–58] to predict HTM regions. Other methods include the use of neural networks,[59,60] Hidden Markov Models (HMM)[61] as well as the incorporation of evolutionary information.[62]

There have been efforts at trying to utilize wavelet transforms in order to predict HTM regions. Lio et al.[63] used a change-point based wavelet shrinkage along with a new propensity scale to predict the location and topology of HTM segments. Their algorithm consists of four main steps: creation of a propensity profile based on the frequency of residues in the HTM segments, decomposition of the profile into wavelet coefficients, data-dependent thresholding of the wavelet coefficients in order to detect abrupt changes in the profile followed by reconstruction of the de-noised profile from wavelet coefficients. The algorithm makes use of the nature of HTM regions that are composed of HTM segments of 15–30 predominantly hydrophobic residues followed by polar connecting loops. Thus propensity profiles based on hydrophobic or other properties tend to show sharp changes in correspondence of HTM regions.[63]

Pashou et al.[64] use a different approach in identifying HTM regions. The authors first created an averaged hydrophobicity profile followed by a wavelet deconstruction and denoising by thresholding. They then used a dynamic programming (DP) algorithm on the denoised signal to predict HTM spanning regions. The DP algorithm produces models of the number, the length and the location of the membrane-spanning segments. The model with the optimal score is selected among those satisfying the determined constraints.

## 4. Hydrophobic Cores

Proteins folding occurs by the burial of hydrophobic residues in the interior with the hydrophilic residues predominantly residing on the surface of the proteins. This structure is stabilized by the interaction between the hydrophobic residues. As a result, the buried hydrophobic residues associate and form solvent-shielded hydrophobic cores, defined by the degree of solvent accessibility to the side-chain of each amino acid. There are essentially two prominent methods for prediction of hydrophobic cores; those that perform in the absence of a sequence alignment and those that require a sequence alignment.

The first category consists of algorithms that utilize hydropathy plots[65] but this approach has the drawback that site-specific information is lost due to averaging of individual residue hydrophobicity. Methods such as neural networks[66] or those using homologous proteins[67–69] are however site-specific and lead to more accurate predictions. On the other hand, for methods that require a sequence alignment, the query protein is compared with similar sequences with known 3D structures.[70] The buried residues are predicted accurately in cases of high sequence similarity but prediction accuracy falls in the case of low similarity of sequences.

Hirakawa et al.[71] utilized wavelet transforms to decompose hydropathy plots into high and low frequencies. They used Daubechies wavelet with tap 4 as a mother wavelet. They utilized only the low frequency wavelet coefficients. After thresholding the extract low frequencies, the residues above the threshold were predicted as buried residues to form hydrophobic cores.

## 5. Protein Repeat Motifs

Protein repeats are ubiquitous and of different types ranging from single residue repeats to heptad repeats in coiled coils, motif repeats as well as to the repetition of homologous domains of 100 or more residues. Motif repeats are defined as secondary or supersecondary structural units such as $\alpha$-helices, $\beta$-strands, $\beta$-sheets, Rosmann folds etc. connected together by short lengths of peptide in a repeating pattern.[30] Traditional repeat detection methods utilize standard sequence comparison algorithms adapted to find repeats.[72–74] Such methods work well for perfect repeats; however, in the absence of significant sequence similarity due to insertions, deletions and mutations, repeat detection becomes very difficult. Additionally repeats may be incomplete, widely spaced and of multiple types interspersed throughout a sequence. Thus there is a need to develop algorithms to identify such repeats.

Murray et al.[30] provide an alternative approach by using continuous wavelet transforms (CWT) to detect and characterize repeating protein motifs from sequence and structural data. They first obtain the CWT of sequences transformed using accessible surface area and Kyte-Dolittle[54] hydrophobicity scales. The same is also done for secondary structural information by means of a simple ternary encoding of $\{-1, 0, 1\}$ to represent helix, sheet and turn. In order to characterize the relationship between wavelet scale and frequency, the Fourier transform of each wavelet scale is calculated. The normalized Fourier power spectrum of the wavelet coefficients, indexed by scale is thus calculated. The position of extrema across scales is used to denoise the wavelet coefficients and thereby obtain dominant features of interest by using the wavelet transform modulus maxima.

Murray et al. applied their technique to different types of proteins known to contain repeating motifs such as four-bladed propeller domain of rabbit serum haemopexin, coiled coil dimerization domain from cortexillin I, TIM barrel domain from the enzyme triosephosphate isomerase and a $\beta\alpha$ repeat from the leucine rich repeat ribonuclease inhibitor. Their results indicate that hydrophobicity can provide useful information on repeating motifs and their topology with the caveat that wavelet transforms cannot easily detect insertions and deletions.

## 6. Sequence Comparison

Sequence comparison and alignment, as described in an earlier part of this review, represent one of the most important areas of protein sequence analysis. Here the objective is to determine the similarities between two or more protein sequences. Based on a wavelet decomposition of protein sequences,[75] people have demonstrated that a protein sequence can be examined at different spatial resolutions. Two sequences can be compared at different spatial resolutions using similarity vectors. This new similarity concept is an expansion of the conventional sequence similarity, which only takes into account the local pairwise amino acid match and ignores the information contained in coarser spatial resolutions.

Two protein sequences with low sequential identity may show similarities in their physicochemical properties, tertiary structure, resonance recognition model (RRM) spectra[76] and biological functions. The RRM multiple-cross spectral function can be regarded as a measurement of the similarity among different protein sequences in the frequency domain when each protein sequence is treated as a numerical series. When wavelet transform is introduced to a protein sequence, the similarity can be measured at different resolution scales. The sequences being compared are initially converted into numerical series using RRM. These numerical series are normalized to zero mean and unit standard deviation and zero-padded to have an identical sequence length. Subsequently they are decomposed to $M$ levels with details from level 1 to level $M$ and an approximation at level $M$ by discrete wavelet transform using Bior3.3 biorthogonal wavelets.[77] Because a correlation function quantifies the degree of interdependence of one process upon another, the cross-correlation coefficients are calculated at each level to establish and quantify the similarity between the two compared protein sequences. There are a total of $M + 1$ correlation coefficients.

The maximum absolute value of the correlation coefficient at each decomposition level is regarded as the similarity score for these two proteins at that level. Therefore, a total of $M + 1$ maximum values are taken out to form a sequence-scale similarity vector. This vector describes the correlation with a multiresolution point of view.

As an example, although having different lengths and sharing no significant similarity (measured by BLAST), the 83 amino acids chymotrypsin and 56 amino acids subtilison do have a common proteolytic function and catalytic mechanism. The sequence-scale similarity analysis[75] shows a week correlation at D4 for these two distantly related proteins.

In a separate study,[78] a novel metric based on discrete wavelet transform and various protein substitution models has been proposed to fund functionally similar proteins. The authors adopted two types of substitution models to transform a protein sequence into a numerical one. The first relies on the commonly used amino acid substitution matrices such as PAM and BLOSUM, the other is based on amino acid physicochemical properties. Once converted, discrete wavelet transform is applied to the sequences and cross-correlation coefficients can be calculated as before at each scale to quantify the similarity between the two compared protein sequences. In the paper, however, a new metric was proposed where similarities on different decomposition levels are calculated separately and a weighted sum is taken using each level's energy as weight. The new metric was shown to be able to detect functional similarity of protein sequence with low identity and can even complement the existing sequence alignment methods.

## 7. Prediction of Protein Secondary Structures

The secondary structure of proteins is dominated by $\alpha$-helices, $\beta$-sheets and short peptides connecting $\alpha$-helices and $\beta$-strands (such as $\beta$-turns, $\gamma$-turns, $\pi$-turns, $\Omega$-turns and specific coils). The prediction of secondary structure from amino acid sequence is a well-defined problem in bioinformatics, and is often regarded as the first step in understanding the protein folding problem.

It has been known that hydrophobicity of amino acid sequences is the most important factor that affects the stability of a protein. Mandell et al.[79] have demonstrated that by representing a protein as a sequence of hydrophobic free energies per residue, the distributions of the amplitudes of the sequential fluctuations in hydrophobic free energy could be examined with respect to location and scale using Morlet wavelet transforms. This in turn gives good hints about the secondary structure of that protein.

A later paper[80] adopted continuous wavelet transform to predict $\alpha$-helices and connecting peptides simultaneously by utilizing the hydrophobicity of amino acid sequence. Using Morlet wavelet, the authors found that the local minima of wavelet coefficients under the appropriate scales corresponding well to $\alpha$-helices and connecting peptides after the amino acid sequence of protein is transformed into hydrophobicity sequence per residue. The best prediction accuracy obtained was 92.8% for $\alpha$-helices and 92.3% for connecting peptides.

## 8. Disease Related Studies

It has been shown that wavelet and other signal processing techniques can be used to characterize disease related proteins.[81] Here the above described Resonant Recognition Model (RRM) was adopted to study proteins related to Alzheimer's disease: beta amyloid protein, amyloid protein precursor and Nerve Growth Factor (NGF). The author found that both BGF and beta amyloid precursors have a common frequency ($f = 0.387 \pm 0.018$ and $0.404 \pm 0.017$) that is missing from beta amyloid protein. This observation supports experimental evidence[82] that NGF could be used to protect neurons and brain cells against degeneration.

## 9. Other Functional Prediction

Mass spectrometry is another area where wavelet techniques have been shown to be useful. The latest example is a high-accuracy peak picking program for proteomics data.[83] The authors described an algorithm that addresses the main difficulties faced by peak picking algorithms: (1) the considerable asymmetry in the mass spectrum peaks which confounds a correct mass to charge computation; (2) the convolution of isotopic peaks makes it hard to distinguish individual peaks. Since mass spectra are of inherently multiscale nature, different effects, typically localized in different frequency ranges, would add up to result in the final signal. This makes for an ideal problem for applying wavelet techniques.

The algorithm uses a continuous wavelet transform to split the signal into different frequency ranges or length scales that can be regarded independently of each other. This decomposition allows us to determine each feature of a peak in the domain from which it can be computed best. The first stage of the algorithm determines the positions of putative peaks in the wavelet-transformed signal. In the second stage, an analytically given peak function is fit to the data in that peak region. The major contribution in applying the wavelet transform is the ability to determine the end points of a peak even if it overlaps heavily with another one.

## 10. Conclusion

The strength of wavelet techniques comes from its ability for capturing hidden components from biological data. We have described a few bioinformatics applications using wavelet techniques, mainly in the fields of protein sequence analysis. Certainly there are more problems in biology that can be explored by wavelet techniques.

There are two main aims of this review. Firstly, we would like to increase the familiarity of wavelet techniques in the bioinformatics community. Secondly, for the mathematics and engineering communities, this review provides general ideas about existing applications of wavelet based techniques in bioinformatics. We hope that this review serves as a link for the two types of people; the biologists and the signal processing specialists. Furthermore, we have no doubt that wavelet applications in biology will continue to be seen in abundance the years to come.

## References

1. A. M. Lesk, Computational molecular biology, A. Kent and J. G. Williams (eds), *Encyclopedia of Computer Science and Technology*, Marcel Dekker, New York **31**, 101–165 (1994).
2. S. F. Altschul, W. Gish, W. Miller, E. W. Myers and D. J. Lipman, Basic local alignment search tool, *J. Mol. Biol.* **215**(3), 403–410 (1990).
3. S. F. Altschul, T. L. Madden, A. A. Schaffer, J. Zhang, Z. Zhang, W. Miller and D. J. Lipman, Gapped BLAST and PSI-BLAST: a new generation of protein database search programs, *Nucleic Acids Res.* **25**(17), 3389–3402 (1997).
4. W. Gish. WU BLAST. http://blast.wustl.edu (1996–2003).
5. B. Ma, J. Tromp and M. Li, PatternHunter: faster and more sensitive homology search, *Bioinformatics* **18**(3), 440–445 (2002).
6. S. Schwartz, W. J. Kent, A. Smit, Z. Zhang, R. Baertsch, R. C. Hardison, D. Haussler and W. Miller, Human-mouse alignments with BLASTZ, *Genome Res.* **13**(1), 103–107 (2003).
7. S. B. Needleman and C. D. Wunsch, A general method applicable to the search for similarities in the amino acid sequence of two proteins, *J. Mol. Biol.* **48**(3), 443–453 (1970).
8. T. F. Smith and M. S. Waterman, Identification of common molecular subsequences, *J. Mol. Biol.* **147**(1), 195–197 (1981).
9. X. Q. Huang and W. Miller, A time-efficient, linear-space local similarity algorithm, *Adv. Appl. Math.* **12**(3), 337–357 (1991).
10. Z. Zhang, S. Schwartz, L. Wagner and W. Miller, A greedy algorithm for aligning DNA sequences, *J. Comput. Biol.* **7**(1–2), 203–214 (2000).
11. Z. Ning, A. J. Cox and J. C. Mullikin, SSAHA: a fast search method for large DNA databases, *Genome Res.* **11**(10), 1725–1729 (2001).
12. W. J. Kent, BLAT — the BLAST-like alignment tool, *Genome Res.* **12**(4), 656–664 (2002).
13. S. Mallat, A theory for the multiresolution signal decomposition: the wavelet representation, *IEEE Pattern Analysis and Machine Intelligence* **11**(7), 676–693 (1989).

14. I. Daubechies, Orthonormal bases of compactly supported wavelets, *Commun. Pure Appl. Math.* **41**, 906–966 (1988).
15. I. Daubechies, Ten lectures on wavelets, *Cbms-Nsf Regional Conference Series in Applied Mathematics* **61** (1992).
16. B. Jawerth and W. Sweldens, An overview of wavelet based multiresolution analyses, *SIAM Review* **36**(3), 377–412 (1994).
17. G. Strang. Wavelet transforms versus fourier transforms, *Bulletin of the American Mathematical Society* **28**(2), 288–305 (1993).
18. A. S. Willsky, Modeling and estimation of multiresolution stochastic processes, *Special issue of the IEEE Trans. on Inf. Theory on Wavelet Transforms and Multiresolution Signal Analysis* **38**(2), 766–784 (1992).
19. C. K. Chui, *An Introduction to Wavelets*, Academic Press (1992).
20. B. Rost and A. Valencia, Pitfalls of protein sequence analysis, *Curr. Opin. Biotechnol.* **7**(4), 457–461 (1996).
21. L. Falquet, M. Pagni, P. Bucher, N. Hulo, C. J. A. Sigrist, K. Hofmann and A. Bairoch, The PROSITE database, its status in 2002, *Nucleic Acids Res.* **30**(1), 235–238 (2002).
22. C. O. Pabo and R. T. Sauer, Transcription factors: structural families and principles of DNA recognition, *Annu. Rev. Biochem.* **61**, 1053–1095 (1992).
23. C. G. Nevill-Manning, T. D. Wu and D. L. Brutlag, Highly specific protein sequence motifs for genome analysis, *Proc. Natl. Acad. Sci. USA* **95**(11), 5865–5871 (1998).
24. M. Gribskov, A. D. McLachlan and D. Eisenberg, Profile analysis: detection of distantly related proteins, *Proc. Natl. Acad. Sci. USA* **84**(13), 4355–4358 (1987).
25. I. B. Dodd and J. B. Egan, Improved detection of helix-turn-helix DNA-binding motifs in protein sequences, *Nucleic Acids Res.* **18**(17), 5019–5026 (1990).
26. D. A. Parry, Coiled-coils in alpha-helix-containing proteins: analysis of the residue types within the heptad repeat and the use of these data in the prediction of coiled-coils in other proteins, *Biosci. Rep.* **2**(12), 1017–1024 (1982).
27. A. Lupas, M. Van Dyke and J. Stock, Predicting coiled coils from protein sequences, *Science* **252**(5010), 1162–1164 (1991).
28. B. Berger, D. B. Wilson, E. Wolf, T. Tonchev, M. Milla and P. S. Kim, Predicting coiled coils by use of pairwise residue correlations, *Proc. Natl. Acad. Sci. USA* **92**(18), 8259–8263 (1995).
29. N. Leibowitz, R. Nussinov and H. J. Wolfson, MUSTA — a general, efficient, automated method for multiple structure alignment and detection of common motifs: application to proteins, *J. Comput. Biol.* **8**(2), 93–121 (2001).
30. K. B. Murray, D. Gorse and J. M. Thornton, Wavelet transforms for the characterization and detection of repeating motifs, *J. Mol. Biol.* **316**(2), 341–363 (2002).
31. T. L. Bailey and C. P. Elkan, Fitting a mixture model by expectation-maximization to discover motifs in biopolymers, *Proceedings of the 2nd International Conference on Intelligent Systems for Molecular Biology*, Menlo Park, CA, AAAI Press, pp. 28–36 (1994).
32. T. L. Bailey, M. E. Baker and C. P. Elkan, An artificial intelligence approach to motif discovery in protein sequences: application to steriod dehydrogenases, *J. Steroid Biochem. Mol. Biol.* **62**(1), 29–44 (1997).
33. A. P. Dempster, N. M. Laird and D. B. Rubin, Maximum likelihood from complete data vis the EM algorithm, *J. R. Statist. Soc.* **39**, 1–38 (1977).

34. G. Narasimhan, C. Bu, Y. Gao, X. Wang, N. Xu, and K. Mathee, Mining protein sequences for motifs, *J. Comput. Biol.* **9**(5), 707–720 (2002).
35. A. Krishnan, K.-B. Li and P. Issac, Rapid detection of conserved regions in protein sequences using wavelets, *Silico Biology* **4**, 0013 (2004).
36. C. H. De Trad, Q. Fang and I. Cosic, An overview of protein sequence comparisons using wavelets, *Proceedings of the IEEE-EMBS*, pp. 115–119 (2001).
37. R. Agrawal, C. Faloutsos and A. N. Swami, Efficient similarity search in sequence databases, D. Lomet, (ed.), *Proceedings of the 4th International Conference of Foundations of Data Organization and Algorithms (FODO)*, Chicago, Illinois, Springer Verlag, pp. 69–84 (1993).
38. C. Faloutsos, M. Ranganathan and Y. Manolopoulos, Fast subsequence matching in time-series databases, *Proceedings 1994 ACM SIGMOD Conference*, Minneapolis, MN, pp. 419–429 (1994).
39. R. Agrawal, K.-I. Lin, H. S. Sawhney and K. Shim, Fast similarity search in the presence of noise, scaling, and translation in time-series databases, *Twenty-First International Conference on Very Large Data Bases*, Zurich, Switzerland, Morgan Kaufmann, pp. 490–501 (1995).
40. D. Rafiei and A. Mendelzon, Similarity-based queries for time series data, *ACM Special Interest Group on Management of Data*, pp. 13–25 (1997).
41. H. V. Jagadish, A. O. Mendelzon and T. Milo, Similarity-based queries, *Proc. 14th Symp. on Principles of Database Systems (PODS'95)*, pp. 36–45 (1995).
42. J. D. Thompson, F. Plewniak and O. Poch, BAliBASE: a benchmark alignment database for the evaluation of multiple alignment programs, *Bioinformatics* **15**(1), 87–88 (1999).
43. S. J. Ono, Molecular genetics of allergic diseases, *Annu. Rev. Immunol.* **18**, 347–366 (2000).
44. R. E. Hileman, A. Silvanovich, R. E. Goodman, E. A. Rice, G. Holleschak, J. D. Astwood and S. L. Hefle, Bioinformatic methods for allergenicity assessment using a comprehensive allergen database, *Int. Arch. Allergy Immunol.* **128**(4), 280–291 (2002).
45. S. M. Gendel, Sequence analysis for assessing potential allergenicity, *Ann N.Y. Acad. Sci.* **964**, 87–98 (2002).
46. C. Bindslev-Jensen, E. Sten, L. K. Earl, R. W. R. Crevel, U. Bindslev-Jensen, T. K. Hansen, P. S. Skov and L. K. Poulsen, Assessment of the potential allergenicity of ice structuring protein type III HPLC 12 using the FAO/WHO 2001 decision tree for novel foods, *Food Chem. Toxicol.* **41**(1), 81–87 (2003).
47. R. C. Aalberse, Structural biology of allergens, *J. Allergy Clin. Immunol.* **106**(2), 228–238 (2000).
48. M. B. Stadler and B. M. Stadler, Allergenicity prediction by protein sequence, *FASEB J.* **17**(9), 1141–1143 (2003).
49. L. Kaufman and P. J. Rousseeuw, *Finding Groups in Data: An Introduction to Cluster Analysis*, John Wiley & Sons, Brussels, Belgium (1990).
50. R. Ihaka and R. Gentleman, R: a language for data analysis and graphics, *J. Comput. Graph. Stat.* **5**(3), 299–314 (1996).
51. J. D. Thompson, D. G. Higgins and T. J. Gibson, CLUSTAL W: improving the sensitivity of progressive multiple sequence alignment through sequence weighting, position-specific gap penalties and weight matrix choice, *Nucleic Acids Res.* **22**(22), 4673–4680 (1994).

52. C. Notredame, D. G. Higgins and J. Heringa, T-Coffee: a novel method for fast and accurate multiple sequence alignment, *J. Mol. Biol.* **302**(1), 205–217 (2000).
53. S. R. Eddy, Profile hidden Markov models, *Bioinformatics* **14**(9), 755–763 (1998).
54. J. Kyte and R. F. Doolittle, A simple method for displaying the hydropathic character of a protein, *J. Mol. Biol.* **157**(1), 105–132 (1982).
55. G. von Heijne, Membrane protein structure prediction, hydrophobicity analysis and the positive-inside rule, *J. Mol. Biol.* **225**(2), 487–494 (1992).
56. D. T. Jones, W. R. Taylor and J. M. Thornton, A model recognition approach to the prediction of all-helical membrane protein structure and topology, *Biochemistry* **33**(10), 3038–3049 (1994).
57. D. M. Engelman, T. A. Steitz and A. Goldman, Identifying nonpolar transbilayer helices in amino acid sequences of membrane proteins, *Annu. Rev. Biophys. Biophys. Chem.* **15**, 321–353 (1986).
58. P. K. Ponnuswamy, Hydrophobic characteristics of folded proteins, *Prog. Biophys. Mol. Biol.* **59**(1), 57–103 (1993).
59. B. Rost, R. Casadio, P. Fariselli and C. Sander, Transmembrane helices predicted at 95% accuracy, *Protein Sci.* **4**(3), 521–533 (1995).
60. B. Rost, P. Fariselli and R. Casadio, Topology prediction for helical transmembrane proteins at 86% accuracy, *Protein Sci.* **5**(8), 1704–1718 (1996).
61. E. L. Sonnhammer, G. von Heijne and A. Krogh, A hidden Markov model for predicting transmembrane helices in protein sequences, *Proc. Int. Conf. Intell. Syst. Mol. Biol.* **6**, 175–182 (1998).
62. P. Lio and N. Goldman, Using protein structural information in evolutionary inference: transmembrane proteins, *Mol. Biol. Evol.* **16**(12), 1696–1710 (1999).
63. P. Lio and M. Vannucci, Wavelet change-point prediction of transmembrane proteins, *Bioinformatics* **16**(4), 376–382 (2000).
64. E. E. Pashou, Z. I. Litou, T. D. Liakopoulos and S. J. Hamodrakas, wavetm: wavelet-based transmembrane segment prediction, *Silico Biol.* **4**(2), 127–131 (2004).
65. G. D. Rose and S. Roy, Hydrophobic basis of packing in globular proteins, *Proc. Nat. Acad. Sci. USA* **77**(8), 4643–4647 (1980).
66. S. R. Holbrook, S. M. Muskal and S. H. Kim, Predicting surface exposure of amino acids from protein sequence, *Protein Eng.* **3**(8), 659–665 (1990).
67. T. J. Hubbard and T. L. Blundell, Comparison of solvent-inaccessible cores of homologous proteins: definitions useful for protein modelling, *Protein Eng.* **1**(3), 159–171 (1987).
68. B. Rost and C. Sander, Conservation and prediction of solvent accessibility in protein families, *Proteins* **20**(3), 216–226 (1994).
69. M. B. Swindells, A procedure for the automatic determination of hydrophobic cores in protein structures, *Protein Sci.* **4**(1), 93–102 (1995).
70. H. Wako and T. L. Blundell, Use of amino acid environment-dependent substitution tables and conformational propensities in structure prediction from aligned sequences of homologous proteins. I. solvent accessibility classes. *J. Mol. Biol.* **238**(5), 682–692 (1994).
71. H. Hirakawa, S. Muta and S. Kuhara, The hydrophobic cores of proteins predicted by wavelet analysis, *Bioinformatics* **15**(2), 141–148 (1999).

72. M. A. Andrade, C. P. Ponting, T. J. Gibson and P. Bork, Homology-based method for identification of protein repeats using statistical significance estimates, *J. Mol. Biol.* **298**(3), 521–537 (2000).
73. M. Pellegrini, E. M. Marcotte and T. O. Yeates, A fast algorithm for genome-wide analysis of proteins with repeated sequences, *Proteins* **35**(4), 440–446 (1999).
74. A. Heger and L. Holm. Rapid automatic detection and alignment of repeats in protein sequences, *Proteins* **41**(2), 224–237 (2000).
75. C. H. de Trad, Q. Fang and I. Cosic, Protein sequence comparison based on the wavelet transform approach, *Protein Eng.* **15**(3), 193–203 (2002).
76. I. Cosic, *The Resonant Recognition Model of Macromolecular Bioactivity*, Birkhäuser Verlag, Basel, Switzerland (1997).
77. A. Cohen, I. Daubechies and J. C. Feauveau, Bi-orthogonal bases of compactly supported wavelets, *Commun. Pure Appl. Math.* **45**, 485–560 (1992).
78. Z.-N. Wen, K.-L. Wang, M.-L. Li, F.-S. Nie and Y. Yang, Analyzing functional similarity of protein sequences with discrete wavelet transform, *Comput. Biol. Chem.* **29**(3), 220–228 (2005).
79. A. J. Mandell, K. A. Selz and M. F. Shlesinger, Wavelet transformation of protein hydrophobicity sequences suggests their memberships in structural families, *Physica A* **244**(1–4), 254–262 (1997).
80. J. Qiu, R. Liang, X. Zou and J. Mo, Prediction of protein secondary structure based on continuous wavelet transform, *Talanta* **61**(3), 285–293 (2003).
81. C. H. de Trad, Prediction of characteristic frequencies of Alzheimers disease key proteins with signal processing techniques, *Proceedings of the 4th Annual UAE Research Conference* (2003).
82. D. E. Smith, J. Roberts, F. H. Gage and M. H. Tuszynski, Age-associated neuronal atrophy occurs in the primate brain and is reversible by growth factor gene therapy, *Proc. Nat. Acad. Sci. USA* **96**(19), 10893–10898 (1999).
83. E. Lange, C. Gröpl, K. Reinert, O. Kohlbacher and A. Hilderbrandt, High-accuracy peak picking of proteomics data using wavelet techniques, R. B. Altman (ed), *Proceedings of the 11th Pacific Symposium on Biocomputing*, CRPIT, Hawaii, USA, **11**, 243–254 (2006).

# CHAPTER 6

# FILTERING PROTEIN SURFACE MOTIFS USING NEGATIVE INSTANCES OF ACTIVE SITES CANDIDATES

Nripendra L. Shrestha

*Department of Multimedia Engineering*
*Graduate School of Information Science and Technology*
*Osaka University, Japan*

Takenao Ohkawa

*Department of Computer Science and Systems Engineering*
*Graduate School of Engineering*
*Kobe University, Japan*

Protein surface motifs are defined as patterns of shape and physical properties that commonly appear in protein molecular surface data. We have proposed SUMOMO, a protein surface motif mining tool, which provide filtering process based on the fact that active sites from proteins having a particular function have similar shape and physical properties. However SUMOMO still tends to extract a large amount of surface motifs, making it difficult to distinguish whether they correspond to active sites. In this chapter, we propose a method of filtering surface motifs to further reduce the number of active sites candidates by using the negative instances of active sites. The filtering of active sites candidates is based on the concept of ranking them in the order of their possibility of being an active site. The possibility of an active sites candidate as an active site is determined by the count of negatives instances which are similar to it at a given value of dissimilarity. The proposed method of filtering was applied to protein surface motif data extracted by SUMOMO from two different sets of protein data. Precision-recall curves were drawn for each set of data verifying the effectiveness of the proposed method.

## 1. Introduction

The function of a protein depends on its structure. Since protein functional sites tend to be conserved in evolutionary process, they occur frequently as common local structures, which we call motifs, in different proteins with similar functions. The concept of motifs appeared in 1990s, and a lot of motif databases have been developed. PROSITE,[1] which is one of the most popular motif databases, collects common amino acids sequence patterns. Besides PROSITE, many other works to identify functional sites from the viewpoint of amino acids sequences have been explored.[2,3] On the other hand, identification of protein 3D structural motifs is relatively new in comparison with the protein sequence motifs. Wang et al. have used geometrical hashing to discover protein 3D motifs by focusing on mining motifs from protein 3D structures and $\alpha$-surfaces as sub graphs.[4] Nussinov et al. have also used geometrical hashing paradigms for mining protein structures and surfaces for pharmacophoric patterns.[5] de Rinaldis et al. have used 3D profiles, which are generated by alignment of protein 3D structures, for identification of protein surface similarities.[6]

We have proposed SUMOMO (SUrface MOtif mining in protein MOlecular surface data), a method of mining protein molecular surface database for extracting surface motifs, which are patterns of shape and physical properties that commonly appear in protein molecular surfaces.[7] SUMOMO focuses on spatial, geometrical and physical attributes at each point on a protein molecular surface since the functions of a protein exhibit strong dependency on the clefs and cavities on its surface, which act like interface for protein-ligand and protein-protein interactions.[8,9] Surface motifs are discovered by using normal vectors with attributes to abstract the spatial, geometrical, and physical properties of protein molecular surfaces, and the use of flexible buckets to absorb small structural aberrations.

If surface motifs are extracted only by focusing on similarity of local structures from different combinations of proteins, all of the extracted surfaces cannot be related to function. Consequently, a large number of insignificant structures get extracted. For example, 3183 surface motifs are extracted from 15 proteins, and motifs corresponding to all four type of active sites are recognized. In order to address the issue of filtering significant surface motifs out of a larger number of motifs extracted during the extraction phase of SUMOMO, we have proposed a method of filtering surface motifs by explicitly relying on structural data without any prior information regarding the function of proteins.[7] This approach of filtering is based on popular TF-IDF concept used for classification of documents, and uses the analogy of proteins to documents and surface motif extracted to terms. After the proteins are grouped based on structural similarity of extracted surface motifs, the motifs which occur within a certain group of structurally similar proteins are considered active sites candidates. The surface motifs which occur in proteins from structurally different groups are filtered out as negative instances of active sites candidates. This method of filtering was able to reduce 3183 active site candidates extracted by SUMOMO to 448(14%) without any loss in number of functions.[7]

Nevertheless, to distinguish surface motifs associated four different functions out of 448 active site candidates is still requires an amount of time and effort. To cope with this problem, this chapter presents an improved method to filter surface motifs further by using the negative instances of active site candidates which were filtered out in previous works. The fundamental principle behind the proposed method of filtering is to filter the active sites candidates which are similar to the negative instances. As such the proposed method can have implications in ranking the active sites candidates such that the most likely active site candidates can be selected for *in vitro* experiments.

The next section presents descriptions of protein structure, protein structural data and protein surface motifs. Section 3 overviews the whole process of SUMOMO, which include surface motif extraction and surface motif filtering. The following sections present a new method of filtering surface motifs and details of implementation as well as experiments related with tuning of parameters and evaluations.

## 2. Protein Structural Data and Surface Motifs

### 2.1. *Protein structural data*

There are a number of freely accessible databases which store and maintain protein structural data for public use. The Protein Data Bank (PDB),[10] maintained by Research Collaboratory for Structural Bioinformatics (RCSB), is a worldwide repository for the processing and distribution of three dimensional structure data of proteins and nucleic acids. Data in PDB are typically obtained by X-ray crystallography or NMR spectroscopy, and is submitted from around the world. As of 5 June 2007, the database contained over

43 000 released atomic coordinate entries, and it is increasing at the rate of about 5000 new entries each year.

A number of different kinds of information, including those referring to the protein's function, can be derived from three dimensional structure data stored in PDB. The information regarding relative juxtaposition of various groups within protein structure helps in identifying catalytic clusters, structural motifs, etc. Shape and electrostatic properties of structural data could reveal active sites.

Each structure in PDB receives a four-character alphanumeric identifier, its PDB ID. This should not be used as an identifier for biomolecules, since often several structures for the same molecule (in different environments or conformations) are contained in PDB with different PDB IDs. PDB ID is used throughout this chapter in order to distinguish proteins.

## 2.2. *Protein molecular surface data*

Protein molecule is formed by the complicated folding of a single polypeptide chain. As a result, some of its composing amino acid residues are exposed at the surface while other amino acid residues exist in the interior of its three dimensional structure. The functions of a protein, such as molecular recognition, binding, docking, etc. can be attributed to shape and physical properties of its molecular surface. There are examples of proteins with completely different folding of their main chain giving but identical enzymatic function due to similar shape and physical properties of their local surfaces. In such context, protein molecular surface has been considered an important topic of research.

The source of protein molecular surface data used in this chapter, which is based on the structural data of PDB, is from eF-site (electrostatic-surface of Functional site),[9,a] which is a database for molecular surface of proteins' functional sites maintained by the Protein Data Bank Japan (PDBj).

Protein molecular surface can be detected by tracing the external surface of protein three dimensional structure data. The Connolly surfaces made by using the Molecular Surface Package (MSP) program along with the electrostatic potentials calculated by solving Poisson-Boltzmann equations with the self-consistent boundary method has been used.[11,12]

---

[a] http://ef-site.protein.osaka-u.ac.jp/eF-site

The resulting protein molecular surface data comprises of a set of large number of polygons. Information about physical properties, electrostatic potential and hydrophobic strength, deduced from PDB data, is appended to each point in the polygons forming protein molecular surface.

### 2.3. Functions of a protein and structural motifs

#### 2.3.1. Functions of a protein

The term *function of a protein* can have wide implications. The function of a protein may be described at a variety of levels from biochemical level through to phenotypic function. *Biochemical, or molecular function* refers to the particular catalytic activity, binding properties or conformational changes of a protein. At a higher level, the complex, or metabolic or signal transduction pathway, in which a protein participates describes the *cellular function*. This is always context-dependent with respect to tissue, organ and taxon. Lastly, the phenotypic functions determine the physiological and behavioral properties of an organism. In this chapter, by terms 'function of a protein', we refer to its biochemical function.

#### 2.3.2. Active sites

Active sites are localized portions in protein structures which are responsible for expressing functions. The structural and physical properties of the active site is responsible for the expression of its function.[13]

Similarity in structural and physical properties of active sites will give rise to similarity in functions in different proteins. These structures could be cavities in protein structure which may serve as an interface for bonding with ligands as well as other proteins.

#### 2.3.3. Protein structural motifs

Protein structural motifs are conserved patterns of small fragments/local structures which commonly appear in protein, and are conjectured to have biological significance such as being responsible for structural stability of protein or in expressing functions of the protein. Depending upon the type of protein structure there can be different levels of structural motifs, ranging from sequence motifs, which are patterns of highly conserved amino acid

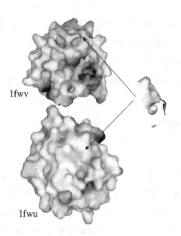

**Fig. 1.** An example of surface motif extracted.

residues found in primary structures, to higher level motifs which could be spatial patterns of atomic arrangement in protein three dimensional structure data.

Protein surface motifs can be defined as patterns of shape and physical properties which commonly appear in protein molecular surfaces. The term *shape* refers to the geometric attributes such as curvatures of irregular surfaces, relative spatial positions, etc. while *physical properties* refers to physical attributes like electrostatic potential and hydrophobic strength. In our research, surface motifs are basically represented as a set of normal vectors with attributes. In addition, a surface motif contains information about proteins, from which it has been extracted, as well as its location in those proteins. Figure 1 shows an example of surface motifs extracted from different proteins.

## 3. Overview of SUMOMO

SUMOMO (SUrface MOtif extraction in protein MOlecular surface data) consists of an extraction phase to extract all possible surface motifs from molecular surface data, and a filtering phase to classify and filter the surface motifs extracted by the extraction phase. See Ref. 7 for a detailed description of SUMOMO procedures.

## 3.1. Surface motif extraction

Since surface motifs cannot be assigned predetermined shapes and sizes, to extract surface motifs of different sizes, a given set of protein molecular surfaces is divided into several small surfaces collectively called *unit surfaces*. Commonly occurring unit surfaces and each of the resulting surfaces formed by merging adjacently located unit surfaces are potential surface motifs, and are termed *candidate motifs*.

### 3.1.1. Unit surfaces construction

The projections and depressions of the protein molecular surface are represented as normal vectors with attributes. To construct vectors, vertices with high local mean curvature $H$, and Gaussian curvature $K$, respectively the sum and the product of principal curvatures, are extracted. These curvatures specify the shape of a surface.[14] Each of these vertices represents normal vectors with attributes with information about its coordinates, normal vectors, as well as its physical attributes of electrostatic potential (*elct*), and hydrophobic strength (*hydr*) with values of $H$ and $K$ added.

Figure 2 illustrates the relative attributes of a pair of normal vectors: the relative distance between starting points ($\delta$), the dihedral angle between the vectors ($\phi$), and the angles made of each vector with the line bonding the starting points of the vectors ($\theta_1, \theta_2$).

To construct the vector pairs, a threshold value $T_R$ is determined experimentally, and each vector is paired only with those vectors that are within radial distance $T_R$ from it.

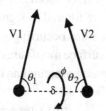

relative attributes
 $\delta$ : distance
 $\phi$ : dihedral angle
 $\theta_1, \theta_2$ : bond angles

individual attributes
 elct : electrostatic potential
 hydr : hydrophobicity
 H : Mean curvature
 K : Gaussian curvature

**Fig. 2.** A pair of normal vectors with attributes.

### 3.1.2. *Sorting of vector pairs using buckets*

Vector pairs are sorted according to their relative attributes and the individual attributes of the composing vectors. Since vectors whose relative attributes are dissimilar cannot be extracted as candidate motifs, vector pairs are grouped according to the value of their relative attributes. The extracted vector pairs are further classified into groups according to the individual attributes of the composing vectors.

### 3.1.3. *Merging of candidate motifs*

Repetitive merging of adjacent candidate motifs, including the unit surfaces, generates surface motifs. Only candidate motifs with two or more vector pairs satisfying the following conditions can be merged.

- Vector pairs from a given protein should have at least one common vector. The vector pairs in other proteins corresponding to these vector pairs should also have corresponding vector(s) common to them.
- All vectors to be merged in a protein should also have vectors with similar relative attributes in other proteins.
- At least two of the input proteins must satisfy the above conditions.

## 3.2. *Filtering using similarity between local surfaces*

As local molecular surface structures of active sites are highly conserved, they occur commonly in some proteins. Nevertheless, active sites require specific structures to express functions, and instead of occurring as merely abundant ones, active sites occur in proteins with similar functions as structures with common shapes and properties. Therefore, by grouping proteins based on the similarity of local structures, surface motifs, which occur frequently within the same group of proteins, are filtered as significant motifs.

Considering surface motifs as local structure occurring in a number of proteins, the proteins from which particular surface motifs are extracted have similar local structures. Hence, similarity among proteins can be determined by focusing on surface motifs common to both.

Moreover, the structures corresponding to active sites tend to be composed of a variable number of surface motifs, as active sites may comprise of several discontinuous but adjacently located surface motifs. Therefore, proteins can be clustered by focusing on the similarity among *adjacent motif sets*, which are the sets of surface motifs close to each other around the same area on the molecular surface. Using adjacent motif sets, proteins can be compared more precisely at local structures comparable to active sites.

### 3.2.1. *Similarity focused on adjacent motif sets*

Since surface motifs are found concentrated around active sites, motifs belonging to adjacent motif sets are considered significant. While evaluating similarity between two proteins, a surface motif can exist as a member of an adjacent motif set in a protein, and it also has an isolated existence in another. Therefore, such surface motifs cannot be considered when evaluating the similarity between two proteins.

Each protein can be expressed as a protein vector defined as:

$$P = (p_1, p_2, \ldots, p_i, \ldots, p_n),$$

where $n$ is the total number of motifs extracted and $p_i (1 \leq i \leq n)$ is the number of composing normal vectors for a surface motif. Each element in the vector represents a surface motif extracted.

Proteins are expressed as protein vectors whose similarity is calculated using the cosine of the vectors. Furthermore, the distance between proteins is defined based on their similarity for clustering proteins and evaluating significance of surface motifs.

### 3.2.2. *Significance of surface motifs*

The significance of the surface motif can be evaluated with an analogy of term weighting in the field of document analysis. Term Frequency (TF) is a measure of how often a term is found in a collection of documents. Inverse Document Frequency (IDF) is a measure of how rare a term is in a collection, calculated by total collection size divided by the number of documents containing the term.[15] Coupling the concepts of TF and IDF, based on the popular TF-IDF, the degree of significance $E_i$ of a surface motif $i$ can be defined as:

$$E_i = \max_j \left( TF_{ij} \times \left( \log \frac{N}{DF_i} + 1 \right) \right),$$

where $N$ is the number of groups of proteins, $TF_{ij}$ is the frequency of surface motif $i$ in protein group $j$, and $DF_i$ is the number of groups in which surface motif $i$ occurs.

The motifs with significance more than or equal to threshold value $T_e$ are referred as active sites candidates, while the rest are filtered out as the negative instances of active sites candidates.

### 3.3. *Problems with SUMOMO*

The filtering phase in SUMOMO was able to filter active sites candidates to 14% of the original number of surface motifs extracted from 15 different proteins without any loss of function.[7] For example, it was able to reduce 3183 surface motifs extracted to 448. Notwithstanding this reduction, to distinguish surface motifs associated four different functions out of 448 active site candidates still requires an amount of time and effort. Moreover, considering that there could only be a few common active sites in a group of proteins with similar functions, 448 active sites candidates is quite a large number to experimentally verify their biological significance.

## 4. Filtering Surface Motifs using Negative Instances of Protein Active Sites Candidates

The surface motifs which have been discriminated as insignificant candidates for protein actives-sites by SUMOMO at the filtering phase is termed as *negative instances of active-sites candidates*. The approach to the proposed method of filtering is to effectively use the information extracted from these negative instances of active-sites candidates to filter the remaining active-sites candidates. The characteristics of these negative instances, despite being absolutely structure based, have negative functional (do not have functional) implications. In other words, the active sites candidates similar to the negatives instances cannot possibly be active sites. Filtering of active sites candidates in the proposed method is achieved by the removal of active sites candidates whose features are similar to those of the negative instance, leaving behind the active sites candidates whose features are similar to those of real active sites. The main issue in filtering active sites candidates using negative instances is to discover and recognize the features which can distinguish active sites candidates which correspond to active sites from the rest of active sites candidates.

## 4.1. Survey on the features to distinguish real active sites from the active sites candidates

A survey was conducted to discover the features to distinguish filter real active-sites from the rest of active sites candidates. *Dissimilarity* between two protein molecular surfaces was used as a simple measure to quantitatively evaluate similarities.

### 4.1.1. Dissimilarity

Dissimilarity $D$ is defined as:

$$D = \frac{1}{N_{min}} \sum_{i,j} d_{i,j},$$

where $N_{min}$ is the minimum of the number of vertices composing the surfaces being compared. $d_{i,j}$ is the difference between the two vertices $(i, j)$, and is given by:

$$d_{i,j} = W_l \cdot d_{l_{i,j}} + W_e \cdot d_{e_{i,j}} + W_h \cdot d_{h_{i,j}},$$

where $d_{l_{i,j}}$, $d_{e_{i,j}}$ and $d_{h_{i,j}}$ are the differences of coordinates, electrostatic potential and hydrophobic property between the nearest pair $(i, j)$ on the compared surfaces.[14]

### 4.1.2. Procedure

The pilot survey was conducted in the following steps:

(1) As a preliminary step for the survey, proteins surface data with prior knowledge of active-sites were subjected to surface motif extraction using SUMOMO. The PDB ID of protein data used in this pilot survey group according to the interacting ligand are shown in Table 1.
(2) The filtering phase of SUMOMO distinguished 257 surface motifs as active-sites candidates, while rest (2152 surface motifs) were distinguished as the negative instances of active-sites candidates. Furthermore, out of the 257 active-sites candidates, only 57 surface motifs were recognized to be the parts of actual active-sites.

Table 1. Protein data used in the pilot survey to verify proposed approach.

| Ligand | PDB ID |
|--------|--------|
| 13P | 1ado, 1j4e, 1fdj |
| 2AM | 1rgk, 6rnt, 7rnt |
| 3GP | 1rls, 1rms, 1rgc, 2gsp, 4gsp |
| 2FG | 1jz0, 1jz1, 1jz2 |

(3) Surface data of each active-sites candidates were compared with 2152 negative instances and dissimilarity was calculated for each pair. During this surface motif comparison, 57 surface motifs recognized to be the parts of actual active-sites and rest of active-sites candidates were grouped separately so as to isolate the patterns of similarity which distinguishing features of active-sites candidates along with the negative instances of active-sites candidates.

(4) Graphs of several statistical quantities such as minimum, average, frequencies of various ranges of dissimilarity were drawn for each active-site candidate based on the results of the experiment.

### 4.1.3. Results

The features which could be used to distinguish known active sites from other active site candidates are not so obvious, as shown in Figs. 3 and 4, each of which shows minimum dissimilarities and average dissimilarities in the $y$-axis for a sample of 20 active sites candidates shown in the $x$-axis. Therefore, instead of relying on individual and simple statistical features, such as minimum dissimilarity, average dissimilarity, etc. of active sites candidates when compared to negative instances, it is necessary to use different feature which is able to discriminate real active sites from rest of the active sites candidates.

A method to assess whether an active sites candidate is similar or different from the negative instances on the whole is to *count the number of similar the negative instances taking some value of dissimilarity as a reference.*

The graphs of frequencies of the negative instances to the candidates corresponding to active sites and others at various intervals of dissimilarities

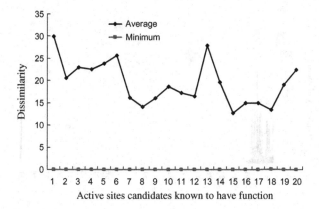

**Fig. 3.** Average and minimum of dissimilarities between the negative instances and active sites candidates corresponding to real active sites.

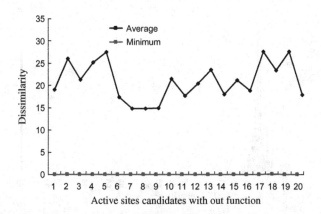

**Fig. 4.** Average and minimum dissimilarities between the negative instances and active sites candidates other than real active sites.

for active sites candidates which correspond to real active sites, and the rest, ranging from 0 to 100 and above are shown in Figs. 5 and 6 respectively. Careful observations of these graphs focusing on the distribution of the dissimilar motifs will reveal that the number of negative instances which are similar to the candidates which correspond to real active sites is fewer than the number of the candidates which do not.

This can be elucidated further by taking cumulative frequencies at each point of reference in dissimilarity, shown in Fig. 7, is remarkably different

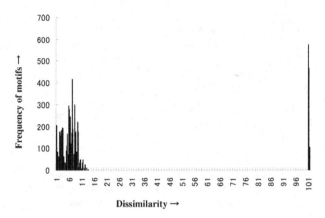

**Fig. 5.** Frequency of dissimilarities between the negative instances and active sites candidates corresponding to real active sites.

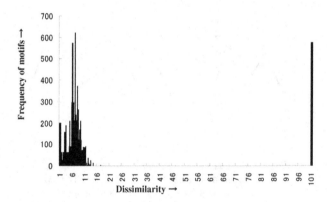

**Fig. 6.** Frequency of dissimilarities between the negative instances and active sites candidates other than real active sites.

for the candidates which correspond to active sites and those which do not. The cumulative frequency of active sites candidates known to be real active sites are quite lower at higher dissimilarities. On the contrary, the candidates which do not correspond to real active sites end up at higher dissimilarities compared to the motif candidates which correspond to real active sites. This phenomenon can be thought in accordance with the fact that surface motifs are extracted by merging of different combinations of proteins and normal vectors, and are labeled as important significant and insignificant

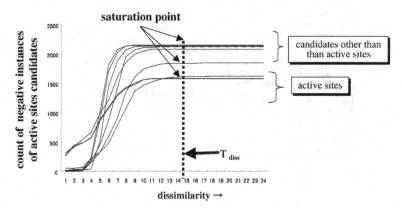

**Fig. 7.** Cumulative frequency of dissimilarity between the negative instances and the active sites candidates.

just because of small variations in composing normal vectors or/and source proteins.

Thus, it can be concluded that the active sites candidates which are most likely to be active sites have lesser number of negative instances which are similar to it.

### 4.2. Ranking active sites candidates

Figure 8 shows an outline of the proposed method to filter surface motifs. The similarity of each members in the input set of active site candidates and set of negative instances are assessed using the molecular surface comparison method using dissimilarity.[14] As a method of filtering, the active sites candidates are ranked according to their similarity with the negative instances of active sites candidates.[16]

As it was concluded from the survey, the active sites candidates which do not correspond to real active sites have comparatively a larger number of the negative instances compared to the ones which do. Therefore, within a given dissimilarity, the more an active sites candidate has negative instances similar to it, the less likely it is to correspond to an active site. On the contrary, the less a candidate has negative instances similar to it, the more likely it is to correspond to an active site. Hence, active sites candidates are ranked by using the count of negative instances similar to an active sites candidate below a threshold value $T_{diss}$ of dissimilarity, as shown in

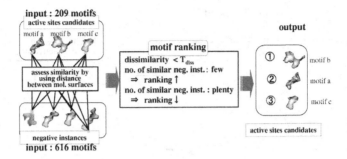

**Fig. 8.** An outline of the proposed method.

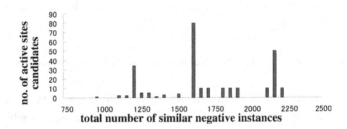

**Fig. 9.** Ranking active sites candidates.

Fig. 7. Threshold dissimilarity $T_{diss}$ is set at the point where the cumulative frequency of each of the candidates becomes constant. Figure 9 shows the graph of count of similar negative instances and the number of motifs at dissimilarity is less than $T_{diss}$. The active sites candidates are ranked in ascending order of the count of similar negative instances, i.e. from the left to the right of the graph.

## 5. Evaluations

The proposed method of filtering is used, in combination of SUMOMO to mine surface motifs from a set of proteins with unknown functions. However, for the purpose of evaluation, we prepared a set of protein molecular surfaces data in Table 2 that has been classified into three groups based on ligands (functions) information and respective lists of interacting proteins.

Table 2. Proteins dataset1.

| Ligands | Proteins |
|---|---|
| MTX | 1bdo, 1stp, 2izi |
| DAN | 1eus, 1ivf, 1nnb, 1sli, 2sim, 2qwe |
| BTN | 1dls, 1dre, 1ra3, 1rx3 |

SUMOMO, followed by proposed method of filtering was applied to the evaluation data. 825 surface motifs extracted by SUMOMO during the extraction process was reduced to 209 active sites candidates by the filtering stage of SUMOMO. Also, at the same time 616 motifs were discriminated as negative instances of active sites candidates. Out of these 209 active site candidates, 36 are known to be the parts of active sites.

In the next stage, the surface of active sites candidates were compared with those of negative instances, calculating the dissimilarity between each member of both active sites candidates and negative instances. For each active sites candidate, cumulative frequency was calculated. At threshold of $T_{diss} = 12$, the sum of similar negative instance were calculated for each active site candidate.

Likewise, evaluation experiment was also carried out with another set of protein molecular surfaces data in Table 3.

SUMOMO extracted and discriminated 257 surface motifs as active sites candidates and 2152 surface motifs as negative instances from the proteins in Table 3. Out of these 257 active site candidates, 57 are known to be the parts of real active sites. At threshold of $T_{diss} = 14$, the sum of similar negative instance were calculated for each active site candidate.

The active sites candidates were placed in ascending order of their sum of similar negative instances below $T_{diss}$. As such the active sites candidates

Table 3. Proteins dataset2.

| Ligands | Proteins |
|---|---|
| 13P | 1ado, 1j4e, 1fdj |
| 2AM | 1rgk, 6rnt, 7rnt |
| 3GP | 1rls, 1rms, 1rgc, 2gsp, 4gsp |
| 2FG | 1jz0, 1jz1, 1jz2 |

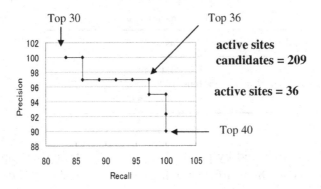

**Fig. 10.** Precision-recall curve for dataset1.

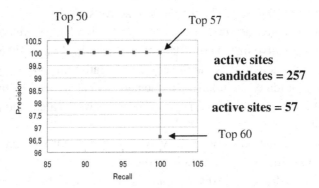

**Fig. 11.** Precision-recall curve for dataset2.

were ranked. Precision-recall curves, shown in Figs. 10 and 11 were drawn for the values of precision and recall, where

$$precision = \frac{no.\ of\ active\ sites\ present\ in\ the\ selected\ set}{total\ no.\ of\ active\ sites\ candidates\ selected},$$

$$recall = \frac{no.\ of\ active\ sites\ present\ in\ the\ selected\ set}{total\ no.\ of\ active\ sites\ present\ in\ the\ whole\ data\ set}.$$

From the Precision-recall curves obtained from the evaluation experiments, we can conclude that the proposed method of filtering active sites candidates using ranking of candidates in order of their likeliness to be an active site is very effective. Precision-recall curve normally is used in balancing the trade off between precision and recall. However, in our evaluation experiment the active sites candidates were selected with high values of precision at the higher ranks in the ranking table.

As seen from the results of the evaluation the proposed method of filtering surface motifs by using negative instances of active sites candidates has been extremely good. One of the reason for such a good result is the use of ranking while filtering the active sites candidates. In experiments where we used some threshold value instead of ranking to filter the active sites candidates yielded poor precision values. Therefore, it can be said that the method of filtering actives sites candidates by raning is better in the sense that the active sites candidates can be selected with a considerably high precision such that their function can be verified *in vitro*.

## 6. Conclusions and Future Works

In this chapter, a method of filtering surface motifs by using negative instances of active sites candidates was proposed. The propose method is an empirically formulated method based on surface motifs data extracted from SUMOMO, and discriminated as significant and insignificant by the filtering stage of SUMOMO. The precision-recall curve from the experiment, besides confirming the validity of the proposed method, pointed at its ability to distinguish surface motifs as possible candidates of active sites.

The proposed method was based on a simple concept of using total of count of the similar negative instances active sites candidates in order to rank the active sites candidates. However, there still remains some gray region in which it is very difficult to distinguish whether an active site candidate can be considered true active site or false active site. Solving such complicacy may require features other than sum of dissimilar negative candidates. In combination with basic features such as average, or minimum of dissimilarities, the total of count of the similar negative instances active sites candidates may be quite useful. In addition, the selection of threshold $T_{diss}$ which depends upon input data set, could be a key point in determining the precision and recall of active sites candidates. Works remain to be done in merging the closely located active sites candidates extracted, which may reduce the number of active sites candidates actually retrieved, and improve the ranking of important candidates even within the resulting filtered active sites candidates.

## References

1. A. Bairoch and P. Bucher, PROSITE: recent developments, *Nucl. Acids Res.* **19**, 3583–3589 (1994).
2. T. K. Attwood, M. E. Beck, A. J. Bleasby, K. Degtyarenko and D. J. Parry-Smith, Progress with the PRINTS Protein fingerprint database, *Nucleic Acids Research* **24**(1), 182–188 (1996).
3. F. Servant, C. Bru, S. Carrere, E. Courcelle, J. Gouzy, D. Peyruc and D. Kahn, ProDom: automated clustering of homologous domains, *Briefings in Bioinformatics* **3**(3), 246–251 (2002).
4. X. Wang, J. T. L. Wang, D. Shasha, B. Shapiro, S. Dikshitulu, I. Rigoutsos and K. Zhang, Finding patterns in three dimensional graphs: algorithms and applications to scientific data mining, *IEEE Transactions on Knowledge and Data Engineering* **14**(4), 731–749 (2002).
5. M. Rosen, S. L. Lin, H. Wolfson and R. Nussinov, Molecular shape comparisons in searches for active sites and functional similarity, *Protein Engineering* **11**(4), 263–277 (1998).
6. M. de Rinaldis, G. Ausiello, G. Cesareni and M. Helmer-Citterich, Three-dimensional profiles: a new tool to identify protein surface similarities, *J. Mol. Biol.* **284**(4), 1211–1221 (1998).
7. N. L. Shrestha, Y. Kawaguchi and T. Ohkawa, SUMOMO: a surface motif mining module, *International Journal of Computational Intelligence and Applications* **4**(4), (2004).
8. G. J. Barlett, C. T. Porter, N. Borakakoti and J. M. Thornton, Analysis of catalytic residues in enzyme active sites, *Journal of Molecular Biology* **324**, 105–121 (2002).
9. K. Kinoshita, J. Furui and H. Nakamura, Identification of protein functions from a molecular surface database, eF-site, *J. Struct. Funct. Genomics* **2**, 9–22 (2001).
10. H. M. Berman, J. Westbrook, Z. Feng, G. Gilliland, T. N. Bhat, H. Weissig, I. N. Shindyalov and P. E. Bourne, the protein data bank, *Nucleic Acids Research* **28**, 235–242 (2000).
11. M. L. Connolly, Solvent-accessible surfaces of proteins and nucleic acids, *Science* **221**, 709–713 (1983).
12. H. Nakamura and S. Nishida, Numerical calculations of electrostatic potentials of protein-solvent systems by the self consistent boundary method, *J. Phys. Soc. Jpn.* **56**, 1609–1622 (1987).
13. A. E. Todd, C. A. Orengo and J. M. Thornton, Evolution of protein function, from a structural perspective, *Curr. Op. Chem. Biol.* **3**, 548–556 (1999).
14. Y. Kaneta, N. Shoji, T. Ohkawa and H. Nakamura, A method of comparing protein molecular surface based on normal vectors with attributes and its application to function identification, *Information Sciences* **146**(1–4), 41–54 (2002).
15. K. S. Jones, A statistical interpretation of term specificity and its application in retrieval, *J. Documentation* **28**(1), 11–20 (1972).
16. N. J. Belkin W. B. Croft, Information filtering and information retrieval: two sides of the same coin? *Communications of the ACM* **35**(12), 29–38 (1992).

# CHAPTER 7

# DISTILL: A MACHINE LEARNING APPROACH TO *AB INITIO* PROTEIN STRUCTURE PREDICTION

Gianluca Pollastri[*], Davide Baú[†] and Alessandro Vullo[‡]

*School of Computer Science and Informatics*
*UCD Dublin, Belfield, Dublin 4, Ireland*
[*]*gianluca.pollastri@ucd.ie*
[†]*davide.bau@ucd.ie*
[‡]*alessandro.vullo@ucd.ie*

We present Distill, a simple and effective scalable architecture designed for modeling protein $C_\alpha$ traces based on predicted structural features. Distill targets those chains for which no significant sequential or structural resemblance to any entry of the Protein Data Bank (PDB) can be detected. Distill is composed of: (1) a set of state-of-the-art predictors of protein structural features based on statistical learning techniques and trained on large, non-redundant subsets of the PDB; (2) a simple and fast 3D reconstruction algorithm guided by a pseudo-energy defined according to these predicted features.

At CASP6, a preliminary implementation of the system was ranked in the top 20 predictors in the Novel Fold hard target category. Here we test an improved version on a non-redundant set of 258 protein structures showing no homology to the sets employed to train the machine learning modules. Results show that the proposed method can generate topologically correct predictions, especially for relatively short (up to 100–150 residues) proteins. Moreover, we show how our approach makes genomic scale structural modeling tractable by solving hundreds of thousands of protein coordinates in the order of days.

## 1. Introduction

Of the nearly two million protein sequences currently known, only about 10% are human-annotated, while for fewer than 2% has the three-dimensional (3D) structure been experimentally determined. Attempts to

---

[*]Corresponding author.

predict protein structure from primary sequence have been carried out for decades by an increasingly large number of research groups.

Experiments of blind prediction such as the CASP series[1–4] demonstrate that the goal is far from being achieved, especially for those proteins for which no resemblance exists, or can be found, to any structure in the PDB[5] — the field known as *ab initio* prediction. In fact, as reported in the last CASP competition results[4] for the New Fold (NF) category, even the best predicted models have only fragments of the structure correctly modeled and poor average quality. Reliable identification of the correct native fold is still a long-term goal. Nevertheless, improvements observed over the last few years suggest that *ab initio* generated low-resolution models may prove to be useful for other tasks of interest. For instance, efficient *ab initio* genomic scale predictions can be exploited to quickly identify similarity in structure and functions of evolutionary distant proteins.[6,7]

Here we describe Distill, a fully automated computational system for *ab initio* prediction of protein $C_\alpha$ traces. Distill's modular architecture is composed of: (1) a set of state-of-the-art predictors of protein features (secondary structure, relative solvent accessibility, contact density, residue contact maps, contact maps between secondary structure elements) based on machine learning techniques and trained on large, non-redundant subsets of the PDB; (2) a simple and fast 3D reconstruction algorithm guided by a pseudo-energy defined according to these predicted features.

A preliminary implementation of Distill showed encouraging results at CASP6, with model 1 in the top 20 predictors out of 181 for GDT_TS on Novel Fold hard targets, and for Z-score for all Novel Fold and Near Novel Fold targets.[6] Here we test a largely revised and improved version of Distill on a non-redundant set of 258 protein structures showing no homology to the sets employed to train the machine learning modules. Results show that Distill can generate topologically correct predictions for a significant fraction of short proteins (150 residues or fewer).

This paper is organized as follows: in Sec. 2, we describe the various structural features predicted; in Sec. 3, we describe in detail the statistical learning methods adopted in all the feature predictors; in Sec. 4, we discuss the overall architecture of the predictive pipeline and the implementation and performances of the individual predictors; in Sec. 5, we introduce the 3D reconstruction algorithm; finally in Sec. 6, we describe the results of benchmarking Distill on a non-redundant set of 258 protein structures.

## 2. Structural Features

We call protein one-dimensional structural features (1D) those aspects of a protein structure that can be represented as a sequence. For instance, it is known that a large fraction of proteins is composed by a few well defined kinds of local regularities maintained by hydrogen bonds: helices and strands are the most common ones. These regularities, collectively known as protein secondary structure, can be represented as a string out of an alphabet of 3 (helix, strand, the rest) or more symbols, and of the same length of the primary sequence. Predicting 1D features is a very appealing problem, partly because it can be formalized as the translation of a string into another string of the same length, for which a vast machinery of tools for sequence processing is available, partly because 1D features are considered a valuable aid to the prediction of the full 3D structure. Several public web servers for the prediction of 1D features are available today, almost all based on machine learning techniques. The most popular of these servers[8-11] process hundreds of queries daily. Less work has been carried out on protein two-dimensional structural features (2D), i.e. those aspects of the structure that can be represented as two-dimensional matrices. Among these features are contact maps, strand pairings, cysteine-cysteine bonding patterns. There is intrinsic appeal in these features since they are simpler than the full 3D structure, but retain very substantial structural information. For example, it has been shown[12] that correct residue contact maps generally lead to correct 3D structures.

In the remainder of this section we will describe the structural features predicted by our systems.

### 2.1. *One-dimensional structural features*

#### 2.1.1. *Secondary structure*

Protein secondary structure is the complex of local regularities in a protein fold that are maintained by hydrogen bonds. Protein secondary structure prediction is an important stage for the prediction of protein structure and function. Accurate secondary structure information has been shown to improve the sensitivity of threading methods (e.g. Ref. 13) and is at the core of most *ab initio* methods (e.g. see Ref. 14) for the prediction of protein structure. Virtually all modern methods for protein secondary structure prediction are based on machine learning techniques,[8,10] and exploit evolutionary information in the form of profiles extracted from alignments of multiple homologous sequences. The progress of these methods over the last 10 years has been slow, but steady, and is due to numerous factors: the ever-increasing size of training sets; more sensitive methods for the detection of homologues, such as PSI-BLAST;[15] the use of ensembles of multiple predictors trained independently, sometimes tens of them;[16] more sophisticated machine learning techniques (e.g. Ref. 10). Distill contains the state-of-the-art secondary structure predictor Porter,[11] described in Sec. 4.

### 2.1.2. *Solvent accessibility*

Solvent accessibility represents the degree to which amino acids in a protein structure interact with solvent molecules. The accessible surface of each residue is normalized between a minimum and a maximum value for each type of amino acid, and then reassigned to a number of classes (e.g. buried versus exposed), or considered as such. A number of methods have been developed for solvent accessibility prediction, the most successful of which based on statistical learning algorithms.[17–21] Within DISTILL we have developed a novel state-of-the-art predictor of solvent accessibility in 4 classes (buried, partly buried, partly exposed, exposed), described in Sec. 4.

### 2.1.3. *Contact density*

The contact map of a protein with $N$ amino acids is a symmetric $N \times N$ matrix $C$, with elements $C_{ij}$ defined as:

$$C_{ij} = \begin{cases} 1 & \text{if amino acid } i \text{ and } j \text{ are in contact} \\ 0 & \text{otherwise} \end{cases}. \qquad (1)$$

We define two amino acids as being in contact if their mutual distance is less than a given threshold. Alternative definitions are possible, for instance, based on different mutual $C_\alpha$ distances (normally in the 7–12 Å range), or on $C_\beta - C_\beta$ atom distances (normally 6.5–8 Å), or on the minimal distance between two atoms belonging to the side-chain or backbone of the two residues (commonly 4.5 Å).

Let $\lambda(C) = \{\lambda : Cx = \lambda x\}$ be the spectrum of $C$, $S_\lambda = \{x : Cx = \lambda x\}$ the corresponding eigenspace and $\bar\lambda = \max\{\lambda \in \lambda(C)\}$ the largest eigenvalue of $C$. The principal eigenvector of $C$, $\bar{x}$, is the eigenvector corresponding to $\bar{\lambda}$. $\bar{x}$ can also be expressed as the argument which maximizes the Rayleigh quotient:

$$\forall x \in \mathcal{S}_\lambda : \frac{x^T C x}{x^T x} \leq \frac{\bar{x}^T C \bar{x}}{\bar{x}^T \bar{x}}. \qquad (2)$$

Eigenvectors are usually normalized by requiring their norm to be 1, e.g. $\|x\|_2 = 1 \; \forall x \in \mathcal{S}_\lambda$. Since $C$ is an adjacency (real, symmetric) matrix, its eigenvalues are real. Since it is a normal matrix ($A^H A = AA^H$), its eigenvectors are orthogonal. Other basic properties can also be proven: the principal eigenvalue is positive; non-zero components of $\bar{x}$ have all the same sign.[22] Without loss of generality, we can assume they are positive, as in Ref. 23. We define a protein's Contact Density as the principal eigenvector of its residue contact map, multiplied by its corresponding eigenvalue: $\bar{\lambda}\bar{x}$.

Contact Density is a sequence of the same length as a protein's primary sequence. Recently[23] a branch-and-bound algorithm was described that is capable of reconstructing the contact map from the exact PE, at least for single domain proteins of up to 120 amino acids. Predicting Contact Densities is thus interesting: as one-dimensional features, they are significantly more tractable than full contact maps; nonetheless a number of ways to obtain contact maps from contact densities may be devised, including modifying the reconstruction algorithm in Ref. 23 to deal with noise, or adding Contact Densities as an additional input feature to systems for the direct prediction of contact maps (such as Ref. 24). Moreover, contact densities are informative in their own right and may be used to guide the search for optimal 3D configurations, or to identify protein domains.[25,26] Contacts among residues, in fact, constrain protein folding and characterize different protein structures (see Fig. 1), constituting a structural fingerprint of the given protein.[27]

Distill contains a state-of-the-art Contact Density predictor.[28]

## 2.2. Two-dimensional structural features

### 2.2.1. Contact maps

Contact maps (see definition above), or similar distance restraints have been proposed as intermediate steps between the primary sequence and the 3D structure (e.g. in Refs. 24, 29, 30), for various reasons: unlike 3D coordinates, they are invariant to rotations and translations, hence less challenging to

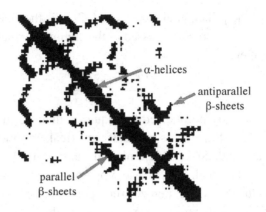

**Fig. 1.** Different secondary structure elements like helices (thick bands along the main diagonal) and parallel — or anti-parallel — β-sheets (thin bands parallel — or anti-parallel — to the main diagonal) are easily detected from the contact map.

predict by machine learning systems;[24,31] quick, effective algorithms exist to derive 3D structures from them, for instance, stochastic optimization methods,[12,32] distance geometry,[33,34] or algorithms derived from the NMR literature and elsewhere.[35-37] Numerous methods have been developed for protein residue contact map prediction[24,29,30,38] and coarse (secondary structure element level) contact map prediction,[31] and some improvements are slowly occurring (e.g. in Ref. 38, as shown by the CASP6 experiment[39]).

Accurate prediction of residue contact maps is far from being achieved and limitations of existing prediction methods have again emerged at CASP6 and from automatic evaluation of structure prediction servers such as EVA.[40] There are various reasons for this: the number of positive and negative examples (contacts versus non-contacts) is strongly unbalanced; the number of examples grows with the squared length of the protein making this a tough computational challenge; capturing long ranged interactions in the primary sequence is difficult, hence grasping an adequate global picture of the map is a formidable problem.

The Contact Map predictor included in Distill relies on a combination of one-dimensional features as inputs and is state-of-the-art.[28]

### 2.2.2. Coarse topologies

We define the coarse structure of a protein as the set of three-dimensional coordinates of the N- and C-terminus of its secondary structure segments (helices, strands). By doing so, we: ignore coil regions, which are normally more flexible than helices and strands; assume that both strands and helices can be represented as rigid rods.

The actual coarse topology of a protein may be represented in a number of alternative ways: the map of distances, thresholded distances (contacts), or multi-class discretized distances between the centers of secondary structures;[31,41] the map of angles between the vectors representing secondary structure elements, or some discretization thereof.[41] In each of these cases, if a protein contains $M$ secondary structure elements, its coarse representation will be a matrix of $M \times M$ elements.

Although coarse maps are simpler, less informative representations of a protein structure than residue- or atom-level contact maps, they nonetheless can be exploited for a number of tasks, such as the fast reconstruction of coarse structures[41] and the rapid comparison and classification of proteins into structural classes.[42]

Coarse contact maps represent compact sets of constraints and hence clear and synthetic pictures of the shape of a fold. For this reason, it is much less challenging to observe, and predict, long-range interactions between elements of a protein structure within a coarse model than in a finer one: a typical coarse map is composed by only hundreds of elements on a grid of tens by tens of secondary structure elements, while a residue-level contact map can contain hundreds of thousands or millions of elements and can typically be modeled only locally by statistical learning techniques.

For this reason, coarse maps cannot only yield a substantial information compression with respect to residue maps, but can also assist in detecting interactions that would normally be difficult to observe at a finer scale, and contribute to improving residue maps, and structure predictions.

Distill contains predictors of coarse contact, multi-class distance and multi-class angle maps.[41]

## 3. Review of Statistical Learning Methods Applied

### 3.1. *RNNs for undirected graphs*

A data structure is a graph whose nodes are marked by sets of domain variables, called labels. A skeleton class, denoted by the symbol #, is a set of unlabelled graphs that satisfy some topological conditions. Let $\mathcal{I}$ and $\mathcal{O}$ denote two label spaces: $\mathcal{I}^{\#}$ (resp. $\mathcal{O}^{\#}$) refers to the space of data structures with vertex labels in $\mathcal{I}$ (resp. $\mathcal{O}$) and topology #. Recursive models such as RNNs[43] can be employed to compute functions $\mathcal{T} : \mathcal{I}^\# \to \mathcal{O}^\#$ which map a structure into another structure of the same form but possibly different labels. In the classical framework, # is contained in the class of bounded DPAGs, i.e. Directed Acyclic Graphs (DAGs) where each vertex has bounded outdegree (number of outgoing edges) and whose children are ordered. Recursive models normally impose causality on data processing: the state variables (and outputs) associated to a node depend only on the nodes upstream (i.e. from which a path leads to the node in question). The above assumption is restrictive in some domains and extensions of these models for dealing with more general undirected structures have been proposed.[24,44,45]

A more general assumption is considered here: # is contained in the class of bounded-degree undirected graphs. In this case, there is no concept of causality and the computational scheme described in Ref. 43 cannot be directly applied. The strategy consists in splitting graphical processing into a *set* of causal "dynamics", each one computed over a plausible orientation of $U$.

More formally, assume $U = (V, E) \in \mathcal{I}^\#$ has one connected component. We identify a set of spanning DAGs $G_1, \ldots, G_m$ with $G_i = (V, E_i)$ such that:

- the undirected version of $G_i$ is $U$,
- $\forall v, u \in V \; v \neq u \; \exists i : (v, u) \in E_i^*$ being $E_i^*$ the transitive closure of $E_i$,

and for each $G_i$, introduce a state variable $X_i$ computed in the usual way. Figure 2 (left) shows a compact description of the set of dependencies among the input, state and output variables.

Connections run from vertices of the input structure (layer $I$) to vertices of the spanning DAGs and from these nodes to nodes of the output structure (layer $O$).

Using weight-sharing, the overall model can be summarized by $m + 1$ distinct neural networks implementing the output function $O(v) = g(X_1(v), \ldots, X_m(v), I(v))$ and $m$ state transition functions $X_i(v) = f_i(X_i(ch_1[v]), \ldots, X_i(ch_k[v]), I(v))$. Learning can proceed by gradient-descent (back-propagation) due to the acyclic nature of the underlying graph. Within this framework, we can easily describe all contextual RNNs

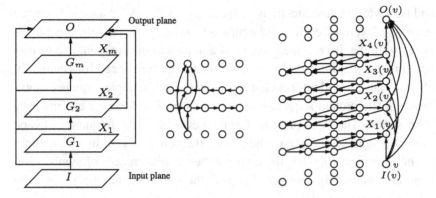

**Fig. 2.** (left): Contextual RNNs, dependencies among input, state and output variables. (center and right): processing of undirected sequences and grids with contextual RNNs (only a subset of connections are shown).

architecture developed so far. Figure 2 (center) shows that an undirected sequence is spanned by two sequences oriented in opposite directions. We then obtain bi-directional recurrent neural networks[44] or 1D DAG-RNNs if we consider a straightforward generalization from sequences to undirected graphs. For the case of two-dimensional objects (e.g. contact maps), they can be seen as two-dimensional grids spanned by four directed grids oriented from each cardinal corner (Fig. 2, right). The corresponding model is called 2D DAG-RNNs.[24] The 1D and 2D DAG-RNNs adopted in our architectures are described in more detail below.

### 3.2. *1D DAG-RNN*

In the 1D DAG-RNNs we adopt, connections along the forward and backward hidden chains span more than 1-residue intervals, creating shorter paths between inputs and outputs. These networks take the form:

$$o_j = \mathcal{N}^{(O)}(i_j, h_j^{(F)}, h_j^{(B)})$$
$$h_j^{(F)} = \mathcal{N}^{(F)}(i_j, h_{j-1}^{(F)}, \ldots, h_{j-S}^{(F)})$$
$$h_j^{(B)} = \mathcal{N}^{(B)}(i_j, h_{j+1}^{(B)}, \ldots, h_{j+S}^{(B)})$$
$$j = 1, \ldots, N$$

where $h_j^{(F)}$ and $h_j^{(B)}$ are forward and backward chains of hidden vectors with $h_0^{(F)} = h_{N+1}^{(B)} = 0$. We parametrize the output update, forward update and backward update functions (respectively $\mathcal{N}^{(O)}$, $\mathcal{N}^{(F)}$ and $\mathcal{N}^{(B)}$) using three two-layered feed-forward neural networks. In our tests the input associated with the $j$th residue $i_j$ contains amino acid information, and further one-dimensional information in some predictors (see Sec. 4 for details). In all cases amino acid information is obtained from multiple sequence alignments of the protein sequence to its homologues to leverage evolutionary information. The input presented to the networks is the frequency of each of the non-gap symbols, plus the overall frequency of gaps in each column of the alignment. If $n_{jk}$ is the total number of occurrences of symbol $j$ in column $k$, and $g_k$ the number of gaps in the same column, the $j$th input to the networks in position $k$ is:

$$\frac{n_{jk}}{\sum_{v=1}^{u} n_{vk}} \qquad (3)$$

for $j = 1, \ldots, u$, where $u$ is the number of non-gap symbols while the $u + 1$th input is:

$$\frac{g_k}{g_k + \sum_{v=1}^{u} n_{vk}}. \qquad (4)$$

In some of our predictors we also adopt a second filtering 1D DAG-RNN.[11] The network is trained to predict the structural feature given first-layer structural feature predictions. The $i$th input to this second network includes the first-layer predictions in position $i$ augmented by first stage predictions averaged over multiple contiguous windows. If $c_{j1}, \ldots, c_{jm}$ are the outputs in position $j$ of the first stage network corresponding to estimated probability of residue $j$ being labelled in class $m$, the input to the second stage network in position $j$ is the array $I_j$:

$$I_j = \left( c_{j1}, \ldots, c_{jm}, \sum_{h=k_{-p}-w}^{k_{-p}+w} c_{h1}, \ldots, \sum_{h=k_{-p}-w}^{k_{-p}+w} c_{hm}, \ldots, \right.$$
$$\left. \sum_{h=k_p-w}^{k_p+w} c_{h1}, \ldots, \sum_{h=k_p-w}^{k_p+w} c_{hm} \right), \qquad (5)$$

where $k_f = j + f(2w + 1)$, $2w + 1$ is the size of the window over which first-stage predictions are averaged and $2p + 1$ is the number of windows considered. In the tests we use $w = 7$ and $p = 7$. This means that 15 contiguous, non-overlapping windows of 15 residues each are considered, i.e. first-stage outputs between position $j - 112$ and $j + 112$, for a total of

225 contiguous residues, are taken into account to generate the input to the filtering network in position $j$.

### 3.2.1. *Ensembling 1D DAG-RNNs*

A few two-stage 1D DAG-RNN models are trained independently and ensemble averaged to build each final predictor. Differences among models are introduced by two factors: stochastic elements in the training protocol, such as different initial weights of the networks and different shuffling of the examples; different architecture and number of free parameters of the models.

In Ref. 16, a slight improvement in secondary structure prediction accuracy was obtained by "brute ensembling" of several tens of different models trained independently. Here we adopt a less expensive technique: a copy of each of the models is saved at regular intervals (100 epochs) during training. Stochastic elements in the training protocol (similar to that described in Ref. 10) guarantee that differences during training are non-trivial.

### 3.3. *2D DAG-RNN*

All systems for the prediction of two-dimensional structural features are based on 2D DAG-RNN, described in Refs. 24 and 31. This is a family of adaptive models for mapping two-dimensional matrices of variable size into matrices of the same size.

We adopt 2D DAG-RNNs with *shortcut connections*, i.e. where lateral memory connections span $N$-residue intervals, where $N > 1$. If $o_{j,k}$ is the entry in the $j$th row and $k$th column of the output matrix, and $i_{j,k}$ is the input in the same position, the input-output mapping is modeled as:

$$o_{j,k} = \mathcal{N}^{(O)}\left(i_{j,k}, h_{j,k}^{(1)}, h_{j,k}^{(2)}, h_{j,k}^{(3)}, h_{j,k}^{(4)}\right)$$

$$h_{j,k}^{(1)} = \mathcal{N}^{(1)}\left(i_{j,k}, h_{j-1,k}^{(1)}, \ldots, h_{j-S,k}^{(1)}, h_{j,k-1}^{(1)}, \ldots, h_{j,k-S}^{(1)}\right)$$

$$h_{j,k}^{(2)} = \mathcal{N}^{(2)}\left(i_{j,k}, h_{j+1,k}^{(2)}, \ldots, h_{j+S,k}^{(2)}, h_{j,k-1}^{(2)}, \ldots, h_{j,k-S}^{(2)}\right)$$

$$h_{j,k}^{(3)} = \mathcal{N}^{(3)}\left(i_{j,k}, h_{j+1,k}^{(3)}, \ldots, h_{j+S,k}^{(3)}, h_{j,k+1}^{(3)}, \ldots, h_{j,k+S}^{(3)}\right)$$

$$h_{j,k}^{(4)} = \mathcal{N}^{(4)}\left(i_{j,k}, h_{j-1,k}^{(4)}, \ldots, h_{j-S,k}^{(4)}, h_{j,k+1}^{(4)}, \ldots, h_{j,k+S}^{(4)}\right)$$

$$j, k = 1, \ldots, N,$$

where $h_{j,k}^{(n)}$ for $n = 1, \ldots, 4$ are planes of hidden vectors transmitting contextual information from each corner of the matrix to the opposite corner. We parametrize the output update, and the four lateral update functions (respectively $\mathcal{N}^{(O)}$ and $\mathcal{N}^{(n)}$ for $n = 1, \ldots, 4$) using five two-layered feedforward neural networks, as in Ref. 31.

In our tests the input $i_{j,k}$ contains amino acid information, and structural information from one-dimensional feature predictors. Amino acid information is again obtained from multiple sequence alignments.

## 4. Predictive Architecture

In this section we briefly describe the individual predictors composing Distill. Currently we adopt three predictors of one-dimensional features: Porter (secondary structure), PaleAle (solvent accessibility), BrownAle (contact density); and two predictors of two-dimensional features: XStout (coarse contact maps/topologies); XXStout (residue contact maps). The overall pipeline is highlighted in Fig. 3.

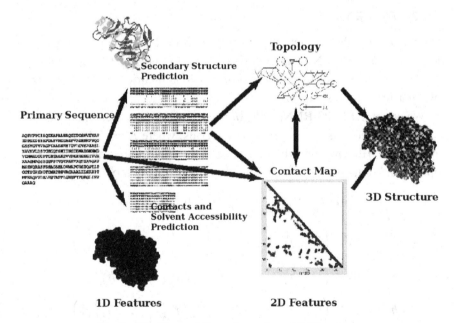

Fig. 3. Distill's modeling scheme (http://distill.ucd.ie).

### 4.1. Data set generation

All predictors are trained on dataset extracted from the December 2003, 25% pdb_select list.[a] We use the DSSP program[46] (CMBI version) to assign target structural features and remove sequences for which DSSP does not produce an output due, for instance, to missing entries or format errors. After processing by DSSP, the set contains 2171 protein and 344 653 amino acids (S2171).

We extract three distinct training/test protocols from S2171:

- Five-fold cross validation splits (5FOLD), in which test sequences are selected in an interleaved fashion from the whole set sorted alphabetically by PDB code (every fifth $+k$ sequence is picked). In this case the training sets contain 1736 or 1737 proteins and the test sets 435 or 434. The performances given on 5FOLD are effectively measured on the whole S2171, as each of its proteins appears once and only once in the test sets.
- The first fold of the above containing a training set of 1736 proteins (S1736) and a test set of 435 (S435).
- The same as the above, but containing only sequences of length at most 200 residues, leaving 1275 proteins in the training set (S1275) and 327 (S327) proteins in the test set.

Multiple sequence alignments for S2171 are extracted from the NR database as available on March 3, 2004 containing over 1.4 million sequences. The database is first redundancy reduced at a 98% threshold, leading to a final 1.05 million sequences. The alignments are generated by three runs of PSI-BLAST[15] with parameters $b = 3000$, $e = 10^{-3}$ and $h = 10^{-10}$.

### 4.2. Training protocols

All RNNs are trained by minimizing the cross-entropy error between the output and target probability distributions, using gradient descent with no momentum term or weight decay. The gradient is computed using the Back-propagation through structure (BPTS) algorithm (for which, see e.g. Ref. 43). We use a hybrid between online and batch training, with 200–600

---

[a] http://homepages.fh-giessen.de/~hg12640/pdbselect

(depending on the set) batch blocks (roughly 3 proteins each) per training set. Thus, the weights are updated 200–600 times per epoch. The training set is also shuffled at each epoch, so that the error does not decrease monotonically. When the error does not decrease for 50 consecutive epochs, the learning rate is divided by 2. Training stops after 1000 epochs for one-dimensional systems, and 300 epochs for two-dimensional ones.

### 4.3. *One-dimensional feature predictors*

#### 4.3.1. *Porter*

Porter[11] is a system for protein secondary structure prediction based on an ensemble of 45 two-layered 1D DAG-RNNs. Porter is an evolution of the popular SSpro[10] server. Porter's improvements include:

- Efficient input coding. In Porter the input at each residue is coded as a letter out of an alphabet of 25. Beside the 20 standard amino acids, B (aspartic acid or asparagine), U (selenocysteine), X (unknown), Z (glutamic acid or glutamine) and . (gap) are considered. The input presented to the networks is the frequency of each of the 24 non-gap symbols, plus the overall proportion of gaps in each column of the alignment.
- Output filtering and incorporation of predicted long-range information. In Porter the first-stage predictions are filtered by a second network. The input to this network includes the predictions of the first stage network averaged over multiple contiguous windows, covering 225 residues.
- Up-to-date training sets. Porter is trained on the S2171 set.
- Large ensembles (45) of models.

Porter, tested by a rigorous 5-fold cross validation procedure (set 5FOLD), achieves 79% correct classification on the "hard" CASP 3-class assignment (DSSP H, G, I $\rightarrow$ helix; E, B $\rightarrow$ strand; S, T, . $\rightarrow$ coil), and currently has the highest performance (over 80%) of all servers tested by assessor EVA.[40]

#### 4.3.2. *Pale Ale*

PaleAle is a system for the prediction of protein relative solvent accessibility. Each amino acid is classified as being in one of 4 (approximately equally frequent) classes: B = completely buried (0–4% exposed);

b = partly buried (4–25% exposed); e = partly exposed (25–50% exposed); E = completely exposed (more than 50% exposed).

The architecture of PaleAle's classifier is an exact copy of Porter's (described above). PaleAle's accuracy, measured on the same large, non-redundant set adopted to train Porter (5FOLD) exceeds 55% correct 4-class classification, and roughly 80% 2-class classification (Buried versus Exposed, at 25% threshold).

### 4.3.3. Brown Ale

BrownAle is a system for the prediction of protein contact density. We define Contact Density as the Principal Eigenvector (PE) of a protein's residue contact map at 8 Å, multiplied by the principal eigenvalue. Contact Density is useful for the *ab initio* the prediction of protein structures for many reasons:

- algorithms exist to reconstruct the full contact maps from the PE for short proteins,[23] and correct contact maps lead to correct 3D structures;
- Contact Density may be used directly, in combination with other constraints, to guide the search for optimal 3D configurations;
- Contact Density may be adopted as an extra input feature to systems for the direct prediction of contact maps, as in the XXStout server described below;
- predicted PE may be used to identify protein domains.[25]

BrownAle predicts Contact Density in 4 classes. The class thresholds are assigned so that the classes are approximately equally numerous, as follows: N = very low contact density (0, 0.04); n = medium-low contact density (0.04, 0.18); c = medium-high contact density (0.18, 0.54); C = very high contact density (greater than 0.54).

BrownAle's architecture is an exact copy of Porter's (described above). The accuracy of BrownAle, measured on the S1736/S435 datasets is 46.5% for the 4-class problem, and roughly 73% if the 4 classes are mapped into 2 (dense versus non-dense).

We have shown[28] that these performance levels for Contact Density prediction yield sizeable gains to residue contact map prediction, and that these gains are especially significant for long-ranged contacts, which are known to be both harder to predict and critical for accurate 3D reconstruction.

### 4.4. Two-dimensional feature predictors

#### 4.4.1. XXStout

XXStout is a system for the prediction of protein residue contact maps. Two residues are considered in contact if their C-$\alpha$s are closer than a given threshold. XXStout predicts contacts at three different thresholds: 6 Å, 8 Å and 12 Å. The contact maps are predicted as follows: protein secondary structure, solvent accessibility and contact density are predicted from the sequence using, respectively, Porter, PaleAle and BrownAle; ensembles of two-dimensional Recursive Neural Networks predict the contact maps based on the sequence, a 2-dimensional profile of amino-acid frequencies obtained from a PSI-BLAST alignment of the sequence against the NR, and predicted secondary structure, solvent accessibility and contact density. The introduction of contact density as an intermediate representation improves significantly the performances of the system. XXStout is trained the S1275 set and tested on S327. Tables 1 and 2 summarize the performances of XXStout on S327. Performances are given for the protein length/5 and protein length/2 contacts with the highest probability, for sequence separations of at least 6, at least 12, and at least 24, in CASP style.[3] These performances compare favorably with the best predictors at the latest CASP competition.[28]

Table 1. XXStout. Top protein length/5 contacts classification performance as: precision% (recall%).

| Separation | ≥ 6 | ≥ 12 | ≥ 24 |
|---|---|---|---|
| 8 Å | 46.4% (5.9%) | 35.4% (5.7%) | 19.8% (4.6%) |
| 12 Å | 89.9% (2.3%) | 62.5% (2.0%) | 49.9% (2.2%) |

Table 2. XXStout. Top protein length/2 contacts classification performance as: precision% (recall%).

| Separation | ≥ 6 | ≥ 12 | ≥ 24 |
|---|---|---|---|
| 8 Å | 36.6% (11.8%) | 27.0% (11.0%) | 15.7% (9.3%) |
| 12 Å | 85.5% (5.5%) | 55.6% (4.6%) | 43.8% (4.9%) |

### 4.4.2. XStout

XStout is a system for the prediction of coarse protein topologies. A protein is represented by a set of rigid rods associated with its secondary structure elements ($\alpha$-helices and $\beta$-strands, as predicted by Porter). First, we employ cascades of recursive neural networks derived from graphical models to predict the relative placements of segments. These are represented as distance maps discretized into 4 classes. The discretization levels ((0 Å,10 Å), (10 Å,18 Å), (18 Å,29 Å), (29 Å,$\infty$)) are statistically inferred from a large and curated data set. Coarse 3D folds of proteins are then assembled starting from topological information predicted in the first stage. Reconstruction is carried out by minimizing a cost function taking the form of a purely geometrical potential. The reconstruction procedure is fast and often leads to topologically correct coarse structures, that could be exploited as a starting point for various protein modeling strategies.[41] Both coarse distance maps and a number of coarse reconstructions are produced by XStout.

## 5. Modeling Protein Backbones

The architecture of Fig. 3 is designed with the intent of making large scale (i.e. genomic level) structure prediction of proteins of moderate ($length \leq 200$ AA) and possibly larger sizes. Our design relies on the inductive learning components described in the previous sections and is based on a pipeline involving stages of computation organized hierarchically. For a given input sequence, first a set of flattened structural representations (1D features) is predicted. These 1D features, together with the sequence are then used as an input to infer the shape of 2D features. In the last stage, we predict protein structures by means of an optimization algorithm searching the 3D conformational space for a configuration that minimizes a cost. The cost is modeled as a function of geometric constraints (pseudo energy) inferred from the underlying set of 1D and 2D predictions (see Sec. 4).

Inference of the contact map is a core component of the pipeline and is performed in $O(|w|n^2)$ time, where $n$ is the length of the input sequence and $|w|$ is the number of weights of our trained 2D-DAG RNNs (see Sec. 4.4.1). In Sec. 5.3, we illustrate a 3D reconstruction algorithm with

$O(n^2)$ time complexity. All the steps are then fully automated and fast enough to make the approach suitable to be applied to multi-genomic scale predictions.

### 5.1. *Protein representation*

To avoid the computational burden of full-atom models, proteins are coarsely described by their main chain alpha carbon ($C_\alpha$) atoms without any explicit side-chain modeling. The bond length of adjacent $C_\alpha$ atoms is restricted to lie in the interval 3.803 Å ± 0.07 in agreement with the experimental range ($D_B = 3.803$ is the average observed distance). To mimic the minimal observed distance between atoms of different amino acids, the excluded volume of each $C_\alpha$ is modeled as a hard sphere of radius $D_{HC} = 5.0$ Å (distance threshold for hard core repulsion). Helices predicted by Porter are modeled directly as ideal helices.

### 5.2. *Constraints-based pseudo energy*

The pseudo-energy function used to guide the search is shaped to encode the constraints represented by the contact map and by the particular protein representation (as described above).

Let $\mathcal{S}_n = \{r_i\}_{i=1...n}$ be a sequence of $n$ 3D coordinates, with $r_i = (x_i, y_i, z_i)$ the coordinates of the $i$th $C_\alpha$ atom of a given conformation related to a protein $p$. Let $\mathcal{D}_{\mathcal{S}_n} = \{d_{ij}\}_{i<j}, d_{ij} = \|r_i - r_j\|_2$, be the corresponding set of $n(n-1)/2$ mutual distances between $C_\alpha$ atoms. A first set of constraints comes from the (predicted) contact map which can be represented as a matrix $C = \{c_{ij}\} \in \{0, 1\}^{n^2}$. The representation of protein models discussed in the previous paragraph induces the constraints $\mathcal{B} = \{d_{ij} \in [3.733, 3.873], |i - j| = 1\}$, encoding bond lengths, and another set $\mathcal{C} = \{d_{ij} \geq D_{HC}, i \neq j\}$ for clashes. The set $\mathcal{M} = C \cup \mathcal{B} \cup \mathcal{C}$ defines the configurational space of physically realizable protein models.

The cost function measures the degree of structural matching of a given conformation $\mathcal{S}_n$ to the available constraints. Let $\mathcal{F}_0 = \{(i, j) \mid d_{ij} > d_T \land c_{ij} = 1\}$ denote the pairs of amino acid in contact according to $C$ but not in $\mathcal{S}_n$ ("false negatives"). Similarly, define $\mathcal{F}_1 = \{(i, j) \mid d_{ij} \leq d_T \land c_{ij} = 0\}$ as the pairs of amino acids in contact in $\mathcal{S}_n$ but not according to $C$ ("false positives"). The objective function is then defined as:

$$C(\mathcal{S}_n, \mathcal{M}) = \alpha_0 \left\{ 1 + \sum_{(i,j) \in \mathcal{F}_0} (d_{ij}/D_T)^2 + \sum_{(i,j):d_{ij} \notin \mathcal{B}} (d_{ij} - D_B)^2 \right\} \\ + \alpha_1 |\mathcal{F}_1| + \alpha_2 \sum_{(i,j):d_{ij} \notin \mathcal{C}} e^{(D_{HC} - d_{ij})}. \tag{6}$$

Note how the cost function is based only on simple geometric terms. The combination of this function with a set of moves allows the exploration of the configurational space.

### 5.3. *Optimization algorithm*

The algorithm we used for the reconstruction of the coordinates of protein $C_\alpha$ traces is organized in two sequential phases, *bootstrap* and *search*.

The function of the first phase is to *bootstrap* an initial physically realizable configuration with a self-avoiding random walk and explicit modeling of predicted helices. A random structure is generated by adding $C_\alpha$ positions one after the other until a draft of the whole backbone is produced. More specifically, this part runs through a sequence of $n$ steps, where $n$ is the length of the input chain. At stage $i$, the position of the $i$th $C_\alpha$ is computed as $r_i = r_{i-1} + d \frac{r}{|r|}$ where $d \in [3.733, 3.873]$ and $r$ is a random direction vector. Both $d$ and $r$ are uniformly sampled. If the $i$th residue is predicted at the beginning of an helix all the following residues in the same segment are modelled as an ideal helix with random orientation.

In the *search* step, the algorithm refines the initial bootstrapped structure by global optimization of the pseudo-potential function of Eq. (6) using local moves and a simulated annealing protocol. Simulated annealing is a good choice in this case, since the constraints obtained from various predictions are in general not realizable and contradictory. Hence the need for using a "soft" method that tries to enforce as many constraints as possible never terminating with failure, and is robust with respect to local minima caused by contradictions. The search strategy is similar to that in Ref. 12, but with a number of modifications. At step $t$ of the search, a randomly chosen $C_\alpha$ atom at position $r_i^{(t)}$ is displaced to the new position $r_i^{(t+1)}$ by a crankshaft move, leaving all the others $C_\alpha$ atoms of the protein in their original position (see Fig. 4). Secondary structure elements are displaced

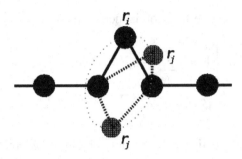

**Fig. 4.** Crankshaft move: the $i$th $C_\alpha$ at position $r_i$ is displaced to postition $r_j$ without moving the other atoms of the protein.

**Fig. 5.** Secondary structure elements are displaced as a whole, without modifying their geometry.

as a whole, without modifying their geometry (see Fig. 5). The move in this case has one further degree of freedom in the helix rotation around its axis. This is assigned randomly, and uniformly distributed. A new set of coordinates $\mathcal{S}^{(t+1)}$ is accepted as the best next candidate with probability $p = min(1, e^{\Delta C/T^{(t)}})$ defined by the annealing protocol, where $\Delta C = C(\mathcal{S}^{(t)}, \mathcal{M}) - C(\mathcal{S}^{(t+1)}, \mathcal{M})$ and $T^{(t)}$ is the temperature at stage $t$ of the schedule.

The computational complexity of the above procedure depends on the maximum number of available steps for the annealing schedule and the number of operations required to compute the potential of Eq. (6) at each step. The computation of this function is dominated by the $O(n^2)$ steps required for the comparison of all pairwise mutual distances with the entries of the given contact map. Note however that the types of move adopted allow to explicitly avoid the evaluation of the potential for each pair of positions. In the case of a residue crankshaft move, since only one position of the structure is affected by it, $\Delta C$ can be directly computed in $O(n)$ time by summing only the terms of Eq. (6) that change. For instance, in the case of the terms taking into account the contact map, the displacement of one $C_\alpha$ changes only a column and a row of the map induced by the configuration, hence the effect of the displacement can be computed by evaluating the $O(n)$ contacts on the row and column affected. The complexity of evaluating the energy after moving rigidly a whole helix is the same as moving all the amino acids on the helix independently. Hence, the overall cost of a search is $O(ns)$ where $n$ is the protein length and $s$ is the number of residues moved during the search. In practice, the number $s$ necessary to achieve convergence is proportional to the protein length, which makes the search complexity quadratic in $n$. A normal search run for a protein of length 100 or less takes a few tens of seconds on a single state-of-the-art CPU, roughly the same as computing the complex of 1D and 2D feature predictions.

## 6. Reconstruction Results

The protein data set used in reconstruction simulations consists of a non-redundant set of 258 protein structures showing no homology to the sequences employed to train the underlying predictive systems. This set includes proteins of moderate size (51 to 200 amino acids) and diverse topology as classified by SCOP (all-$\alpha$, all-$\beta$, $\alpha/\beta$, $\alpha + \beta$, surface, coiled-coil and small).

In all the experiments, we run the annealing protocol using a nonlinear (exponential decay) schedule with initial (resp. final) temperature proportional to the protein size (resp. 0). Pseudo-energy parameters are set to $\alpha_0 = 0.2$ (false non-contacts), $\alpha_1 = 0.02$ (false contacts) and $\alpha_2 = 0.05$ (clashes), so that the conformational search is biased towards the generation of compact clash-free structures and with as many of the predicted contacts realized.

Our algorithm is first benchmarked against a simple baseline that consists in predicting models where all the amino acids in the chain are collapsed into the same point (center of mass).

We run two sets of simulations: one where the reconstructions are based on native contact maps and structural features; one where all the structural features including contact maps are predicted. Using contact maps of native folds allows us to validate the algorithm and to estimate the expected upper bound of reconstruction performance. In order to assess the quality of predictions, two measures are considered here: root mean square deviation (RMSD); longest common sequence (LCS) averaged over four RMSD thresholds (1, 2, 4 and 8 Å) and normalized by the sequence length.

For each protein in the test set, we run 10 folding simulations and average the distance measures obtained over all 258 proteins. In Fig. 6, the average RMSD versus sequence length is shown for models derived from true contact maps (red crosses) and from predicted contact maps (green crosses), together with the baseline computed for sequences of the same length of the query. With true (resp. predicted) contact maps, the RMSD averaged over all test chains is 5.11 Å (resp. 13.63 Å), whereas the LCS1248 measure is 0.57 (resp. 0.29). For sequences of length up to 100 amino acids,

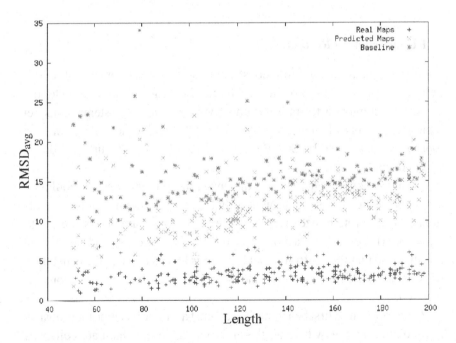

**Fig. 6.** Average RMSD versus sequence length.

the reconstruction algorithm using predicted contact maps obtains an an average LCS1248 of 0.43.

In a second and more detailed suite of experiments, for each protein in the test set, we run 200 folding simulations and cluster the corresponding optimal structures at final temperature after 10 000 iterations. The clustering step aims at finding regions of the configurational space that are more densely populated, hence are likely to represent tight minima of the pseudo-energy function. The centroid of the $n$th cluster is the configuration whose $q$th closest neighbor is at the smallest distance. After identifying a cluster, its centroid and $q$ closest neighbors are removed from the set of configurations. Distances are measured by RMSD (see below), and a typical value of $q$ is between 2 and 20. In our experiments, the centroids of the top 5 clusters are on average slightly more accurate than the average reconstruction. A protein is considered to be predicted with the correct topology if at least one of the five cluster centroids is within 6.5 Å RMSD to the native structure for at least 80% of the whole protein length (LCS(6.5) $\geq$ 0.8).

A first set of results is summarized in Table 3, where the proteins are divided into separate categories based on the number of models in each cluster (2, 3, 10 and 20) and length: from 51 to 100 amino acids (small), between 100 and 150 amino acids (medium) and from 150 to 200 amino acids (long). Table 3 shows for each combination of length and cluster size the number of proteins in the test set for which at least one of the five cluster centroids is within 6.5 Å of the native structure over 80% of the structure. From the table it is evident that correctly predicted topologies are restricted to proteins of limited size (up to 100–150 amino acids). Distill is able to identify the correct fold for short proteins in almost half of the cases, and for a few further cases in the case of proteins of moderate size (from 100 to 150 residues).

Table 3. Number of topologically correct (LCS(6.5) $\geq$ 0.8) predicted models.

| Cluster size | All | Short | Medium | Long |
| --- | --- | --- | --- | --- |
| 2 | 26/258 | 22/62 | 3/103 | 1/93 |
| 3 | 32/258 | 24/62 | 7/103 | 1/93 |
| 10 | 29/258 | 23/62 | 6/103 | 0/93 |
| 20 | 32/258 | 26/62 | 6/103 | 0/93 |

**Table 4.** Percentage of correctly predicted topologies with respect to sequence length and structural class.

| Length | $\alpha$ | $\beta$ | $\alpha + \beta$ | $\alpha/$ | $\beta$Surface | Coiled-coil | Small |
|---|---|---|---|---|---|---|---|
| All    | 20.3 | 4.0  | 7.3  | 6.3  | 33.3 | 66.7 | 16.7 |
| Short  | 64.7 | 25.0 | 35.7 | 33.3 | 60.0 | 60.0 | 25.0 |
| Medium | 6.3  | 0    | 2.8  | 11.8 | 0    | 100  | 0    |
| Long   | 0    | 0    | 0    | 0    | 0    | —    | —    |

In Table 4, we group the results for 20-dimensional clusters according to the SCOP assigned structural class and sequence length. For each combination of class and length, we report the fraction of proteins where at least one of the five cluster centroids has LCS(6.5) $\geq$ 0.8 to the native structure. These results indicate that a significant fraction of $\alpha$-helical proteins and those lacking significant structural patterns are correctly modeled. Reliable identification of strands and the corresponding patterns of connection is a major source of difficulty. Nevertheless, the reconstruction pipeline identifies almost correct folds for about a third of the cases in which a short protein contains a significant fraction of $\beta$-paired residues.

Figures 7–9 contain examples of predicted protein models from native and predicted contact maps.

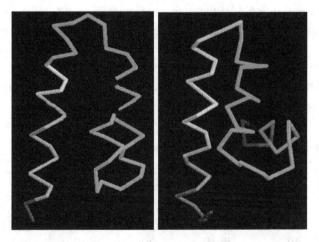

**Fig. 7.** Examples of reconstruction, protein 1OKSA (53 amino acids): real structure (left) and derived protein model from predicted contact map (right, RMSD = 4.24 Å).

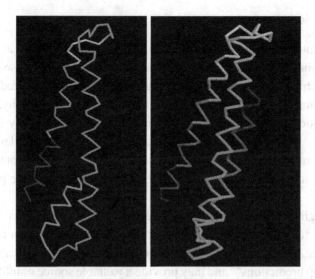

**Fig. 8.** Example of reconstruction, protein 1LVF (106 amino acids): real structure (left) and derived protein model from predicted contact map (right, RMSD = 4.31 Å).

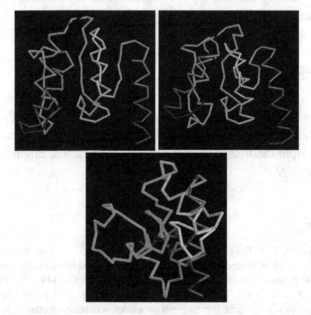

**Fig. 9.** Examples of reconstruction, protein 2RSL (119 amino acids): real structure (top-left), predicted model from true contact map (top-right, RMSD = 2.26 Å) and predicted model from predicted contact map (bottom, RMSD = 11.1 Å).

## 7. Conclusions

In this chapter we have presented Distill, a modular and fully automated computational system for *ab initio* prediction of protein coarse models. Distill's architecture is composed of: (1) a set of state-of-the-art predictors of protein features (secondary structure, relative solvent accessibility, contact density, residue contact maps, contact maps between secondary structure elements) based on machine learning techniques and trained on large, non-redundant subsets of the PDB; (2) a simple and fast 3D reconstruction algorithm guided by a pseudo-energy defined according to these predicted features.

Although Distill's 3D models are often still crude, nonetheless they may yield important information and support other related computational tasks. For instance, they can be effectively used to refine secondary structure and contact map predictions[47] and may provide a valuable source of information to identify protein functions more accurately than it would be possible by sequence alone.[7] Distill's modeling scheme is fast and makes genomic scale structural modeling tractable by solving hundreds of thousands of protein coordinates in the order of days.

## Acknowledgment

This work is supported by Science Foundation Ireland grants 04/BR/CS0353 and 05/RFP/CMS0029, grant RP/2005/219 from the Health Research Board of Ireland, a UCD President's Award 2004, and an Embark Fellowship from the Irish Research Council for Science, Engineering and Technology to AV.

## References

1. C. A. Orengo, J. E. Bray, T. Hubbard, L. Lo Conte and I. I. Sillitoe, Analysis and assessment of *ab initio* three-dimensional prediction, secondary structure, and contacts prediction, *Proteins: Structure, Function and Genetics* **37**(S3), 149–170 (1999).
2. A. M. Lesk, L. Lo Conte and T. J. P. Hubbard, Assessment of novel fold targets in CASP4: predictions of three-dimensional structures, secondary structures, function and genetics, *Proteins: Structure, Function and Genetics* **S5**, 98–118 (2001).
3. J. Moult, K. Fidelis, A. Zemla and T. Hubbard, Critical assessment of methods of protein structure prediction (casp)-round v, *Proteins* **53**(S6), 334–339 (2003).

4. J. Moult, K. Fidelis, A. Tramontano, B. Rost and T. Hubbard, Critical assessment of methods of protein structure prediction (casp)-round vi, *Proteins*, Epub 26 September 2005 (in press).
5. H. M. Berman, J. Westbrook, Z. Feng, G. Gilliland, T. N. Bhat, H. Weissig, I. N. Shindyalov and P. E. Bourne, The protein data bank, *Nucl. Acids Res.* **28**, 235–242 (2000).
6. J. J. Vincent, C. H. Tai, B. K. Sathyanarayana and B. Lee, Assessment of casp6 predictions for new and nearly new fold targets, *Proteins*, Epub 26 September 2005 (in press).
7. R. Bonneau, C. E. Strauss, C. A. Rohl, D. Chivian, P. Bradley, L. Malmstrom, T. Robertson and D. Baker, De novo prediction of three-dimensional structures for major protein families, *Journal of Molecular Biology* **322**(1), 65–78 (2002).
8. D. T. Jones, Protein secondary structure prediction based on position-specific scoring matrices, *J. Mol. Biol.* **292**, 195–202 (1999).
9. B. Rost and C. Sander, Prediction of protein secondary structure at better than 70% accuracy, *J. Mol. Biol.* **232**, 584–599 (1993).
10. G. Pollastri, D. Przybylski, B. Rost and P. Baldi, Improving the prediction of protein secondary structure in three and eight classes using recurrent neural networks and profiles, *Proteins* **47**, 228–235 (2002).
11. G. Pollastri and A. McLysaght, Porter: a new, accurate server for protein secondary structure prediction, *Bioinformatics* **21**(8), 1719–1720 (2005).
12. M. Vendruscolo, E. Kussell and E. Domany, Recovery of protein structure from contact maps, *Folding and Design* **2**, 295–306 (1997).
13. D. T. Jones, Genthreader: an efficient and reliable protein fold recognition method for genomic sequences, *J. Mol. Biol.* **287**, 797–815 (1999).
14. P. Bradley, D. Chivian, J. Meiler, K. M. S. Misura, C. A. Rohl, W. R. Schief, W. J. Wedemeyer, O. Schueler-Furman, P. Murphy, J. Schonbrun, C. E. M. Strauss and D. Baker, Rosetta predictions in casp5: successes, failures, and prospects for complete automation, *Proteins* **53**(S6), 457–468 (2003).
15. S. F. Altschul, T. L. Madden and A. A. Schaffer, Gapped blast and psi-blast: a new generation of protein database search programs, *Nucl. Acids Res.* **25**, 3389–3402 (1997).
16. T. N. Petersen, C. Lundegaard, M. Nielsen, H. Bohr, J. Bohr, S. Brunak, G. P. Gippert and O. Lund, Prediction of protein secondary structure at 80% accuracy, *Proteins: Structure, Function and Genetics* **41**(1), 17–20 (2000).
17. B. Rost and C. Sander, Conservation and prediction of solvent accessibility in protein families, *Proteins: Structure, Function and Genetics* **20**, 216–226 (1994).
18. H. Naderi-Manesh, M. Sadeghi, S. Arab and A. A. Moosavi Movahedi, Prediction of protein surface accessibility with information theory, *Proteins: Structure, Function and Genetics* **42**, 452–459 (2001).
19. M. H. Mucchielli-Giorgi, S. Hazout and P. Tuffery, PredAcc: prediction of solvent accessibility, *Bioinformatics* **15**, 176–177 (1999).
20. J. A. Cuff and G. J. Barton, Application of multiple sequence alignments profiles to improve protein secondary structure prediction, *Proteins: Structure, Function and Genetics* **40**, 502–511 (2000).
21. G. Pollastri, P. Fariselli, R. Casadio and P. Baldi, Prediction of coordination number and relative solvent accessibility in proteins, *Proteins* **47**, 142–235 (2002).

22. N. Biggs, Algebraic graph theory, Second Edition (1994).
23. M. Porto, U. Bastolla, H. E. Roman and M. Vendruscolo, Reconstruction of protein structures from a vectorial representation, *Phys. Rev. Lett.* **92**, 218101 (2004).
24. G. Pollastri and P. Baldi, Prediction of contact maps by recurrent neural network architectures and hidden context propagation from all four cardinal corners, *Bioinformatics* **18** (Suppl.1), S62–S70 (2002).
25. L. Holm and C. Sander, Parser for protein folding units, *Proteins* **19**, 256–268 (1994).
26. U. Bastolla, M. Porto, H. E. Roman and M. Vendruscolo, Principal eigenvector of contact matrices and hydrophobicity profiles in proteins, *Proteins: Structure, Function, and Bioinformatics* **58**, 22–30 (2005).
27. P. Fariselli, O. Olmea, A. Valencia and R. Casadio, Progress in predicting inter-residue contacts of proteins with neural networks and correlated mutations, *Proteins: Structure, Function and Genetics* **S5**, 157–162 (2001).
28. A. Vullo, I. Walsh and G. Pollastri, A two-stage approach for improved prediction of residue contact maps, *BMC Bioinformatics* **7**, 180 (2006).
29. P. Fariselli and R. Casadio, Neural network based predictor of residue contacts in proteins, *Protein Engineering* **12**, 15–21 (1999).
30. P. Fariselli, O. Olmea, A. Valencia and R. Casadio, Prediction of contact maps with neural networks and correlated mutations, *Protein Engineering* **14**(11), 835–439 (2001).
31. P. Baldi and G. Pollastri, The principled design of large-scale recursive neural network architectures — DAG-RNNS and the protein structure prediction problem. *Journal of Machine Learning Research* **4**, 575–602 (2003).
32. D. A. Debe, M. J. Carlson, J. Sadanobu, S. I. Chan and W. A. Goddard, Protein fold determination from sparse distance restraints: the restrained generic protein direct Monte Carlo method, *J. Phys. Chem.* **103**, 3001–3008 (1999).
33. A. Aszodi, M. J. Gradwell and W. R. Taylor, Global fold determination from a small number of distance restraints, *J. Mol. Biol.* **251**, 308–326 (1995).
34. E. S. Huang, R. Samudrala and J. W. Ponder, Ab initio fold prediction of small helical proteins using distance geometry and knowledge-based scoring functions, *J. Mol. Biol.* **290**, 267–281 (1999).
35. J. Skolnick, A. Kolinski and A. R. Ortiz, Monsster: a method for folding globular proteins with a small number of distance restraints, *J. Mol. Biol.* **265** (1997) 217–241.
36. P. M. Bowers, C. E. Strauss and D. Baker, De novo protein structure determination using sparse NMR data. *J. Biomol. NMR* **18**, 311–318 (2000).
37. W. Li, Y. Zhang, D. Kihara, Y. J. Huang, D. Zheng, G. T. Montelione, A. Kolinski and J. Skolnick, Touchstonex: protein structure prediction with sparse NMR data, *Proteins: Structure, Function, and Genetics* **53**, 290–306 (2003).
38. R. M. McCallum, Striped sheets and protein contact prediction. *Bioinformatics* **20** (Suppl. 1), 224–231 (2004).
39. Casp6 home page.
40. V. A. Eyrich, M. A. Marti-Renom, D. Przybylski, M. S. Madhusudan, A. Fiser, F. Pazos, A. Valencia, A. Sali and B. Rost, Eva: continuous automatic evaluation of protein structure prediction servers, *Bioinformatics* **17**, 1242–1251 (2001).
41. G. Pollastri, A. Vullo, P. Frasconi and P. Baldi, Modular DAG-RNN architectures for assembling coarse protein structures, *Journal of Computational Biology* **13**(3), 631–650 (2006).

42. C. A. Orengo, A. D. Michie, S. Jones, D. T. Jones, M. B. Swindells and J. M. Thornton, Cath — a hierarchic classification of protein domain structures, *Structure* **5**, 1093–1108 (1997).
43. P. Frasconi, M. Gori and A. Sperduti, A general framework for adaptive processing of data structures. *IEEE Trans. on Neural Networks* **9**, 768–786 (1998).
44. P. Baldi, S. Brunak, P. Frasconi, G. Soda and G. Pollastri, Exploiting the past and the future in protein secondary structure prediction, *Bioinformatics* **15**, 937–946 (1999).
45. A. Vullo and P. Frasconi, Disulfide connectivity prediction using recursive neural networks and evolutionary information, *Bioinformatics* **20**(5), 653–659 (2004).
46. W. Kabsch and C. Sander, Dictionary of protein secondary structure: pattern recognition of hydrogen-bonded and geometrical features, *Biopolymers* **22**, 2577–2637 (1983).
47. A. Ceroni, P. Frasconi and G. Pollastri, Learning protein secondary structure from sequential and relational data, *Neural Networks* **18**(8), 1029–1039 (2005).

# CHAPTER 8

# IN SILICO DESIGN OF LIGANDS USING PROPERTIES OF TARGET ACTIVE SITES

Sanghamitra Bandyopadhyay* and Santanu Santra[†]

*Machine Intelligence Unit, Indian Statistical Institute
Kolkata 700108, India
*sanghami@isical.ac.in
†santanu_t@isical.ac.in*

Ujjwal Maulik

*Department of Computer Science and Engineering
Jadavpur University, Kolkata 700 032, India
drumaulik@cse.jdvu.ac.in*

Heinz Muehlenbein

*Fraunhofer Institute, AiS, Sankt Augustin, Germany
heinz.muehlenbein@online.de*

In this chapter we present an evolutionary approach for designing a ligand that can bind to the active site of a target protein. A tree like structure of the ligand is assumed, whose nodes are to be filled up from a library of functional units. Variable string length genetic algorithm (VGA) is used for evolving an appropriate arrangement of the basic functional units of the molecule to be designed. Since only the outer groups (i.e. those on the surface) take a major part in the van der Waals energy computation, a local search is first used to design the core part of the ligand by optimizing its internal energy. Once the geometry of the core part of the molecule is obtained, VGA is used for finding the appropriate arrangement for the outer layer of the ligand. The crossover and mutation operators are appropriately redesigned in order to tackle the concept of variable length chromosomes in VGA. Results are demonstrated for different target proteins both numerically and pictorially using Insight II (MSI/Accelrys, San Diego, CA, USA). Comparative results in form of ligand energy, docking energy and computation time shows a significant improvement over an earlier attempt.

## 1. Introduction

The task of drug design has traditionally been dependent on nature, with some of the most effective drugs, like morphine and penicillin, having been obtained from natural sources. Their power stems from their unique structures that have evolved over millions of years of random variation and natural selection. However a potential drawback of the natural processes is that they are extremely slow. Therefore, for developing and commercializing drugs within a reasonable amount of time, pharmaceutical companies have devised proactive drug discovery methods, including a recent innovation called rational drug design. In this approach, researchers build and test small drug-like molecules based on prior knowledge about the three-dimensional structures of known drug molecules. Quantitative structure-activity relationships (QSAR)[13,14] studies form an integral part of rational drug design. However, this approach has its own limitations, since it is not always clear which variations on known molecules are worth testing.

Identification or design of drug molecules, without assuming any similarity with some known structure, that can target proteins crucial for the proliferation of microbial organisms, cancer cells or viruses is one of the important approaches in drug design. Such molecules can disrupt the action of the target protein, that sustain viral proliferation, by binding to its active site; thereby nullifying its activity which can be lethal. Therefore the task of accurately predicting the structure of the potential inhibitors, while utilizing the knowledge about the structure of a target protein, is another important area of research.

Genetic algorithms (GAs) are randomized search and optimization techniques guided by the principles of evolution and natural genetics, and have a large amount of implicit parallelism. GAs perform multi-model search in complex landscapes and provide near optimal solutions for objective or fitness function of an optimization problem. They have diverse application in the fields as diverse as pattern recognition, image processing, VLSI design, neural networks etc.[17,23] GAs have also been applied to the domain of bioinformatics[4,6] including that of drug design.[2,9,10,14,18,19] The approach adopted earlier[2] is based on the use of genetic algorithms for evolving small molecules represented using a graphical structure composed of atoms as the vertices and the bonds as the edges. The task here is to determine the effectiveness of GAs in evolving a molecule that is similar to a target molecule.

Thus knowledge about the target molecule is assumed, which may not be readily available in many situations.

Another approach for ligand design, that is based on the presence of a fixed pharmacophore and that uses the search capabilities of genetic algorithms, was studied by Goh and Foster,[10] where the harmful protein human Rhinovirus strain14 was used as the target. This pioneering work assumed a fixed 2-dimensional tree structure representation of the molecule on both sides of the pharmacophore. Evidently, an *a priori* knowledge of the size of the tree is difficult to obtain. Moreover, it is known that no unique ligand structure is best for a given active site geometry. Therefore, in Ref. 16 a variable length representation of the trees in 2-dimensions on both sides of the pharmacophore for designing the ligand molecule was proposed. Variable string length genetic algorithm (VGA)[1] was used for this purpose. In contrast to Ref. 10, the chromosome, that encodes the ligand tree, can be of any size in Refs. 1 and 16. The VGA searches for an appropriate arrangement of the functional groups in the designed ligand such that the van der Waals interaction energy of the target and the ligand is minimized. This is a more natural representation since the size of the active site will itself be different for different proteins.

Note that even with the tree based representation, the task of drug design remains extremely difficult because of the huge number of possibilities. In fact by allowing the size of the tree structure to vary, the search space becomes even larger. Moreover, all possible arrangements of functional groups are biologically not possible since they may not satisfy several electrochemical requirements. This issue was not adequately tackled in Ref. 16 (though a penalty factor was imposed for certain conformations). Additionally, in Ref. 16 only the van der Waals interaction energy was optimized using the VGA; the internal energy of the ligand was not considered. The present article is an attempt to address the above mentioned limitations in Ref. 16. For restricting the search to plausible structures, a more detailed set of rules and penalty factors are now incorporated. These are mentioned in Sec. 3 of this article. To make the search faster and more focused, a local search technique is incorporated in the VGA in two steps. Firstly, local search is used to generate the core portion of the ligand by optimizing its internal energy. Thereafter VGA considers the generated core for population initialization which is then modified appropriately (that can both increase or decrease the size of the tree) so that the interaction energy between the ligand and the target is now minimized. Once this is done, again a local search is employed to optimize the structure of the designed ligand as far as possible. Finally, rather than evolving the left and right tree separately as in Ref. 16, the VGA now evolves both of them simultaneously. This has been done in view of the fact that by evolving the left and right tree separately, any interaction between them, if they are possible, have been disregarded in Ref. 16. Therefore to make the model more realistic, the entire small ligand is designed here in one shot.

Two-dimensional molecules designed using genetic algorithms are visualized in three dimensions using a tool Insight II (MSI/Accelrys, San Diego, CA, USA) which also has an in built optimizer to evolve the best 3D structure given a 2D structure. Experimental results are reported for four proteins with different characteristics of the active site. Comparative study with the VGA based approach[16] is provided.

## 2. Relevance of Genetic Algorithm for Drug Design

One of the main approaches in the field of drug design is the identification of proper ligands so that they can be used to generate the structure of a drug molecule. In general one or more proteins are typically involved in the bio-chemical pathway of a disease. The treatment aims to appropriately reduce the effect of such proteins, by designing a ligand molecule that can bind to its active site. In a rational approach to drug design, the structure of a ligand molecule is evolved from a set of groups in close proximity to crucial residues of the protein; thereby a molecule is designed that fits the protein target receptor such that a criterion (for example, van der Waals interaction energy) is optimized. Such an approach is adopted in this article.

The design of ligands for a particular protein target with respect to some of its active sites can be viewed as a problem of searching for a particular arrangement of several functional groups so that the protein ligand interaction becomes favorable. It has been shown in Ref. 16 that the search space is very large and complex. Consequently the application of GAs for solving this problem appears to be natural and appropriate. In Ref. 10, a tree like representation of the ligand was considered, with the size of the tree being fixed on both side of the pharmacophore, and GA was used for evolving a best arrangement of a given set of functional group. Inspired by this approach, in Ref. 16, a variable length GA (VGA) was used for evolving the tree like ligand, composed of variable sized left and right subtrees. Its advantage with respect to the earlier fixed length version was demonstrated in Ref. 16.

The present article further extends the work in Ref. 16. Though the concept of VGA was found to improve the performance, there is still room for improvization for exploiting the full potential of the method. In this article the GA is augmented by incorporating local search in the process. Note that the primary objective is to maximize the energy of interaction between the ligand and the target protein, and generally only the surface groups are involved in this computation. Consequently for appropriately generating the core part of the ligand, a local search is employed starting from a configuration provided by the GA.

In the VGA based approach for ligand design by incorporating local search, subsequently referred to as LVGALD, the main issues involve the formation of the core of the ligand, chromosome representation and local search for refining the final solution. These are first described in detail in Sec. 3 followed by an overall description of the algorithm in Sec. 4.

## 3. Basic Issues

In this section we describe the basic issues concerning the LVGALD algorithm.

### 3.1. Core formation

The tree structure of the ligand is shown in Fig. 1. The tree grows along the major axis of the active site, on both sides of the pharmacophore $P$. The major axis is determined by the coordinates of the active site. The whole structure follows a lower leaf $(L)$, base $(B)$, upper leaf $(U)$ strategy. The nodes of the tree are numbered 1 for L, 2 for B and 3 for U as shown in Fig. 1. These nodes are to be filled in from a library of functional groups shown in Table 1 in such a way that the ligand as well as its interaction with the protein are stable. This indicates that the corresponding energies are the minimum. The van der Waals energy is used for this purpose. The van der

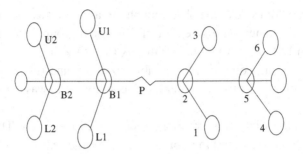

**Fig. 1.** LBU structure.

Waals energy between two groups $x$ and $y$ is defined as

$$E(x, y) = \left(\frac{C_n}{r_{xy}^6}\right) - \left(\frac{C_m}{r_{xy}^{12}}\right), \quad (1)$$

where $C_n$ and $C_m$ are constants,[28] and $r_{xy}$ is the distance between $x$ and $y$. The value of $r_{xy}$ is calculated as follows:

$$r_{xy} = \sqrt{(x_x - y_x)^2 - (x_y - y_y)^2}, \quad (2)$$

where $(x_x, x_y)$ and $(y_x, y_y)$ are the 2D coordinates of the groups $x$ and $y$ respectively. The way of computing the coordinates is explained later. The total energy is the sum of all the individual energy terms computed using Eq. (1). This is defined as

$$E = \sum_x \sum_y E(x, y) \text{ such that } 0.65\,\text{Å} \leq r_{xy} \leq 2.7\,\text{Å}. \quad (3)$$

Note that $r_{xy} < 0.65\,\text{Å}$ indicates steric contact between $x$ and $y$, and $r_{xy} > 2.7\,\text{Å}$ is too large a distance for the weak van der Waal's force to have any effect.

The way of computing $(x_x, x_y)$, the 2D coordinates of a functional group $x$, is now explained using Fig. 2. Let us assume that the coordinates of BASE in Fig. 2 is known, and is denoted by $(Base_x, Base_y)$. Then the coordinates of $L$, $B$ and $U$, denoted by $(L_x, L_y)$, $(B_x, B_y)$ and $(U_x, U_y)$ respectively, for the left subtree are computed as

$$L_x = BASE_x + r \times \cos(-109°47')$$
$$L_y = BASE_y + r \times \sin(-109°47')$$
$$B_x = BASE_x - r \times \cos(0)$$

# In Silico Design of Ligands

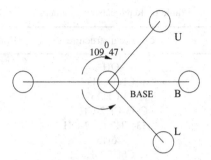

**Fig. 2.** Basic structure for designing the tree

$$B_y = BASE_y - r \times \sin(0)$$
$$U_x = BASE_x + r \times \cos(109°47')$$
$$U_y = BASE_y + r \times \sin(109°47'). \quad (4)$$

For the right sub-tree this is calculated as

$$\begin{aligned}
L_x &= BASE_x + r \times \cos(\pi + 109°47') \\
L_y &= BASE_y + r \times \sin(\pi + 109°47') \\
B_x &= BASE_x + r \times \cos(0) \\
B_y &= BASE_y + r \times \sin(0) \\
U_x &= BASE_x + r \times \cos(\pi - 109°47') \\
U_y &= BASE_y + r \times \sin(\pi - 109°47').
\end{aligned} \quad (5)$$

Here $r$ is the bond length of the group whose coordinates is being determined. The bond length values for the different functional groups are provided in Table 1. The bond angle for carbon is known to be $109°47'$.[21]

The core part of the tree is formed in a greedy manner. The length of the core tree ($c_T$) is then determined randomly in between $lmin$ and $lmax - 1$, where $lmax$ and $lmin$ are the maximum and minimum possible levels that a right or left tree respectively may have. These are as defined below:

$$lmax = \frac{\text{length of the major axis}}{\text{minimum bond length}} \quad \text{and}$$
$$lmin = \frac{\text{length of the major axis}}{\text{maximum bond length}}. \quad (6)$$

**Table 1.** Representation of groups

| Group name | Chemical structure | Bond length Å |
|---|---|---|
| Polar | OH | 0.01 |
| Alkyl 1C | CH3-CH3 | 0.65 |
| Alkyl 1C Polar | CH3-CH2-OH | 1.10 |
| Alkyl 3C | CH3-CH2-CH3 | 1.75 |
| Alkyl 3C Polar | CH3-CH2-CH2-OH | 1.20 |
| Aromatic | C6H6 | 1.9 |
| Aromatic Polar | C6H5-OH | 2.7 |

The length of the major axis is computed as the Euclidean distance between the two farthest points of the active site. A core tree with length $l_{c_T}$ needs $1 + 3(l_{c_T} - 1)$ positions (nodes) of the tree to be filled up for designing the core. As an example, let us consider the process of forming the right subtree of the core (refer to Fig. 1). First position ($B$) is filled with the first functional group, say $Group_1$. Thereafter for filling up the next positions 2 to $3(l_{c_T} - 1)$, each group is considered in turn, and the one that results in the minimum energy of the core designed so far is accepted. The sum of the van der Waals energy between all the interacting groups of the core tree is taken as its internal energy. This is denoted by $Int\_E(Core, Group_1)$, which represents the internal energy of the core, with $Group_1$ as the starting group in the subtree. Similarly $Int\_E(Core, Group_2), \ldots, Int\_E(Core, Group_7)$ are computed by starting from each of the remaining six groups (see Table 1). The core corresponding to $min_{i=1,2,\ldots,7} Int\_E(Core, Group_i)$ is selected as the tree to initiate the genetic process.

### 3.2. Chromosome representation

A chromosome represents an entire tree structure on both sides of the pharmacophore which are separated by a #, with the nodes being filled up by the functional groups as demonstrated graphically in Fig. 3. A number in any position of the chromosome represents the presence of the corresponding group in that position. For example, consider a chromosome $C = 3426\#512444$. This is decoded as the tree shown in Fig. 3. Both the trees start from a position within the active site that is centrally located within it with respect to the major axis.

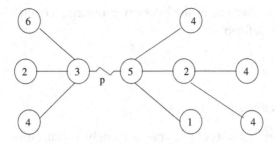

**Fig. 3.** Structure of the ligand corresponding to the chromosome $\mathcal{C} = 3426\#512444$.

## 3.3. Fitness computation

In GA, fitness of a chromosome is so defined that it reflects the relative quality of the solutions. In LVGALD the objective to be minimized is the energy. Firstly the internal energy of the ligand, which is used for local search, should be minimized for better stability of the ligand. This energy calculation is based on the calculation of interaction energy of the different functional groups within the ligand and is considered during the core formation. The fitness value, which is thereafter optimized by the VGA, is defined as the interaction energy between the ligand and the active site, and is termed as the docking energy. Again this should also be minimized for better interaction of the ligand and the protein. This energy calculation is based on the proximity of the different residues in the active site to the closest functional groups in the ligand and their chemical properties. For both the cases (internal and interaction energies), the distance of the interacting groups should be at most 2.7 Å for the molecules to interact and should not be closer than 0.65 Å for avoiding steric effect. A penalty value is incorporated in the fitness value to penalize the chromosome (and therefore the corresponding tree like ligand) that fails to fulfill the above criteria. Again, a chromosome having the OH group in the non-leaf base position of the corresponding tree is penalized, since for this group it is not possible to have any further outgoing branches due to valency saturation of the atoms in the group.

The van der Waals energy value is used to compute the interaction energy between the ligand and the residues of the active site, using Eq. (3). In this case one of the functional groups belong to the ligand (and its coordinate is generated in a manner described earlier), while the other is one of the residues of the active site of the interacting protein. The fitness value of a chromosome is defined as

$$F = \frac{1}{E}. \qquad (7)$$

## 4. Main Algorithm

The main process consists of five phases, namely, initialization, fitness computation, selection, crossover and mutation. These phases are described below in detail.

Initialization: as mentioned in Sec. 3.1, initially chromosomes are designed through a local search. In order to create only feasible chromosomes, restriction is imposed during initialization so that group with feasible valency only can be considered as base (B). For example, a base with -OH functional group cannot have any child. So -OH group can only be placed in a leaf node (i.e. L and U, or B if it is a leaf). For each chromosome $C_i$, $1 = 1, \ldots, P$, where $P$ is the population size, a random number between $lmin$ and $lmax$ is selected. This corresponds to the length of the core. Once the core is formed using the local search technique mentioned in Sec. 3.1 (which optimizes the internal energy), this is encoded in a string like structure to yield the $i$th chromosome.

Fitness computation: each chromosome is assigned a fitness value ($F$) computed using the van der Waals equation as mentioned in Sec. 3.3.

Selection: several alternatives are possible for implementing selection in a GA. Here truncation selection is used as the selection procedure. In this process, the chromosome are first ordered according to descending value of their fitness function. Then a mating pool is created by selecting the top $\alpha$ percent of the population. These form the parents for the next generation.

Crossover: in crossover, information is exchanged between two solutions in order to yield potentially better solutions. A VGA compatible crossover technique is used in the LVGALD method. In this crossover technique, performed with probability $cprob$, two different portions of the parent chromosomes (which may be of different lengths) are selected so that the exchange of these parts does not produce any offspring below length $lmin$ and above length $lmax$. Moreover, care is taken to extract the parts from the central (or, base) position. This crossover is capable of producing offspring of different length. The following example demonstrates the crossover process.

```
Parents                                Length
C1 →    4  3  2  4 | 6  5 | 4  2         8
C2 →    1 | 5  4  2  1 | 4  4            7
Children              ⇓
C'1 →   4  3  2  4  5  4  2  1  4  2    10
C'2 →   1  6  5  4  4                     5
```

Mutation: mutation is used to increase the diversity of the population, and to recover lost information. In mutation, a position of the chromosome is selected, and its value is replaced by another group randomly selected from [$Group_1, Group_2, \ldots, Group_7$]. The mutation operation is so defined that it ensures the feasibility of the resultant chromosome. Here each position of a chromosome is subjected to mutation with probability *mprob*, which is generally kept low.

Elitism: an elite population is maintained after each generation. Let the size of the elite population be $P_E$. Thus, it contains the $P_E$ best chromosomes (those with the lowest energy values) seen so far. On termination, all these $P_E$ chromosomes are tested. The reason for retaining an elite set of good chromosomes is as follows. Note that, since the chromosome provided by the LVGALD are only in 2 dimensions, their energy values change substantially when the structures are subjected to further optimization by InsightII. Although it is generally true that if two LVGALD designed chromosomes be $C_1$ and $C_2$, and energy of $C_1$ is less than energy of $C_2$, then after 3D simulation in InsightII, the final energy of $C_1$ will be less than that of $C_2$. However, sometimes it may happen that although $C_1$ is better than $C_2$ as indicated by the energy value provided by LVGALD, the final 3-dimensional interaction energy of $C_1$ is more than that of $C_2$. The reason for this is that $C_1$ may have some structure which is not particularly amenable to too much optimization, while $C_2$ has this scope. Therefore, rather than concentrating attention on only the best chromosome, an elite set of $P_E$ best chromosomes are finally considered.

## 5. Experimental Results

This section presents an experimental evaluation of LVGALD along with its comparison to the VGA method. The main algorithm is implemented in

C++ language on a Sun-Solaris machine. The performance is tested using InsightII software (MSI/Accelrys, San Diego, CA) on a Silicon Graphics O2 Workstation. InsightII is a molecular visualization software to visualize biological macromolecules like DNA, proteins etc. There are several modules incorporated in the software to manipulate the macromolecules. Mainly Builder and Biopolymer are used to design and optimize the ligand, encoded in the chromosomes. To measure the interaction energy between a ligand and a protein (on some defined active site) mainly Docking and Ludi are used.

For the experiments, four different proteins are considered. These are HIV-I Nef,[15] HIV-I Integrase,[20] HIV-I capsid[8] and 1SMT[24] protein. The coordinates of the active sites of these proteins were obtained from Protein Data Bank.

HIV-I Nef protein is a retroviral protease acting on proteins of host cells. HIV-I Integrase accelerates virulent progression of acquired immunodeficiency syndrome (AIDS) by its interaction with specific cellular proteins involved in signal transduction and host cell activation. HIV-I capsid protein is the protein present in the outer covering of the virus. The absence of the protein leads to the disorganization of the viral cellular materials. Note that these three proteins (viz. HIV-I Nef protein, HIV-I Integrase, HIV-I capsid protein) are known to be involved in the proliferation of the virus causing AIDS inside host cells, though each has a separate function. 1SMT is a trans-acting dimeric repressor that is required for Zn(2+)-responsive expression of the metallothionein SmtA. The structure of 1SMT was solved using multiple isomorphous replacement techniques and refined at 2.2 Å resolution by simulated annealing to an R-factor of 0.218. It has an alpha + beta topology, and the arrangement of the three core helices and the beta hairpin is similar to the HNF-3/fork head, CAP and diphtheria toxin repressor proteins.[24]

The active site geometry of these proteins varies from a barrel shaped structure to an ellipsoidal one. The coordinates of an active site of the protein 1SMT is shown in Table 2. The aim is to find non-peptide molecule(s) to fit into the given active sites. From the active site geometries of the molecules it is quite clear that any appropriate drug molecule must be flexible enough to bend in order to fit into protein target site. This flexible backbone is part of a pharmacophore. The essential part of the pharmacophore is found to have more or less the same structural arrangement of groups, i.e. they are

Table 2. Active site coordinates of 1SMT.

| Molecules | x-coordinates | y-coordinates | z-coordinates |
|---|---|---|---|
| A24:N | −14.61 | −1.03 | −9.91 |
| A24:O | −14.36 | −7.13 | −5.35 |
| A25:O | −11.34 | 3.58 | −4.24 |
| B50:O | 0.04 | −3.84 | −2.98 |
| B50:NH2 | 0.31 | −0.24 | 1.16 |
| B51:CD1 | 3.70 | 0.81 | −4.62 |
| B51:0 | −0.29 | −7.20 | 0.18 |
| B54:CD2 | −7.49 | −4.15 | −6.44 |
| B54:CD1 | −5.71 | −2.42 | −6.36 |
| B55:CA2 | −4.62 | −3.49 | −5.12 |
| B60:CA | −8.35 | −0.30 | 14.21 |
| B60:CD1 | −6.66 | −1.73 | −3.46 |
| B60:CD2 | −10.36 | −1.56 | −11.86 |
| B64:NE2 | −13.53 | −4.73 | −5.18 |
| B64:OA1 | −12.24 | 3.23 | −10.93 |
| B92:NH2 | −14.10 | −0.78 | −12.82 |

made up of alkyl chains, which make them hydrophobic enough in order to fit into the hydrophobic core of the protein target sites.

For the sake of comparison, the energy values of the ligands (obtained using the VGA based method and the proposed LVGALD method which are then subjected to optimization using DISCOVER module of InsightII) as well as those of the ligand-protein interaction are computed. These values are presented in Tables 3 and 4 respectively. Lower internal energy value suggests better stability of the ligand. As seen from Table 3, in all the cases LVGALD provides more stable ligands that are associated with lower energy values.

Table 4 shows the interaction energy between the ligand and the protein for the four different targets. As can be seen from the table, three out of four interaction energies are found to be better in LVGALD. Only in the case of 1SMT, LVGALD performs slightly poorer than the earlier VGA method.

Figures 4–7 show the molecules designed using the proposed LVGALD (denoted by (a) in the figure captions) and docked into the protein target (denoted by (b) in the figure captions). As is evident from the figures, the designed molecules are found to fill up the active site reasonably well.

Table 3. Internal energies of the ligands in kcal/mol.

| Process | HIV Nef | HIV Integrase | HIV Capsid | 1SMT |
|---|---|---|---|---|
| VGA | 39.91 | 45.32 | 30.01 | 32.89 |
| LVGALD | 27.94 | 38.25 | 22.69 | 27.56 |

Table 4. Interaction energies of the protein-ligand complexes in kcal/mol.

| Process | HIV Nef | HIV Integrase | HIV Capsid | 1SMT |
|---|---|---|---|---|
| VGA | 11.86 | 19.65 | 9.06 | 2.46 |
| LVGALD | 5.36 | 11.61 | 8.7 | 2.89 |

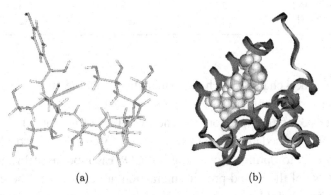

(a)  (b)

**Fig. 4.** Result of LVGALD for HIV-I Nef protein. (a) Structure of the ligand (represented in line), (b) the interaction of the ligand (represented by CPK) with the protein.

In order to compare the designed ligand to an existing one, the protein 1SMT is considered. Figure 8 shows the original ligand, nicotinamide adenine dinucleotide phosphate (NADP). The PDB contains the coordinates of the 1SMT-NADP complex that is available in a docked state in the PDB. It has been calculated in InsightII that the energy of the ligand NADP and the interaction energy of NADP with 1SMT are 146.93 and 62.0 kcal/mol respectively. Comparing with the values in Tables 3 and 4, it is evident that the ligand designed by LVGALD has lower values of both the internal energy and the interaction energy. This indicates that LVGALD is able to

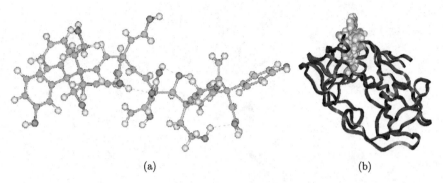

**Fig. 5.** Result of LVGALD for HIV-I Integrase protein. (a) Structure of the ligand (represented in ball-stick), (b) the interaction of the ligand (represented by CPK) with the protein.

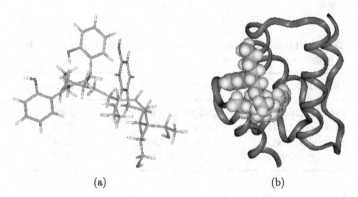

**Fig. 6.** Result of LVGALD for HIV-I Capsid protein. (a) Structure of the ligand (represented in line), (b) the interaction of the ligand (represented by CPK) with the protein.

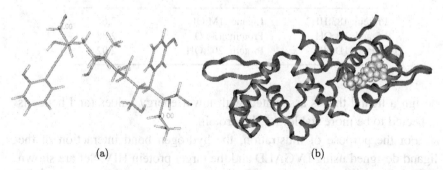

**Fig. 7.** Result of LVGALD for 1SMT protein. (a) Structure of the ligand (represented in line), (b) the interaction of the ligand (represented by CPK) with the protein.

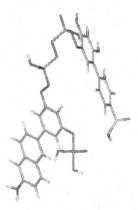

**Fig. 8.** Original structure of ligand nicotinamide adenine dinucleotide phosphate (NADP) of 1SMT protein.

**Table 5.** H-Bond interaction of the designed ligand as obtained by LVGALD in InsightII with HIV-Nef protein.

| H-bond donor | H-bond acceptor | H-bond distance |
|---|---|---|
| Protein: 216:N | Ligand: 1X:OH | 3.46 |
| Protein: 162:NH1 | Ligand: 1Y:OH | 2.76 |
| Protein: 219:NE | Ligand: 1T:OH | 2.24 |
| Protein: 162:NH2 | Ligand: 1B:OH | 2.42 |

**Table 6.** H-Bond interaction of the designed ligand as obtained by VGA in InsightII with HIV-nef protein.

| H-bond donor | H-bond acceptor | H-bond distance |
|---|---|---|
| Protein: 183:HH2 | Ligand: 1M:OH | 2.65 |
| Ligand: 1B:OH | Protein: 136:O | 2.88 |
| Ligand: 1D:HH | Protein: 202:OH | 2.92 |

design a ligand that is associated with lower energy values (and hence is expected to be more stable) for the protein.

For the purpose of illustration, the hydrogen bond interaction of the ligand designed using LVGALD and the target protein HIV-Nef are shown in Table 5. As seen from Tables 5 and 6, more number of hydrogen bonds is formed after docking the LVGALD designed ligand than the VGA designed ligand. Again the hydrogen bond distances are also small for the LVGALD designed ligand which ensure better stability in docking.

## 6. Discussion

A new evolutionary approach for rational design of a drug like molecule has been presented in the present article. This molecule can bind to the active site of a protein target receptor, thereby preventing the proliferation of the microorganism for which the protein is a vital factor. Hence, one of the possible applications of this methodology is the design of inhibitors for a given target protein. The proposed methodology in this article is based on a variable length genetic algorithm. Unlike a previous GA based approach[10] and a more recent one,[16] the proposed method makes no assumption regarding the size of the tree formed on both sides of the pharmacophore. Moreover the present method considers both the internal as well as the interaction energies for optimization in different stages. A local search is also incorporated to speed up the process and improve the performance.

It is found that the ligand molecules designed using the proposed approach are, in general, associated with lower van der Waals energy values as compared to the previous method described in Ref. 16. This improvement is noted in both the internal and the interaction energies, thereby indicating that the designed ligand could be more stable. Moreover, it is found that the structure of the evolved molecule is, in general, such that it is amenable to more stable configurations because of the presence of more hydrogen bonds at a closer distance. A comparison with a ligand that actually binds to a protein 1SMT, the coordinates of which are available in a docked state in the PDB, also indicates the effectiveness of the proposed approach.

The actual conformation of a molecule depends not only on the bond lengths and functional groups, but also on non-covalent intermolecular forces, such as electrostatic interaction, hydrogen bond formation, between the drug and the receptor. These factors have yet to be considered in this model. Moreover, here only seven groups are considered. As a scope for future study more groups as well as the energy contributions due to the other interactions can be taken into account. Another area of further research in this regard is to extend the model to three dimensions and use some other representation to encode the ligand. It is to be noted that in this study we assume that the active site geometries of the receptors are known. Another attempt may be made in the lines of the work done in Ref. 25 such that a representation of the active site may be produced using 3DQSAR[13,14] studies as a first step to the ligand design problem.

## References

1. A. Bagchi, S. Bandyopadhyay and U. Maulik, Determination of molecular structure for drug design using variable string length genetic algorithm, *Workshop on Soft Computing, High Performance Computing (HiPC) Workshops 2003: New Frontiers in High-Performance Computing*, Hyderabad, 145–154 (2003).
2. A. Globus, J. Lawton and T. Wipke, Automatic molecular design using evolutionary techniques, *Sixth Foresight Conference on Molecular Nanotechnology, 1998 and Nanotechnology* **10**(3), 290–299 (1999).
3. A. R. Leach, *Molecular Modeling Principles and Applications*, Pearson, Prentice Hall (2001).
4. A. R. Leach and V. J. Gillet, *An Introduction to Chemoinformatic*, Kluwer Academic Publishers (2003).
5. C. Branden and J. Tooze, *Introduction to Protein Structure*, Garland, New York (1991).
6. D. E. Clark (ed), *Evolutionary Algorithms in Molecular Design*, John Wiley (2000).
7. D. Freifelder, *Molecular Biology*, Narosa Publishing House (1998).
8. D. K. Worthilake, H. Wang, Y. Soo, W. I. Sundquist and C. P. Hill, Structures of the HIV-1 capsid protein dimerization domain at 2.6 A resolution, *Acta Crystallogr. D. Biol. Crystallogr.* **55**, 85–95 (1999).
9. G. B. Fogel and D. W. Corne (eds), *Evolutionary Computation in Bioinformatics*, Morgan Kaufmann (2002).
10. G. Goh and J. A. Foster, Evolving molecules for drug design using genetic algorithm, *Proc. Int. Conf. on Genetic and Evol. Computing*, Morgan Kaufmann, 27–33 (2000).
11. G. Jones, P. Willett, R. C. Glen, A. R. Leach and R. J. Taylor, Development and validation of a genetic algorithm for flexible docking, *Journal of Molecular Biology* **267**(3), 727–748 (1997).
12. J. L. R. Filho and P. C. Treleavan, Genetic algorithm programming environments, *IEEE Comput.*, 28–43 (1994).
13. N. Ghosal and P. K. Mukherjee, 3-D QSAR of N-substituted 4-amino-3, 3-dialkyl-2(3H)-furanone GABA receptor modulators using molecular field analysis and receptor surface modeling study, *Bioinorganic and Medical Chemistry Letters* **14**, 103–109 (2004).
14. O. Nicolotti, V. J. Gillet, P. J. Fleming and D. V. S. Green, Multiobjective optimisation in quantitative structure-activity relationships: deriving accurate and interpretable QSARs, *Journal of Medicinal Chemistry* **45**, 5069–5080 (2002).
15. S. Arold, P. Franken, M. P. Strub, F. Hoh, S. Benichou and R. Dumas, The crystal structure of HIV-1 Nef protein bound to the Fyn Kinase SH3 domain suggests a role for this complex in altered T cell receptor signaling, *Structure* **5**, 1361–1372 (1997).
16. S. Bandyopadhyay, A. Bagchi and U. Maulik, Active site driven ligand design: an evolutionary approach, *Journal of Bioinformatics and Computational Biology* **3**(5), 1053–1070 (2005).

17. S. Bandyopadhyay and U. Maulik, Genetic clustering for automatic evolution of clusters and application to image classification, *Pattern Recognition* **35**(2), 1197–1208 (2002).
18. S. C. Pegg, J. J. Haresco and I. D. Kuntz, A genetic algorithm for structure-based de novo design, *Journal Comput. Aided Molecule Design* **15**(10), 911–933 (2001).
19. S. Kamphausen, N. Holtge, F. Wirsching, C. Morys-Wortmann, D. Riester, R. Goetz, M. Thurk and A. Schwienhorst, Genetic algorithm for the design of molecules with desired properties, *Journal Comput. Aided Molecule Design* **16**(8–9), 551–567 (2002).
20. T. Hinck, A. P. Wang, Y. X. Nicholson, L. K. Torchia, D. A. Wingfield, P. Stahl, S. J. Chang, C. H. Domaillel and P. J. Lam, Three-dimensional solution structure of the HIV-1 protease complexed with DMP323, a novel cyclic urea-type inhibitor, determined by nuclear magnetic resonance spectroscopy, *Protein Science* **5**, 495–510 (1996).
21. R. T. Morrison, R. N. Boyd and R. K. Boyd, Organic Chemistry, Benjamin Cummings, **6** (1992).
22. U. Maulik and S. Bandyopadhyay, Genetic algorithm-based clustering technique, *Pattern Recognition* **33**, 1455–1465 (2000).
23. U. Maulik and S. Bandyopadhyay, Fuzzy partitioning using real coded variable length genetic algorithm for pixel cassification, *IEEE Transactions on Geosciences and Remote Sensing* **41**(5), 1075–1081 (2003).
24. W. J. Cook, S. R. Kar, K. B. Taylor and L. M. Hall, Crystal structure of the cyanobacterial metallothionein repressor SmtB: a model for metalloregulatory proteins, *Journal of Molecular Biology* **275**, 337–346 (1998).
25. P. A. Greenidge, S. A. Merette, R. Beck, G. Dodson, C. A. Goodwin, M. F. Scully, J. Spencer, J. Weiser and J. J. Deadman, Generation of ligand conformations in continuum solvent consistent with protein active site topology: application to thrombin, *Journal of Medicinal Chemistry* **46**, 1293–1305 (2003).

# III.
# GENE EXPRESSION AND MICROARRAY DATA ANALYSIS

# CHAPTER 9

# INFERRING REGULATIONS IN A GENOMIC NETWORK FROM GENE EXPRESSION PROFILES

Nasimul Noman* and Hitoshi Iba[†]

*Department of Frontier Informatics*
*Graduate School of Frontier Sciences*
*The University of Tokyo, Kashiwanoha 5-1-5*
*Kashiwa-shi, Chiba 277-8561, Japan*
*\*noman@iba.k.u-tokyo.ac.jp*
*†iba@iba.k.u-tokyo.ac.jp*

With many revolutionary advancements in technology, biological researchers are now attempting to accumulate the fragments of knowledge to make up the whole living system, or at least to a larger fragment that gives a better understanding of the system implicit in a living organism. Reconstructing genetic networks from expression profiles is an important challenge in systems biology. This chapter presents an improved evolutionary algorithm for inferring the underlying gene network from gene expression data. We used the decoupled S-system formalism for representing genetic interactions and proposed a more effective fitness function for attaining the sparse network architecture in biological systems. The performance of the method was evaluated using artificial genetic networks by varying network size and expression profile characteristics (noise-free and noisy). The results showed the superiority of the method in constructing the network structure and predicting the regulatory parameters. The method was also used to analyze real gene expression data for predicting the interactions among the genes in SOS DNA repair network in *Escherichia coli*.

## 1. Introduction

In last decade, many radical progresses in the field of molecular biology have opened the door of extensive research in many new domains. One such emerging field is systems biology that attempts systemic approach to explain every aspect of biological organisms at system level.[1] In order to design new biological systems or to control the state of existing biological systems, which is the ultimate goal of this field, first we need to understand the organisms at structural and behavioral level. The genome, containing all the necessary information in the form of thousands of genes, and the interacting complex regulatory circuits among these molecular components define the features and behavior of an organism.[2] Therefore, an important challenge in systems biology is to unravel the skeletons- and the mechanisms of these networks which are the regulators of different cellular processes.

Gene regulation, the core of many biological processes, may be thought of as a combination of *cis*-acting regulation by the extended promoter of a gene, and of *trans*-acting regulation by the transcription factor products of other genes. If we simplify the *cis*-action by using a phenomenological model that can be tuned to data, then the full *trans*-acting interaction between multiple genes can be modeled as a network which is commonly known as *gene circuit, gene regulatory network* or *genetic network*.[3] However, the mechanism behind such biological networks, which are dynamic and highly nonlinear, is excessively complicated. Because of poor understanding of the biological components, their dependencies, interaction and nature of regulation grounded on molecular level, the study of such systems had been impeded until recently.

Several cutting-edge technologies such as DNA microarrays, oligonucleotide chips permit rapid and parallel measurement of gene expression as either time series or steady-state data. Monitoring the transcriptomes on a genome-wide scale, scientists are forming global views of the structural and dynamic changes in genome activity during different phases in a cells development and following exposure to external agents. Interpretation of this vast amount of experimental data not only is capable of providing comprehensive understanding of the activity of a particular gene in a specific biochemical process, but can facilitate greater understanding of the regulatory architecture also. Therefore, with the availability of these massive amounts of biological data, the researchers are trying to unravel the underlying transcriptional regulations in gene circuits using model-based identification methods.

Cluster analysis[4–6] of gene expression data can help elucidate the regulation (or co-regulation) of individual genes, which provides valuable information and insights, but often fails to identify system-wide functional properties of a given network.[7] Computational modeling and simulation can provide a much more detailed and better understanding of the regulatory architecture such as network connections, rate constants and biochemical concentrations etc.

Various types of gene regulation models have been proposed, which integrate biochemical pathway information and expression data to trace genetic regulatory interactions.[8-12] The modeling spectrum ranges from abstract Boolean descriptions to detailed Differential Equation based models, where every representation has its advantages and limitations.

At one end of the modeling spectrum are Boolean network models, in which each gene is either fully expressed or not expressed at all.[13] Though it is evident from the real gene expression data that the gene expression levels tend to be continuous rather than binary, such coarse representation of the gene state has certain advantages in terms of complexity reduction and computation. These models are typically used to obtain a first representation of the network organization and dynamics. On the other end of the spectrum lie the continuous models with stochastic kinetics and detailed biochemical interactions. In-depth biochemical models are very useful in representing the precise interactions in the gene circuits, but their complexity and the currently available gene expression data restrict their application to very small systems.

Given a dynamic model of gene interaction, the problem of gene network inference is equivalent to learning the structural and functional parameters from the time series that represents the gene expression kinetics, i.e. the network architecture is reverse engineered from its activity profiles. It is often wondered whether it is at all possible to reverse engineer a genetic network from gene expression data. Though reverse engineering is possible in principle, the success depends on the characteristics of the model involved, the availability of the gene expression data and the level of noise present in the data.

In this chapter, we review the S-system model based inference of gene network from gene expression data and present an improved evolutionary algorithm for estimating the network architecture with greater efficiency. Our proposed algorithm and the modified fitness function are intended to estimate the correct network topology and parameter values. Numerical experiments show that the proposed enhancements attain the network structure and parameter values very accurately and effectively. We also evaluated our algorithm against the experimental data relative to the SOS DNA repair network of the *Escherichia coli* bacterium.

## 2. Modeling Gene Regulatory Networks by S-system

Biochemical System Theory (BST)[14,15] is the mathematical basis of well-established methodological framework for analyzing networks of biochemical reactions and provides a general framework for modeling and analyzing nonlinear systems of genetic networks. It is based on the generic approximation of kinetic rate laws with multivariate power-law functions. This representation results from Taylors theorem in logarithmic coordinates. This type of representation is compatible with the observations of biological systems and has been proven to be capable of describing biological systems adequately.[16] System models in BST consist of sets of ordinary differential equations in which the change in each variable is always represented as sums and differences of multivariate products of power-law functions.[17] Two most important variants within BST are the Generalized Mass Action (GMA) and the S-system form. The latter is actually a more computationally and analytically tractable specialization of GMA.

### 2.1. Canonical model description

The S-system model[18] is organizationally rich enough to reasonably capture various dynamics and mechanisms that could be present in a complex system of genetic regulation. The canonical representation of the model is a set of non-linear differential equations and it has the following systematic structure

$$\frac{dX_i}{dt} = \alpha_i \prod_{j=1}^{N} X_j^{g_{ij}} - \beta_i \prod_{j=1}^{N} X_j^{h_{ij}}, \qquad (1)$$

where $N$ is the number of reactants or network components. The terms $g_{ij}$ and $h_{ij}$ represent interactive affectivity of $X_j$ to $X_i$. For each dependent concentration $X_i$ in this biochemical model there exist an aggregate production function and an aggregate consumption function. The first term in right-hand side of (1) represents all influences that increase $X_i$, whereas the second term represents all influences that decrease $X_i$. An exponent of zero for any $X_j$ means that variable has no direct influence on the rate of the corresponding aggregate process, a positive exponent means that they are positively correlated, and a negative exponent means that they are negatively correlated. The parameters that define the S-system are: $\{\alpha, \beta, g, h\}$. In a biochemical engineering context, the non-negative parameters $\alpha_i$, $\beta_i$ are called *rate constants*, and the real-valued exponents $g_{ij}$ and $h_{ij}$ are referred to as *kinetic orders*. It is known that biological networks are sparse, which means that the number of regulators that have effect on a single gene is relatively small; so many of the *kinetic orders* are zero in real condition.

Since the details of the molecular mechanisms that govern interactions among system components are not substantially known or well understood, the description of these processes requires a representation that is general enough to capture the essence of the experimentally observed response. The strength of the S-system model is its structure which is rich enough to satisfy these requirements and to capture all relevant dynamics; an observed response (dynamic response) may be monotone or oscillatory, it may contain limit cycles or exhibit deterministic chaos.[19] Furthermore, the simple homogeneous structure of S-system has a great advantage in terms of system analysis and control design, because the structure allows analytical and computational methods to be customized specifically for this structure. [20]

## 2.2. Genetic network inference problem by S-system

Reconstruction of gene regulatory networks from gene expression profile using S-system is formalized as an optimization problem in which appropriate system parameters of S-system must be found so that the difference between the time course data calculated by the S-system model and the time course data observed in experiments becomes minimum. Tominaga *et al.*[19] used Genetic Algorithm (GA) for searching the set of parameters for S-system that produces the dynamics closest to experimental results. Since methods for finding analytic solution for this problem is almost impracticable, use of Evolutionary Computation (EC) has become more feasible and popular method among researchers.[21-25] While searching for the appropriate parameters, goodness of each set of estimated parameters is evaluated by taking the sum of squared relative error between the calculated gene expression for the estimated parameter set and measured gene expression. In other words, the fitness of each set of estimated parameters for the target system is evaluated using the following function:[19]

$$f = \sum_{i=1}^{N} \sum_{t=1}^{T} \left\{ \left( \frac{X_i^{cal}(t) - X_i^{exp}(t)}{X_i^{exp}(t)} \right)^2 \right\}, \quad (2)$$

where $X_i^{cal}(t)$ is gene expression level of gene $i$ at time $t$ calculated numerically by solving the system of differential equation of (1) for the estimated parameter set, and $X_i^{exp}(t)$ represents the experimentally observed gene expression level of gene $i$ at time $t$, $T$ is the number of sampling points of the experimental data. In this form of optimization problem the search algorithm tries to find a set of parameters that minimizes $f$.

The problem has the difficulty of high dimensionality, since $2N(N+1)$ S-system parameters must be determined in order to solve the set of differential equations (1). Estimation of parameters for a $2N(N+1)$-dimensional optimization problem often causes bottlenecks and fitting the model to experimentally observed responses (time course of relative state variables or reactants) is never straightforward and is almost always difficult. Therefore, the application of the S-system model has been limited to inference of small-scale gene networks only.

### 2.3. Decoupled S-system model

In order to deal with problem of high dimensionality, decoupling of the original model has been performed.[17,24,26,27] This decoupled S-system model allows its application to larger gene network inference problem. Using the suggested decomposition strategy the original optimization problem is divided into $N$ sub-problems.[24,26] In each of these sub-problems the parameter values of a gene are estimated for realizing the temporal profile of gene expression. In other words, this disassociation technique divides a $2N(N+1)$-dimensional optimization problem into $N$ sub-problems of $2(N+1)$ dimension. In the $i$th sub-problem for gene $i$, $X_i^{cal}(t)$ is calculated by solving the following differential equation instead:

$$\frac{dX_i}{dt} = \alpha_i \prod_{j=1}^{N} Y_j^{g_{ij}} - \beta_i \prod_{j=1}^{N} Y_j^{h_{ij}}. \quad (3)$$

For solving the differential equation (3), we need the concentration levels $Y_j$ ($j = 1, \ldots, N$). In the $i$th sub-problem corresponding to gene $i$ the concentration level $Y_{j=i}$ is obtained by solving the differential equation whereas the other expression levels $Y_{j\neq i}$ to be estimated directly from the observed time-series data. The optimization task for the tightly coupled S-system model is not trivial because Eq. (1) is nonlinear in all relevant cases, requiring iterative optimization in a larger parameter space, where 95% of the total optimization time is expended in numerical integration of the differential equations.[28] Therefore, such disassociation could be very useful in reducing the computational burden which will be focused later. Moreover, the experimental results showed its usefulness in estimating the network parameters.[26,27] In this work, we have applied linear spline interpolation[29] for direct estimation of expression levels $Y_{j\neq i}$.

The sum of squared relative error between the experimental and calculated gene expression levels of gene $i$ is used as a means of evaluating the fitness in sub-problem $i$. So the objective function of the sub-problem corresponding to the $i$th gene becomes

$$f_i = \sum_{t=1}^{T} \left\{ \left( \frac{X_i^{cal}(t) - X_i^{exp}(t)}{X_i^{exp}(t)} \right)^2 \right\}, \qquad (4)$$

and in sub-problem $i$ the parameters $\{\alpha_i, \beta_i, g_{ij}, h_{ij} \ (j = 1, \ldots, N)\}$ for gene $i$ are estimated that minimizes $f_i$.

## 2.4. Fitness function for skeletal network structure

In the study of biological networks, it has been found that each gene or protein interacts with few other genes.[30] Different studies in the field have shown that one major pitfall for S-system based gene regulatory network estimation is identifying the sparse network structure which is more usual for real biological systems. Because of the high degree-of-freedom of the model and the deceptive nature of the problem, solutions often converge to different local minima each of which reproduces almost the same time-course. So any method attempting to reproduce the time dynamics only, fails to obtain the skeletal structure.[22] Use of an additional term called *pruning term* or *penalty term* for augmentation of the fitness equation was very successful to deal with this difficulty.[22,24,31] In the canonical optimization problem, the basic fitness function of (2) was extended by adding a penalty, based on Laplacian regularization term, using all *kinetic orders* in Refs. 22

and 31. Because of high dimensionality these fitness functions have been applied to small scale networks only. Based on the same notion Kimura et al. added another more effective *penalty term* to the objective function of (4) for obtaining sparse network structure in the decoupled form of the problem[24,27]

$$f_i = \sum_{t=1}^{T} \left( \frac{X_i^{cal}(t) - X_i^{exp}(t)}{X_i^{exp}(t)} \right)^2 + c \sum_{j=1}^{N-I} \left( |G_{ij}| + |H_{ij}| \right), \quad (5)$$

where $G_{ij}$ and $H_{ij}$ are given by rearranging $g_{ij}$ and $h_{ij}$, respectively, in ascending order of their absolute values (i.e. $|G_{i1}| \leq |G_{i2}| \leq \cdots \leq |G_{iN}|$ and $|H_{i1}| \leq |H_{i2}| \leq \cdots \leq |H_{iN}|$). $I$ is the maximum cardinality of the network and $c$ is the penalty constant. The superiority of this *penalty* lies in including the maximum cardinality of the network. Therefore, this *pruning term* will penalize only when the number of genes that directly affect the $i$th gene is higher than the maximum in-degree $I$, thereby will cause most of the genes to disconnect when this penalty term is applied. However, very few genes affect both the synthesis process and the degradation process of a specific gene. Therefore, designing the penalty term considering both synthetic and degradative regulations together rather than separately will be more effective. Because, such penalty will penalize whenever total number of regulators (whether synthetic or degradative) is greater than maximum cardinality. So we modified the fitness of (5) as follows:

$$f_i = \sum_{t=1}^{T} \left( \frac{X_i^{cal}(t) - X_i^{exp}(t)}{X_i^{exp}(t)} \right)^2 + c \sum_{j=1}^{2N-I} \left( |K_{ij}| \right), \quad (6)$$

where $K_{ij}$ are the *kinetic orders* of gene $i$ sorted in ascending order of their absolute values. The use of (6) instead of (5) as fitness function helps reducing the number of false predicted regulations (as will be shown later). In other words, we can say that the use of (6) as fitness function can identify the zero valued parameters increasingly and thus obtain the skeletal network structure more precisely.

## 3. Inference Method

In this section, we explain our algorithm for reconstructing the network from the time dynamics. As mentioned earlier, EC has become a very popular approach for the gene network inference problem. Before presenting the methodology, we give a brief review of Trigonometric Differential Evolution (TDE) algorithm that we applied as the optimization engine in our algorithm.

### 3.1. *Trigonometric Differential Evolution (TDE)*

Differential Evolution (DE)[32] is an effective, efficient and robust optimization method capable of handling non-differentiable, nonlinear and multimodal objective functions.[32,33] Recently, Fan and Lampinen have proposed a Trigonometric Mutation Operation (TMO) for DE to accelerate its convergence rate and robustness[34] and this modified DE algorithm is called Trigonometric mutation DE (TDE). In the core of our evolutionary algorithm, we have used TDE to search the set of network parameters that gives the smallest fitness score.

Like other Evolutionary Algorithms (EAs) TDE is a population based search algorithm. Each population consists of a certain number of individuals where each individual represents a candidate solution for the problem. A new generation (an instance of population) is created from the current generation and the new one replaces the current one. Thus producing and replacing new generations in an iterative manner TDE searches for the optimal solution of the problem. We explain the search procedure for subproblem $i$.

For estimating the solution for sub-problem $i$, the initial population of random individuals is created where each individual consists of parameters $\{\alpha_i, \beta_i, g_{ij}, h_{ij} \ (j = 1, \ldots, N)\}$ for gene $i$. Then the fitness of each individual is evaluated using Eq. (6). Subsequently, new individuals are generated by the combination of randomly chosen individuals from the current population. Specifically, for each individual $x_G^i$, $i = 1, \ldots, P$, three other random individuals $x_G^j$, $x_G^k$ and $x_G^l$ (such that $j, k$ and $l \in \{1, \ldots, P\}$ and $i \neq j \neq k \neq l$) are selected from generation $G$; $P$ is the number of individuals in $G$. Then a new trial individual $y_G^i$ (i.e. a new candidate solution) is generated using probabilistic mutation operation according to the following equations

$$y_G^i = x_G^j + F(x_G^k - x_G^l), \tag{7}$$

$$y_{G+1}^i = (x_G^j + x_G^k + x_G^l)/3 + (p_k - p_j)(x_G^j - x_G^k) \\ + (p_l - p_k)(x_G^k - x_G^l) + (p_j - p_l)(x_G^l - x_G^j), \tag{8}$$

where

$$p_j = |f(x_G^j)|/p' \quad p_k = |f(x_G^k)|/p' \quad p_l = |f(x_G^l)|/p'$$
$$\text{and} \quad p' = |f(x_G^j)| + |f(x_G^k)| + |f(x_G^l)|$$

where $F$ is called the *scaling factor* or *amplification factor*. Equation (7) represents the regular mutation operation in DE and Eq. (8) represents TMO proposed by Fan and Lampinen.[34] This TMO is applied with probability $M_t$ and the regular one is applied with probability $(1 - M_t)$. In order to achieve higher diversity, the mutated individual $y_G^i$ is mated with the current population member $x_G^i$ using a *crossover* operation to generate the *offspring* $y_{G+1}^i$. The variables of solution $y_{G+1}^i$ are randomly inherited from $x_G^i$ or $y_G^i$ determined by a parameter called *crossover factor* $CF$, i.e. if $r \leq CF$ (where $r$ is a uniform random number in [0, 1]) then it is inherited from $x_G^i$ otherwise from $y_G^i$. Finally, the offspring is evaluated and replaces its parent $x_G^i$ in the next generation if and only if its fitness is better than that of its parent. This is the *selection* process for producing new generation. And this process is repeated until a solution satisfying our criteria is found or a maximum number of generations have elapsed.

In TDE, the trigonometric mutation operation, a rather greedy search operator, makes it possible to straightforwardly adjust the balance between the convergence rate and the robustness through the newly introduced parameter, $M_t$. The greediness of the algorithm can be tuned conveniently by increasing or decreasing $M_t$. Experimental results have shown that TDE has good convergence properties, outperforms other well known EAs[34] and effective in genetic network inference.[31] Because of these admirable properties we have chosen TDE as optimization tool in our algorithm (explained in next section) for gene network reconstruction problem.

### 3.2. *Proposed algorithm*

In our algorithm for estimating the S-system parameters, we optimized the objective function (6) for each sub-problem $i$ $(i = 1, \ldots, N)$ using a modified TDE algorithm with a local search procedure (described later). For explaining the inference algorithm we consider the case of gene $i$. The same is applied for other sub-problems to obtain a complete set of parameters for the full network.

The genetic network reconstruction problem is strongly nonlinear and highly multimodal that causes the search procedure to stuck in a local optimum. Therefore, in order to avoid premature convergence to some local minima we performed optimization using a two step method. In each phase, the parameter values of the gene are represented as an individual of TDE (as mentioned earlier). In the *first stage*, we perform $\Gamma$ trials of optimization starting with different random initial populations. In each of these trials, at the end of the optimization using TDE augmented with local search procedure we obtain a solution for the subproblem i.e. a set of parameters for the target gene. However, the optimization process may converge to a local optimum and may fail to attain actual parameter set. Also because of local convergence it may lose some essential regulatory interaction among the genes. In other words we can say, due to convergence to local minima some parameter value could go down to zero, which is not actually zero in the target parameter set. Since trials are repeated with different initial values, candidate solutions are obtained as possible different local minima. To obtain a more robust and accurate solution, we perform double optimization using elite individuals from different trials in the *second stage* of our algorithm. Double optimization can automatically detect the essential parameters by optimizing multiple local minima once again and has been found to be useful for genetic network inference problem.[22,35] The solutions obtained from the multiple local minima of the first stage retained some essential parameters. So applying another optimization on these solutions we can identify the correct regulations, accurate strength of the regulation and avoid deletion of any necessary parameter. The complete algorithm is illustrated in Fig. 1.

### 3.2.1. *Mutation phase*

In EAs the mutation operator serves to create random diversity in the population.[36] Traditionally, mutation has always been used as much a secondary mechanism in comparison to crossover. Since DE/TDE does not have a direct mutation operation, we periodically apply a mutation operation to introduce new traits in the population and escape the local minima thereby. Our mutation process works as follows: if the fitness of the elite individual does not improve for $G_m$ generations then we pass all the other individuals in current generation through the mutation phase. We applied the

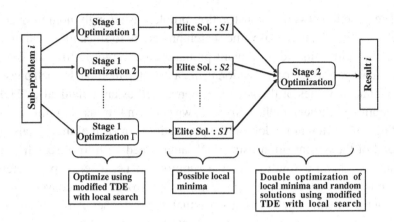

Fig. 1. Optimization procedure for subproblem $i$.

Gaussian mutation with mutation probability $p_m$. Gaussian mutation realizes the mutation operation by adding a random value from the Gaussian distribution. For mutating the *rate constants* of an individual the random numbers are drawn from a Gaussian distribution with mean $\mu_r = 0$ and standard deviation $\sigma_r$ and for mutating the *kinetic orders* the random numbers are drawn from a distribution with mean $\mu_k = 0$ and standard deviation $\sigma_k$.

In each stage of the algorithm, the overall optimization procedure for estimating the model parameters for each subproblem can be summarized as follows:

(1) Prepare initial population $P_G$ with candidate solutions,
(2) Create the new generation $P_{G+1}$ of candidate solutions by applying the recombination/selection operation of TDE,
(3) Apply local search to the best individual and a randomly selected individual of the new generation $P_{G+1}$,
(4) If fitness of the elite individual does not improve for $G_m$ generations then apply mutation to non-elite individuals,
(5) Stop if the termination criteria are satisfied. Otherwise, Set $G = G + 1$ and go to Step 2.

In phase 1, the initial population $P_G$ is created randomly and in Phase 2, the elite individuals from different trials of Phase 1 together with some random individuals are used for initialization.

## 3.3. Local search procedure

In order to provide an effective global optimization method, some meta-heuristics or local searches are often embedded inside the evolutionary algorithm. Genetic Algorithms (GAs) hybridized with local refinement procedures are known as Memetic Algorithm (MAs) which are motivated to take advantage of both the exploration abilities of GA and the exploitation abilities of local search. Therefore, we introduce a local search method inside the TDE algorithm. In our local refinement procedure, we perform a greedy search operation around the best individual and a random individual of each generation. The local search around the best individual and a random individual will accelerate the optimization process as well as maintain the diversity of the population. The local searching is performed as follows: all the *kinetic orders* are sorted in ascending order of their absolute values. Then the *kinetic order* of the lowest absolute value is set to zero, and the fitness of this new individual is evaluated. If this modification improves the fitness of the individual then the new solution is accepted, otherwise its parent solution is restored. This process is repeated for all kinetic orders in the increasing order of their absolute values. This local search process allows us to identify the zero valued parameters by mutating them in the increasing order of their strength and thus helps us to identify the skeletal network structure. The restore capability of the greedy search also allows to recover from wrong elimination of any essential regulation. Hybridizing this greedy local search procedure with the TDE algorithm we can identify the sparse network structure efficiently and estimate the strength of regulations more accurately.

## 4. Simulated Experiment

The performance of our approach was evaluated by simulation. We tested the algorithm using a well-studied small scale network, and a medium scale network under both ideal and noisy conditions. The detailed experiments and results are described in the following sub-sections.

### 4.1. Experiment 1: inferring small scale network in noise free environment

In our first experiment, we reverse engineered an artificial genetic network from noise free gene expression profile. The S-system model of the target

5-gene network is shown in Eq. (9). This small model represents a typical gene interaction system which adequately demonstrates different types of regulations among the genes. Many other researchers have experimented with this network[19,22,24,37,38] hence our results can be easily compared.

$$\left.\begin{array}{l} dx_1/dt = 5.0x_3^{1.0}x_5^{-1.0} - 10.0x_1^{2.0} \\ dx_2/dt = 10.0x_1^{2.0} - 10.0x_2^{2.0} \\ dx_3/dt = 10.0x_2^{-1.0} - 10.0x_2^{-1.0}x_3^{2.0} \\ dx_4/dt = 8.0x_3^{2.0}x_5^{-1.0} - 10.0x_4^{2.0} \\ dx_5/dt = 10.0x_4^{2.0} - 10.0x_5^{2.0} \end{array}\right\}. \qquad (9)$$

If an insufficient amount of time series data is used for estimating the parameters for S-system model, many candidate solutions will evolve due to the high-degree of freedom of the model.[23] This is because it is only one path in a phase diagram and from such a single path no general conclusions about the overall behavior of the dynamic system can be drawn.[37] Therefore, Kikuchi *et al.* and Kimura *et al.* have used 50 and 75 time series data, respectively, for solving this 5-gene network problem.[22,24] We have also used 50 time series data for ensuring sufficient amount of observed gene expression data. The sets of time-series were obtained by solving the set of differential equations of (9). Initial concentration level for each time series was generated randomly in [0.0, 1.0]. Sampling 11 points from each time-course we used $10 \times 11 = 110$ gene expression levels for each gene.

### 4.1.1. *Experimental setup*

The conditions of our experiments were as follows. The search regions of the parameters were [0.0, 20.0] for $\alpha_i$ and $\beta_i$, and [−3.0, 3.0] for $g_{ij}$ and $h_{ij}$. The maximum cardinality $I$ was chosen to be 5, and the penalty coefficient $c$ was 1.0. The parameter values for TDE algorithm were $F = 0.5$, $CF = 0.8$ and $M_t = 0.05$, population size was 60 and the maximum number of generations in each trial of Phase 1 and in Phase 2 was 850. In Phase 1, we evolved 5 ($\Gamma = 1, \ldots, 5$) independent trial solutions from which we selected elite individuals for optimization in Phase 2. The parameter values for the mutation phase were $m_p = 0.01$, $\sigma_r = 3.0$ and $\sigma_k = 1.2$. In Stage 1 of the optimization $G_m = 100$ and in Stage 2,

$G_m = 200$, were used. Our algorithm was implemented in Java language and the time required for solving each subproblem was approximately 10 minutes using a PC with 1700 MHz Intel Pentium processor and 512 MB of RAM.

In order to reduce the computational burden a structure skeletalizing was applied in a similar fashion as used by Tominaga.[19] If the absolute value of a parameter is less than a threshold value $\delta$ then structure skeletalizing resets it to zero. This process reduces the computational cost as well as helps to identify the zero valued parameters. In our experiment $\delta = 0.001$ was used. We performed 5 repetitions of each experiment to assure the soundness of our stochastic search algorithm.

$$\left.\begin{aligned}
dx_1/dt &= 4.741x_1^{-0.059}x_2^{-0.006}x_3^{1.008}x_5^{-1.026} - 9.659x_1^{1.975} \\
dx_2/dt &= 10.206x_1^{1.989}x_2^{0.012} - 10.154x_1^{0.030}x_2^{1.963}x_4^{0.007} \\
dx_3/dt &= 10.156x_2^{-1.002}x_3^{-0.012} - 10.153x_2^{-1.002}x_3^{1.977} \\
dx_4/dt &= 7.905x_1^{-0.074}x_3^{1.969}x_5^{-1.006} - 9.786x_1^{-0.012}x_4^{1.980} \\
dx_5/dt &= 9.763x_3^{0.018}x_4^{2.035}x_5^{-0.028} - 9.732x_4^{0.018}x_5^{2.016}
\end{aligned}\right\}. \quad (10)$$

### 4.1.2. Result

Equation (10) shows the system estimated by our algorithm in a typical run. As shown in Eq. (10), our model was able to attain the overall network structure and the parameter values were also very close to target values. Many of the zero valued parameters were identified correctly and the others are close enough to zero to indicate (possible) false positive interactions.

### 4.2. Experiment 2: inferring small scale network in noisy environment

Noise is inevitable in microarray data and the real challenge lies in designing an inference algorithm capable of constructing the network from noisy data. In this section, we test the performance of our proposed algorithm simulating a noisy real world environment i.e. conducting the experiments with noisy time series data.

### 4.2.1. Experimental setup

As the target model we have selected the same network used in Experiment 1 with the same target parameter set. Data points were generated using the same sets of initial expression levels used in Experiment 1. We added 5% Gaussian noise to the time-series data in order to simulate the measurement error between true expression and observed expression. Eleven sample points from each time course were used for optimization. The rest of the settings were the same as in the previous experiments.

$$\left. \begin{aligned} dx_1/dt &= 4.832 x_2^{0.127} x_3^{1.473} x_5^{-1.507} - 9.533 x_1^{2.617} x_5^{-0.621} \\ dx_2/dt &= 11.925 x_1^{2.271} x_3^{-0.401} x_5^{0.114} - 13.733 x_2^{2.294} x_4^{0.209} \\ dx_3/dt &= 10.012 x_1^{-0.059} x_2^{-0.952} x_3^{0.031} - 10.913 x_2^{-0.873} x_3^{2.057} \\ dx_4/dt &= 6.499 x_3^{2.047} x_4^{-0.145} x_5^{-1.003} - 10.012 x_4^{2.461} x_5^{0.313} \\ dx_5/dt &= 10.653 x_4^{1.674} x_5^{0.026} - 11.172 x_1^{0.271} x_4^{-0.159} x_5^{1.790} \end{aligned} \right\}. \quad (11)$$

### 4.2.2. Results

The network model, inferred by our method in a typical run using the noisy data, is presented in Eq. (11). Even using noisy data the proposed method was successful in identifying all the correct regulations and also predicted a few false regulations. The number of false-negative interactions was zero. In some cases, the estimated parameters were pretty far away from the target values and some false predicted parameter values were too large to ignore.

### 4.3. Experiment 3: inferring medium scale network in noisy environment

In this experiment, we examined the effectiveness of our strategy for medium scale genetic networks. We randomly created a 20-gene network structure as the target model in this simulation. The structure and the S-system parameters for the model are shown in Fig. 2 and Table 1 respectively. This artificial genetic network models several types of regulation (e.g. auto-regulation, cyclic-regulation) commonly found in biological systems. We artificially created 10 sets of gene expression data for each gene of the target network. Initial gene expression level for each time series was randomly selected from [0.0, 1.0]. And using these initial concentrations

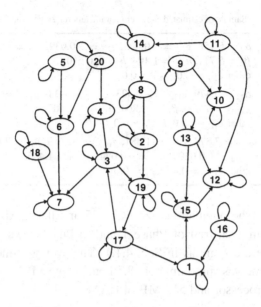

**Fig. 2.** Structure of the 20 gene network.

**Table 1.** Target S-system parameters for 20 gene network.

| | |
|---|---|
| $\alpha_i$ $\beta_i$ | 10.0 |
| $g_{i,j}$ | $g_{1,16} = -1.0$, $g_{1,17} = 1.2$, $g_{2,8} = -2.0$, $g_{3,4} = 1.8$, $g_{3,17} = 1.5$, $g_{4,20} = -0.8$, $g_{6,5} = 1.0$, $g_{6,20} = -1.3$, $g_{7,3} = -1.5$, $g_{7,6} = -0.5$, $g_{7,18} = 1.0$, $g_{8,14} = 1.5$, $g_{10,9} = -1.0$, $g_{10,11} = 1.0$, $g_{12,11} = 1.0$, $g_{12,13} = -2.0$, $g_{12,15} = 1.4$, $g_{14,11} = 2.0$, $g_{15,1} = 0.8$, $g_{15,13} = -1.5$, $g_{17,19} = 2.5$, $g_{19,2} = 1.2$, $g_{19,3} = -2.1$, other $g_{i,j} = 0.0$ |
| $h_{ij}$ | 1.0 if ($i = j$), 0.0 otherwise |

and the network model of Table 1, system dynamics were generated and 11 expression levels were sampled from each time course. Then 5% Gaussian noise was added to each point to simulate noisy real-world condition. We used these $11 \times 10 = 110$ gene expression levels for estimating the transcriptional regulators of each gene.

### 4.3.1. *Experimental setup*

We repeated the experiment five times under the following setup. The search regions of the parameters were [0.0, 15.0] for $\alpha_i$ and $\beta_i$, and [−3.0, 3.0]

Table 2. Sample of estimated S-system parameters for 20 gene network.

| | |
|---|---|
| Gene 2 | $\alpha_2 = 5.158$, $g_{2,2} = -0.403$, $g_{2,7} = -0.039$, $g_{2,8} = -2.492$, $\beta_2 = 5.711$, $h_{2,2} = 1.104$, $h_{2,8} = 0.660$ other $g_{2,j} = 0.0$, $h_{2,j} = 0.0$ |
| Gene 12 | $\alpha_{12} = 9.479$, $g_{12,5} = -0.423$, $g_{12,13} = -3.0$, $g_{12,14} = 0.410$, $g_{12,15} = 1.856$, $\beta_{12} = 8.864$, $h_{12,12} = 1.407$ other $g_{12,j} = 0.0$, $h_{12,j} = 0.0$ |
| Gene 19 | $\alpha_{19} = 7.378$, $g_{19,2} = 1.481$, $g_{19,3} = -2.689$, $\beta_{19} = 7.492$, $h_{19,1} = -0.029$, $h_{19,7} = -0.074$, $g_{19,19} = 1.495$ other $g_{19,j} = 0.0$, $h_{19,j} = 0.0$ |

for $g_{ij}$ and $h_{ij}$. The population size was 125 and the maximum number of generations in each trial of Phase 1 and in Phase 2 was 4000. Other conditions were the same as in Sec. 4.1.1. The average time for solving each sub-problem was approximately 8.5 hours using a PC with 1700 MHz Intel Pentium processor and 512 MB of RAM.

### 4.3.2. Results

Typical regulations estimated for the sub-problems corresponding to gene 2, 12 and 19 are shown in Table 2. Inspecting these results it can be found that the proposed method failed to predict some regulations in the target network and predicted many false interactions. Some of these false regulations can be ignored because of the strength of their affectivity whereas some are too strong to ignore. From 43 true regulations of the target network, the algorithm inferred $38.2 \pm 0.837$ true-positive, $61.8 \pm 0.447$ false-positive and $4.8 \pm 0.837$ false-negative interactions, on an average. The possible reason for such failure is the noise in the experimental data and providing insufficient number of time-series data.

## 5. Analysis of Real Gene Expression Data

After evaluating the proposed method using synthetic data, we tested it using the gene expression data collected from the well-known SOS DNA repair network in *Escherichia coli*. Exposure of *E. coli* to agents or conditions that damage DNA or interfere with DNA replication induces this gene network that produces many defense proteins for repairing the damaged DNA and reactivation of DNA synthesis.[39] The entire system involves more than 30 genes.[7] The functioning of the SOS system seems to be simple and in general operates as proposed by Little et al.[40] Figure 3 diagrams the basic regulatory mechanism of the SOS system schematically. The SOS genes are scattered at different sites and their expression is based on interplay of the two proteins *LexA* and *RecA*. *LexA* is a repressor; in an un-induced cell, it binds to the SOS box of the operator sites of both *lexA* and *recA* and all of the genes belonging to the SOS system and represses their transcription.[39] When a cells DNA is damaged, an inducing signal that leads to the expression of the SOS regulon is generated. One of the SOS proteins, *RecA*, acts as the sensor of DNA damage: by binding to single-stranded DNA, becomes activated (often referred as *RecA* *).[42] The interaction of the activated *RecA* protein with the *LexA* protein results in the proteolytic cleavage of *LexA*. As the pools of intact *LexA* begins to decrease, various SOS genes begin to express. Genes with operators that bind *LexA* weakly are the first to be expressed fully. If the inducing treatment is sufficiently strong, more molecules of *RecA* are

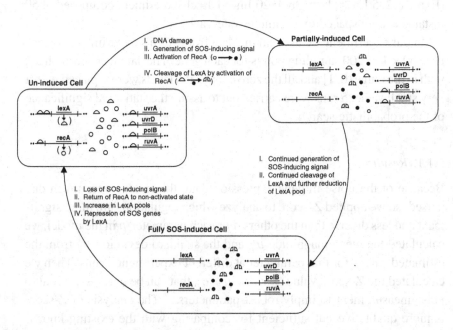

**Fig. 3.** Model of SOS regulatory system (adapted from Ref. 41).

activated, resulting in cleavage of more molecules of *LexA* and genes whose operators bind *LexA* very tightly are expressed.[41]

As the cell begins to recover from the inducing treatment, e.g. by DNA repair, the inducing signal is eliminated, and the *RecA* molecules return to their proteolytically inactive state. In the absence of the *RecA* protease, the continued synthesis of *LexA* molecules now leads to an increase in the *LexA* pools. This in turn leads to repression of the SOS genes and a return to the un-induced state.[41–43]

### 5.1. *Experimental data set*

The experimental data was downloaded from the homepage of Uri Alon Lab.[44] Data are expression kinetics of 8 genes (*uvrD, lexA, umuD, recA, uvrA, uvrY, ruvA* and *polB*) of the SOS DNA repair network. The measurement technology is based on the property of GFPs (green fluorescent proteins). Alon *et al.* have developed a system for obtaining very precise kinetics.[43] Measurements are done after irradiation of the DNA at the initial time with UV light. Four experiments are done for various light intensities (Exp. 1&2: 5 $Jm^{-2}$, Exp. 3&4: 20 $Jm^{-2}$). Each experiment composed of 50 instants evenly spaced by 6 minutes intervals.

In our experiment we used all the data from Alon's experiments, i.e. we used $50 \times 4 = 200$ sampling points for each gene. The data were normalized with in the range (0, 1] and all the zero expression levels were replaced with a very small value. 5 runs were carried out to assure the statistical significance of the probabilistic search.

#### 5.1.1. *Results*

Because of the noise in gene expression data, the results were much dispersed; so we applied Z-score to analyze which regulations are more significant and less diverse than the others. For each parameter $p$ in the model, we calculated the mean magnitude $\mu_p$ and the standard deviation $\sigma_p$ from the estimated values for that parameter at different experimental runs. Then we calculated the Z-score value as $Z_p = \mu_p/\sigma_p$ that can be used as a signal to noise measurement to imply robust parameters.[35] The analysis of Z-score is more qualitative but sufficient for comparing with the existing knowledge or suggesting new regulation.[2] We considered only those regulations

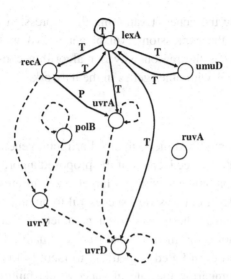

Fig. 4. Estimated SOS network structure.

which have Z-score value above the threshold $Z_{th} = 1.5$, which was set empirically.

Figure 4 shows the structure of the SOS DNA repair network reconstructed by the proposed method. The true/known regulations are drawn using solid arcs labeled with $T$, the unknown regulations or the possible false predictions are drawn using dashed arcs and the regulation drawn in solid arc with label $P$ represents unknown regulation predicted by others. Looking at the inferred structure it can be found that the *lexA* regulation of *lexA*, *umuD*, *recA*, *uvrA* and *uvrD* have been correctly identified. The activation of *lexA* by *recA* was also correctly predicted. Regulation of *lexA* by *umuDC* is also known. An unknown regulation of *uvrA* by *recA* was identified. Perrin *et al.* also identified this regulation in their experiment and hypothesized that this could correspond to the indirect regulation $recA \rightarrow RecA \dashv LexA \dashv uvrA$.[42] Some other regulations were inferred, those are either novel regulatory pathways or false-positive findings. Moreover in the gene expression collecting experiments the UV light shock was not sufficient to lead to the functioning of all SOS genes. From the expression profile it can be found that several genes were not induced sufficiently during the experiment. These genes are activated only when the damage is sufficiently high and their activation would have been useful to identify interrelation among the genes. Examining the expression profiles no significant change in the expression level of gene *ruvA* was observed. We conjecture that this is the reason why our algorithm could not find any relation of gene *ruvA* with other genes in the network.

## 6. Discussion

In our analysis of small and medium-sized artificial genetic networks, we have demonstrated the effectiveness of the proposed algorithm by comparing the estimated parameters with the target ones. In an ideal noise-free environment the algorithm was very successful to predict not only the network topology but also the quasi-correct parameters. But when the data was biased with noise the method could not estimate all the parameters very accurately and even failed sometimes to identify the correct regulations. However, comparing the rate of success and failure, the proposed method can still be considered to be useful for predicting the transcriptional regulations in a genetic circuit.

The number of false positives is directly dependent on the maximum cardinality $I$ used in the fitness function. In our Exp. 1, the number of false positives was 11 where was in Ref. 24, it was 36 using Eq. (5) and 15 sets of time series data. Therefore, modification of the fitness function was obviously useful to decrease the number of false predictions. And the proposed algorithm took approximately 10 minutes (in 1.7 GHz processor) to solve each sub-problem of Exp. 1 whereas the algorithm in Ref. 24 reportedly took 54.5 minutes (in 1 GHz processor). Though in Ref. 24, the average absolute value of the false predicted parameters was smaller than that of our approach, comparing the amount of gene expression data used and the required time for optimization, proposed algorithm can be considered as an efficient alternative for inferring larger networks.

When noisy data is given, the quality of the prediction deteriorates with the increase of the network dimension and complexity. In case of the smaller network, though the parameter values were not predicted accurately but all the regulations were predicted correctly but for the larger network, few true regulations could not be inferred by the method. It is anticipated that by using the proposed algorithm, the number of such true negative predictions will increase with the complexity of the network and the level of noise present in the gene expression data. However, use of additional time courses can decrease the number of false predictions as well as can increase the accuracy of the estimated parameter values. The other possible ways of decreasing the number of false negatives predictions are to estimate the initial gene expression levels (because of noise) and to use the co-evolutionary approach.[24,27]

Due to the level of noise present in the real gene expression data, we did not get consistent results for the estimated parameters in different experimental runs. Therefore, instead of giving the parameter values we perform some qualitative analysis of the results using Z-score, and the estimation gives a partially correct structure for the SOS network.

From the discussions above it is clear that the proposed method is still not adequate to give an exact topology and precise parameter values for large gene networks but it can estimate the overall network structure with the currently available gene expression data, which is sufficient for comparing with the existing knowledge or suggesting new regulation. Moreover, such indication, where most of the network interactions are inferred, may turn out to be very useful for biologist to design additional experiments or to develop conjectures. Using the iterative loop of prediction and experimentation, it is possible to infer larger networks and more complex correlations.

## 7. Conclusion

In this chapter, we presented an evolutionary algorithm for the inverse problem of genetic network identification. We suggested some modification for the existing fitness evaluation criteria proposed by Kimura *et al.*[24] for the decoupled S-system formalism. The proposed enhancement was intended to increase the accuracy of the prediction based on biological facts. Our proposed algorithm was designed taking several issues into consideration, such as: double optimization applied for selecting robust parameter values, local search included for accelerating the identification of the skeletal network topology and a mutation-phase embedded to maintain the diversity in the population for finding the global optimal solution. The performance of the proposed technique was evaluated by varying network size and the level of noise in expression data, and the efficiency and effectiveness was verified. Applicability of the method to real gene expression data was also reviewed. The reconstruction method identified some known regulations and predicted some unknown or false interactions.

The goal of "The Systeome Project" is detailed and comprehensive simulation of cells of different organisms, from yeast to human.[1] A crucial challenge for this ambitious project is to identify the genetic and metabolic networks and their properties. The proposed approach can handle small to medium sized genetic networks. But the reconstruction of large scale convoluted networks of complex organisms is still out of its scope. One way to deal with such large networks is to use different genomic knowledge in the prediction methodology. Integration of such knowledge will be beneficial to the use of currently available gene-expression data as an adequate source of information, to increase the precision of inference and to reduce the effect of the noise in microarray data, limiting the search space and the number of local optima and increasing the efficiency of the search. Our long-term target is to develop a powerful tool, extending the current form of the algorithm by integrating such prior information, for genome-wide gene network inference.

## References

1. H. Kitano, *Science* **295**, 1662 (2002).
2. P. D'haeseleer, S. Liang and R. Somogyi, *Bioinformatics* **16**, 707 (2000).
3. E. Mjolsness, *Computational Modeling of Genetic and Biochemical Networks*, J. M. Bower and H. Bolouri (eds), MIT Press, England, p. 101 (2001).
4. S. Tavazoie, J. D. Hughes, M. J. Campbell, R. J. Cho and G. M. Church, *Nature Genetics* **22**, 281 (1999).
5. P. Tamayo, D. Slonim, J. Mesirov, Q. Zhu, S. Kitareewan, E. Dmitrovsky, E. S. Lander and T. R. Golub, *Proceedings of National Academy of Science USA* **96**, 2907 (1999).
6. X. Wen, S. Fuhrman, G. S. Michaels, D. B. Carr, S. Smith, J. L. Barker and R. Somogyi, *Proceedings of National Academy of Science USA* **95**, 334 (1998).
7. T. S. Gardner, D. di Bernardo, D. Lorenz and J. J. Collins, *Science* **301**, 102 (2003).
8. S. A. Kauffman, *The Origins of Order, Self-Organization and Selection in Evolution*, Oxford University Press (1993).
9. M. A. Savageau, *Conference Record of the World Congress of Nonlinear Analysts*, p. 3323 (1992).
10. A. Arkin, J. Ross and H. H. McAdams, *Genetics* **149**, 1663 (1998).
11. P. D'haeseleer, X. Wen, S. Fuhrman and R. Somogyi, *Conference Record of Pacific Symposium on Biocomputing* **4**, 41 (1999).
12. H. Matsuno, A. Doi, M. Nagasaki and S. Miyano, *Conference Record of the Pacific Symposium on Biocomputing* **5**, 338 (2000).
13. T. Akutsu, S. Miyano and S. Kuhara, *Conference Record of the Pacific Symposium on Biocomputing* **4**, 17 (1999).
14. M. A. Savageau, *Theoretical Biology* **25**, 365 (1969).
15. M. A. Savageau, *Theoretical Biology* **25**, 370 (1969).

16. E. O. Voit and T. Radivoyevitch, *Bioinformatics* **16**, 1023 (2000).
17. J. S. Almeida and E. O. Voit, *Genome Informatics* **14**, 114 (2003).
18. M. A. Savageau, *Biochemical Systems Analysis: A Study of Function and Design in Molecular Biology*, Addison-Wesley, Reading, MA (1976).
19. D. Tominaga, N. Koga and M. Okamoto, *Conference Record of the Genetic and Evolutionary Computation Conference*, 251 (2000).
20. D. Irvine and M. A. Savageau, *SIAM Journal on Numerical Analysis* **27**, 704 (1990).
21. S. Ando, E. Sakamoto and H. Iba, *Information Sciences* **145**, 237 (2002).
22. S. Kikuchi, D. Tominaga, M. Arita, K. Takahashi and M. Tomita, *Bioinformatics* **19**, 643 (2003).
23. R. Morishita, H. Imade, I. Ono, N. Ono and M. Okamoto, *Conference Record of the Congress on Evolutionary Computations*, 615 (2003)
24. S. Kimura, M. Hatakeyama and A. Konagaya, *Chem-Bio Informatics Journal* **4**, 1 (2004).
25. C. Speith, F. Streichert, N. Speer and A. Zell, *Conference Record of the Genetic and Evolutionary Computation Conference*, 461 (2004).
26. Y. Maki, T. Ueda, M. Okamoto, N. Uematsu, K. Inamura, K. Uchida, Y. Takahashi and Y. Eguchi, *Genome Informatics* **13**, 382 (2002).
27. S. Kimura, K. Ide, A. Kashihara, M. Kano, M. Hatakeyama, R. Masui, N. Nakagawa, S. Yokoyama, S. Kuramitsu and A. Konagaya, *Bioinformatics* **21**, 1154 (2005).
28. E. O. Voit and J. S. Almeida, *Bioinformatics* **20**, 1670 (2004).
29. W. Press, S. Teukolsky, W. Vetterling and B. Flannery, *Numerical Recipies in C*, Second Edition, Cambridge University Press (1995).
30. M. I. Arnone and E. H. Davidson, *Development* **124**, 1851 (1997).
31. N. Noman and H. Iba, *Conference Record of the Genetic and Evolutionary Computation Conference*, 439 (2005).
32. R. Storn and K. Price, *Journal of Global Optimization* **11**, 341 (1997).
33. R. Storn, *IEEE Transactions on Evolutionary Computation* **3**, 22 (1999).
34. H.-Y. Fan and J. Lampinen, *Journal of Global Optimization* **27**, 105 (2003).
35. S. Ando and H. Iba, *Genome Informatics* **14**, 94 (2003).
36. W. M. Spears, in *Foundations of Genetic Algorithms 2*, L. D. Whitley (ed), Morgan Kaufmann, 221 (1993).
37. F. Streichert, H. Planatscher, C. Spieth, H. Ulmer and A. Zell, *Conference Record of the Genetic and Evolutionary Computation Conference*, 471 (2004).
38. N. Noman and H. Iba, *Genome Informatics* **16**, 205 (2005).
39. C. Janion, *Acta Biochimica Polonica* **48**, 599 (2001).
40. J. W. Little, S. H, Edmiston, L. Z. Pacelli and D. W. Mount, *Proceedings of National Academy of Science USA* **77**, 3225 (1980).
41. G. C Walker, *Microbiological Reviews* **48**, 60 (1984).
42. B.-E. Perrin, L. Ralaivola, A. Mazurie, S. Bottani, J. Mallet and F. d'Alché-Buc, *Bioinformatics* **19**, ii138 (2003).
43. M. Ronen, R. Rosenberg, B. I. Shraiman and U. Alon, *Proceedings of National Academy of Science USA* **99**, 10555 (2002).
44. http://www.weizmann.ac.il/mcb/UriAlon/

# CHAPTER 10

# A RELIABLE CLASSIFICATION OF GENE CLUSTERS FOR CANCER SAMPLES USING A HYBRID MULTI-OBJECTIVE EVOLUTIONARY PROCEDURE

Kalyanmoy Deb*, A. Raji Reddy and Shamik Chaudhuri

*Kanpur Genetic Algorithms Laboratory (KanGAL)*
*Indian Institute of Technology Kanpur*
*Kanpur, PIN 208016, India*
*\*deb@iitk.ac.in*

In the area of bioinformatics, the identification of gene subsets responsible for classifying available samples to two or more classes (such as 'malignant' or 'benign') based on the gene microarray data is an important task. The main challenges are the availability of only a few samples compared to the number of genes in the samples and the exorbitantly large search space of solutions to search from. Also, in such problems many different gene combinations may provide similar classification accuracy. Thus, researchers are motivated to find a reliable gene classifier which is small in size and capable of producing as accurate a classification as possible. Although there exist a number of studies which use an evolutionary algorithm (EA) for this task, here we treat the problem as a multi-objective optimization problem of minimizing the classifier size and simultaneously minimizing the number of misclassified instances in training and test samples. Optimal classifiers are tested on a set of samples which were not used in the optimization process. The standard weighted voting method is used to design a unified procedure for handling two and multi-class problems. The reliability in the classification process is ensured by using a prediction strength concept. The use of multi-objective EAs here is also unique in finding multiple high-performing classifiers in a single simulation run. Contrary to the past studies, the use of a multi-objective EA (non-dominated sorting GA or NSGA-II) has enabled us to discover a much smaller gene subset size to correctly classify 100% or near 100% samples for three two-class cancer datasets (Leukemia, Lymphoma, and Colon) and a well-studied multi-class dataset (NCI60). In all cases, the classification accuracy is more than that reported earlier. Moreover, an analysis of the multiple high-performing classifiers reveals important commonly-appearing gene combinations which should be of immediate importance to biologists. The flexibility and effectiveness of the proposed multi-objective EAs in tackling the classification task on two and nine class samples amply demonstrate further and immediate use of the technique to other more complex classification problems.

## 1. Introduction

The recent technological improvements in DNA microarray experiments enable to monitor the *expression levels* of thousands of genes simultaneously. Since the cancer cells generally evolve from normal cells due to mutations in genomic DNA, comparison of gene expression profiles from cancerous and normal tissues (e.g. in case of Leukemia cancer,[11] ALL versus AML), or from tissues corresponding to different cancer types (e.g. in case of NCI60 dataset,[17] breast, CNS, melanoma, ovarian, renal etc.) can provide useful insights into genes implicated in different cancer types. For this purpose, different machine learning approaches such as supervised and some unsupervised learning have been previously investigated with varying degrees of success.[1,3,11] However, the important issue in such problems is the availability of only a few samples compared to the large number of genes, a matter which makes the classification task difficult. Furthermore, many of the genes in a dataset are not relevant to the distinction between different tissue types and introduce noise in the classification process. Therefore, identification of small set of genes, sufficient to distinguish between different tissue types under investigation is one of the crucial tasks from cancer diagnostics point of view. This paper goes further in this direction and focuses on the topic for identification of smallest set of informative genes for reliable classification of cancer samples to two or more classes solely based on their expression levels.

The primary objective of the gene subset identification task is to find a classifier with minimum number of genes providing maximum classification accuracy. However, the optimality of a gene subset for a fixed training set can only be guaranteed by an exhaustive search over all possible combinations of gene subsets. In practice, this is computationally infeasible owing to the large number of possible gene combinations or subsets. Several statistical and analytical approaches have been developed to identify key predictive genes for classification purpose. But many of these studies[1,10,11,16] have used simple gene-ranking techniques or correlation matrices, where a set of genes are ranked based on their expression patterns in a set of training samples and a fixed number of top-ranked genes are selected to construct the classifier. Therefore, these ranking-based techniques select the genes which individually provide better classification, but they may not result in meaningful gene combinations for an overall classification task. Hence, approaches which are capable of performing an efficient *search* in high-dimensional spaces, such as evolutionary algorithms (EAs), should prove to be ideal candidates.

The gene subset identification problem is truly a multi-objective optimization problem consisting of a number of objectives.[13,14] Although the optimization problem is multi-objective, all of these previous studies have scalarized multiple objectives into one. In this paper, we have used a multi-objective evolutionary algorithm (MOEA) to find the optimum gene subset in a number of well-studied datasets (Leukemia, Lymphoma, Colon and NCI60 data). A multiple objective optimization should provide a flexible search compared to the scalarized approach. By using three objectives for minimization (gene subset size, number of misclassifications in training samples, and number of misclassifications in test samples) several variants of a particular MOEA (modified non-dominated sorting genetic algorithm or NSGA-II) are applied to investigate if gene subsets exist with 100% correct classifications in both training and test samples. Since the gene subset identification problem may involve multiple gene subsets of the same size causing identical number of misclassifications,[12] in this paper, we use a novel multi-modal NSGA-II for finding multiple gene subsets simultaneously in one single simulation run. One other important matter in the gene subset identification problem is the confidence level with which the samples are being classified. We introduce the classification procedure based on the prediction strength consideration, suggested in Ref. 11.

## 2. Class Prediction Procedure

For the identification task, we begin with the gene expression values available for cancer disease samples obtained from the DNA microarray experiments. In addition to the gene expression values, each sample in the dataset is also labeled to belong to one class or the other. For identifying genes responsible for the classification of different tumor types, the available dataset is divided into three groups: one used for the training purpose and the second used for the testing purpose within NSGA-II and the third used for the final testing purpose. Here, we use a leave-one-out-cross-validation (LOOCV) procedure to estimate the number of class prediction mismatches in the training samples ($\tau_{train}$), in which one sample is excluded from the training set, and rest of the training samples are used to build the classifier. The classifier is used to predict the class of the left-out sample, and the same procedure is repeated for all training samples. Thereafter, we construct a classifier using all training samples and is used to predict the class of independent test samples. We call $\tau_{test}$ as the number of class prediction mismatches in test samples. The class prediction procedure based on weighted voting (WV) approach[11,16] is described in the following.

### 2.1. Two-class classification

For a given gene subset $G$, we can predict the class of any sample $x$ (whether belonging to A or B) with respect to a known set of $S$ samples in the following manner. Let us say that $S$ samples are composed of two subsets $S_A$ and $S_B$, belonging to class A and B, respectively. First, for each gene $g \in G$, we calculate the mean $\mu_A^g$ and standard deviation $\sigma_A^g$ of the normalized gene expression levels $\bar{x}_g$ of all $S_A$ samples. The same procedure is repeated for the class B samples and $\mu_B^g$ and $\sigma_B^g$ are computed. Thereafter, we determine the class of the sample $x$ in the following manner:[11]

$$\text{class}(x) = \text{sign}\left\{\sum_{g \in G}\left(\frac{\mu_A^g - \mu_B^g}{\sigma_A^g + \sigma_B^g}\right)\left(\bar{x}_g - \frac{\mu_A^g + \mu_B^g}{2}\right)\right\}. \quad (1)$$

If the right term of the above equation is positive, the sample belongs to class A and if it is negative, it belongs to class B. One of the difficulties with the above classification procedure is that the sign of the right term in Eq. (1) is checked to identify if a sample belongs to one class or another. For each ($g$) of the 50 genes in a particular Leukemia sample $x$, we have calculated the statistic $S(x, g)$ (the term inside the summation in Eq. (1)). The statistic $S(x, g)$ values are plotted in Fig. 1 for each gene $g$. Although 27 genes cause negative values of $S(x, g)$ (thereby classifying individually that the sample belongs to AML) and the rest 23 genes detects the sample to be ALL, Eq. (1) finds the right side value of Eq. (1) to be 0.01, thereby correctly classifying the sample to be an ALL sample. But it has been argued elsewhere[11] that a correct prediction with such a small strength does not make the classification with any reasonable confidence.

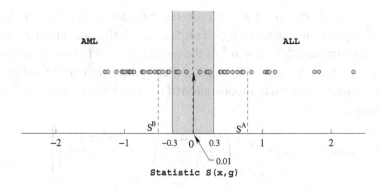

**Fig. 1.** The statistic $S(x, g)$ of a sample for the 50-gene Leukemia data.

For a more confident classification, we may fix a prediction strength threshold $\theta$ ($> 0$) and modify the classification procedure slightly. Let us say that the sum of the positive $S(x, g)$ values is $S^A$ and the sum of the negative $S(x, g)$ values is $S^B$. Then, the prediction strength, as defined in Ref. 11 as $|(S^A - S^B)/(S^A + S^B)|$ is compared with $\theta$. If it is more than $\theta$, the classification is accepted, else the sample is considered to be undetermined for class A or B. Here, we assume these undetermined samples to be identical to mismatched samples and include this sample to increment $\tau_{train}$ or $\tau_{test}$, as the case may be. This way, a 100% correctly classified gene subset will ensure that the prediction strength is outside $(-\theta, \theta)$ for all training and test cases. Figure 1 illustrates this concept with $\theta = 0.3$ (30% threshold) and demands that a match will be scored only when the prediction strength falls outside the shaded area.

### 2.2. Multi-class classification

For predicting the class of a sample belonging to more than two-classes, we use the above WV approach in conjunction with a one-versus-all (OVA) binary pair-wise classification procedure.[2,5,16]

For a given gene subset $G$, the class of a new sample $x$ is predicted in the following manner. Let us say there are total $J$ different tumor types in a dataset such that these $S$ samples are composed of $J$ different subsets $S_1, S_2, \ldots, S_j, \ldots, S_J$, belonging to class 1, class 2, $\ldots$, class $j$, $\ldots$, class $J$, respectively. The class of $x$ is predicted by assuming that the sample $x$ belongs to one particular class $j$ ($j \in J$). For each gene $g \in G$, we first calculate the mean $\mu_j^g$ and the standard deviation $\sigma_j^g$ of the normalized expression values of all $S_j$ samples. Similarly, calculate the mean $\mu_{\bar{j}}^g$ and the standard deviation $\sigma_{\bar{j}}^g$ of the remaining $J-1$ class of samples ($S_1, S_2, \ldots, S_{j-1}, S_{j+1}, \ldots, S_J$) together. Thereafter, the class of sample $x$ can be predicted with slight modification of the above Eq. (1) in the following manner.

$$\text{class}(x_j) = \text{sign}\left\{\sum_{g \in G}\left(\frac{\mu_j^g - \mu_{\bar{j}}^g}{\sigma_j^g + \sigma_{\bar{j}}^g}\right)\left(\bar{x}_g - \frac{\mu_j^g + \mu_{\bar{j}}^g}{2}\right)\right\}. \quad (2)$$

If the sign of the above equation is positive the sample $x$ belongs to class $j$, and if the sign is negative the sample $x$ does not belong to that particular class. The above procedure is repeated for all $j$'s ($j = 1, 2, \ldots, J$). Hence, the multi-class classification problem can be thought of as decomposition of a series of $J$ one-versus-all (OVA) binary classification problems.

One of the difficulties with the above classification procedure is that the sign of the right term of the above equation can be positive for more than one classes while predicting the class of a sample $x$, which ultimately assigns more than one classes for only one sample. One way to overcome this difficulty is that the sample $x$ is assigned to the class with the maximum value of $class(x_j)$ (out of all $j$'s). In this paper, we have used a more stringent classification procedure for generating a generic classifier for reliable classification of multi-class cancer data. For a sample, we calculate the prediction strength $(ps)$[16] of its belonging to each class $j$ and then find the maximum $(ps_{max})$ of them. If $ps_{max}$ is greater than $\theta$ (the chosen threshold), then assign sample $x$ to that particular class, else the sample is added to the mismatch counters $\tau_{train}$ or $\tau_{test}$ as the case may be. Thereafter, the predicted class of a sample is compared with its actual class. If the predicted class matches with the actual class, then the sample is said to be correctly-classified, else the sample is added to the mismatch counter.

## 3. Evolutionary Gene Selection Procedure

In this section, we describe the optimization procedure adopted in this study. First, we describe the optimization problem and then discuss the multi-objective optimization algorithm used in this study and later we describe the overall procedure.

## 3.1. The optimization problem

One of the objectives of the classification task is to identify the smallest size of a gene subset for predicting the class of all samples correctly. Although not obvious, when a too small gene subset is used, the classification procedure becomes erroneous. Thus, minimizations of class prediction mismatches in the training and test samples are also important objectives. From the total available samples, we keep aside $N_\epsilon$ samples for final testing of the developed classifier. These samples are not used in the optimization run. The remaining samples are divided into two parts — training samples used to develop the classifier in the optimization run and testing samples used to test the classifier within the optimization process. Thus, we use three objectives in a multi-objective optimization problem here: (i) the first objective $f_1$ is to minimize the size of gene subset in the classifier, (ii) the second objective $f_2$ is to minimize the number of mismatches in the training samples calculated using the LOOCV procedure described earlier and is equal to $\tau_{train}$ described above and (iii) the third objective $f_3$ is to minimize the number of mismatches $\tau_{test}$ in the test samples. Once an optimization run is complete and a set of Pareto-optimal classifiers are identified, they are applied to classify the final testing $N_\epsilon$ samples.

## 3.2. A multi-objective evolutionary algorithm

Like in a previous study,[13] we use a $\ell$-bit binary string (where $\ell$ is the number of genes in a dataset) to represent a solution. For a particular string, the positions marked with a 1 are included in the gene subset for that solution. For example, in the following example of a 10-bit string (representing a total of 10 genes in a dataset), first, third, and sixth genes are considered in the gene subset (also called the *classifier*):

(1 0 1 0 0 1 0 0 0 0).

The procedure of evaluating a string is as follows. We first collect all genes for which there is a 1 in the string in a gene subset $G$. Thereafter, we calculate $f_1$, $f_2$, and $f_3$ as described above as three objective values associated with the string. Unlike classical methods, an EA uses a number (population) of solutions in each iteration. We initialize each population member by randomly choosing at most 10% of string positions to have a 1. Since the gene subset size is to be minimized, this biasing for 1 in a string allows an EA to start with good population members.

To handle three objectives, we have used a modified non-dominated sorting genetic algorithm or NSGA-II,[7] which we briefly describe here. NSGA-II has the following features:

(1) It uses an elitist principle, meaning that best solution(s) from previous iteration is(are) always included in current iteration.
(2) It uses an explicit diversity preserving mechanism, thereby allowing to maintain multiple trade-off solutions in an iteration.
(3) It emphasizes the *non-dominated* solutions, thereby ensuring convergence close to the *Pareto-optimal* solutions.

In NSGA-II, an offspring population $Q_t$ (of size $N$) is first created from the parent population $P_t$ (of size $N$) by using three genetic operators (reproduction, crossover and mutation operators, described a little later). Thereafter, the two populations are combined together to form $R_t$ of size $2N$. First, all duplicate solutions are deleted from $R_t$. Then, a modified non-dominated sorting procedure[6] (described a little later) is used to classify the entire population $R_t$ according to increasing order of dominance. Once the non-dominated sorting is over, the new parent population $P_{t+1}$ is created by choosing solutions of different non-dominated fronts from the top of the list in $R_t$, one at a time. Figure 2 illustrates this procedure. Since about half the members of $R_t$ can be accommodated to $P_{t+1}$, the above procedure is continued till the last acceptable front $\mathcal{F}_l$. Let us say that solutions remaining to be filled before this last front is considered is $N'$ and the number of non-duplicate solutions in the last front is $N_l$ ($> N'$). Based on a *crowding distance* value,[6] only those $N'$ solutions which will make the diversity of the solutions maximum are chosen from $N_l$.

### 3.3. *A multi-modal NSGA-II*

It is important to realize that in the classification problem there may exist multiple classifiers resulting in an identical set of objective values (having same classifier size and classification accuracy, for example). The following fix-up to the original NSGA-II enables us to find multiple classifiers simultaneously. We compute the number of distinct objective solutions in the set $\mathcal{F}_l$ and let us say it is $n_l$ (obviously, $n_l \leq N_l$). If $n_l \geq N'$ (the top

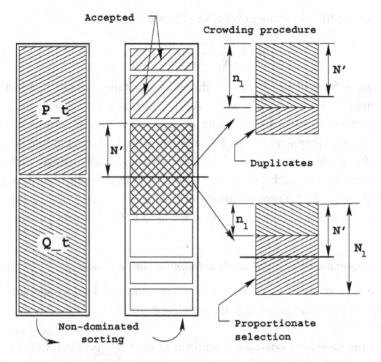

**Fig. 2.** Schematic of the multi-modal NSGA-II procedure is shown.

case shown in Fig. 2), we follow the usual crowding distance procedure to choose $N'$ as the most dispersed and distinct solutions from $n_l$ solutions. The major modification to NSGA-II is made when $n_l < N'$ (bottom case in the figure). This means that although there are fewer distinct solutions than the population slots, the distinct solutions are multi-modal. However, the total number of multi-modal solutions of all distinct solutions ($N_l$) is more than the remaining population slots. Thus, we need to make a decision of choosing a few solutions. The purpose here is to have at least one copy of each distinct objective solution and as many multi-modal copies of them so as to fill up the population. Here, we choose a strategy in which every distinct objective solution is allowed to have a proportionate number of multi-modal solutions as they appear in $\mathcal{F}_l$. To avoid loosing any distinct objective solutions, we first allocate one copy of each distinct objective solution, thereby allocating $n_l$ copies. Thereafter, the proportionate rule is applied to the remaining solutions ($N_l - n_l$) to find the accepted number of solutions for the $i$th distinct objective solution as follows:

$$\alpha_i = \frac{N' - n_l}{N_l - n_l}(m_i - 1), \qquad (3)$$

where $m_i$ is the number of multi-modal solutions of the $i$th distinct objective solution in $\mathcal{F}_l$, such that $\sum_{i=1}^{n_l} m_i = N_l$. The final task is to choose $(\alpha_i + 1)$ random multi-modal solutions from $m_i$ copies for the $i$th distinct objective solution. Along with the duplicate-deletion strategy, the random acceptance of a specified number multi-modal solutions to each distinct objective solution ensures a good spread of solutions in both objective and decision variable space. In the rare occasions of having less than $N$ non-duplicate solutions in $R_t$, new random solutions are used to fill up the population.

### 3.4. *Genetic operators and modified domination operator*

For the reproduction operator, a binary tournament selection operator, in which two solutions are picked at random and the better solution is chosen, is used.

For the crossover operator, a uniform crossover[18] between two parent strings is used. In principle, for an offspring solution, one bit is chosen from each parent with a probability of 0.5. To make the operation less time consuming, the above crossover is restricted only between those bits having a bit difference.

For the mutation operator, a randomly chosen 1 bit position is exchanged with a randomly chosen 0 bit position in the string. It is interesting to note that this operator does not alter the number of 1s in the string.

The usual domination operator described in Ref. 6 compares two solutions and checks if one dominates the other or not. A solution dominating the other solution is considered to be better in the parlance of multi-objective optimization. With NSGA-II, several modifications to the usual domination operator are allowed.[9] Here, we use the following modification. For two solutions, first the sum of the mismatches in training and testing samples $(f_2 + f_3)$ is computed for both solutions under comparison. If they are unequal, the usual domination operator is used. Otherwise, if the difference in the size $(f_1)$ of the classifiers between the two solutions is more than a specified value (10 used here), the usual domination definition is used. Else, they are declared to be *not* dominating to each other. In short, this modification allows two solutions having identical mismatches in $(N - N_\epsilon)$ samples (all samples used for NSGA-II) and having differing classifier sizes (up to a maximum difference of 10) to coexist in the population. Such a modification will enable NSGA-II to find multiple classifiers with differing sizes and having identical accuracy level. Since the final testing on samples which were not used in NSGA-II will reveal the overall accuracy, a search for multiple such high performing solutions is found to be beneficial here.

### 3.5. NSGA-II search using a fixed classifier size

In this procedure, we keep the classifier size fixed to a prescribed value throughout the simulation. We slightly modify the above multi-modal NSGA-II procedure as follows. The initialization procedure creates all classifiers of a certain size (say, $L$). This is achieved by randomly picking $L$ bit-positions and assigning a value 1. The crossover operator first collects the set of all genes having a 1 in either of the two parent and then randomly copies exactly $L$ genes from the set to an offspring solution. The mutation operator exchanges a 1 with a 0, thereby making sure the number of 1s (or genes in the classifier) remains unchanged. The rest of the NSGA-II procedure is as before.

### 3.6. Overall procedure

We suggest and use the following overall procedure:

**Step 1.** Apply the multi-modal NSGA-II described above and obtain the smallest-sized best-performing classifier(s). In this NSGA-II, the classifier size is not restricted to any particular size. Let us say that the smallest size classifier obtained the by this procedure is $L$.

**Step 2.** Apply the multi-modal NSGA-II but fix the classifier size to $L$. This procedure is described earlier and will find multiple best-performing classifiers of size $L$, if available.

The first step makes a global search on the complete problem by evaluating classifiers of different sizes and gene combinations. Thereafter, Step 2 ensures that *multiple* classifiers of a certain size (best dictated by Step 1) are obtained. Step 1 is the main crux of the overall procedure and Step 2 can be considered as a local search procedure to find multiple (and improved, if possible) solutions of a certain size. Such a hybrid evolutionary algorithm procedure are often practiced in solving complex real-world problems[8] for making the overall algorithm more reliable for finding the true optimum of the problem.

## 4. Simulation Results

In this section, we present results obtained by the proposed algorithm on different cancer datasets: Leukemia, Lymphoma, Colon, and NCI60.

### *Leukemia Dataset*

The Leukemia gene expression dataset[11] containing expression profiles of 72 leukemia samples each in 7129 gene was downloaded from http://www.genome.wi.mit.edu/MPR. The dataset was divided into two groups: an initial training set of 27 samples of acute lymphoblastic leukemia (ALL) and 11 samples of acute myeloblastic leukemia (AML), and an independent test set of 20 ALL and 14 AML samples. Here, the expression values are preprocessed by using a threshold of 20 units and a ceiling of 16 000 units, and then exclude genes violating $max(x_g) - min(x_g) > 500$ and $max(x_g)/min(x_g) > 5$ conditions from further consideration, leaving a total of 3859 genes. Based on the suggestion in Ref. 11, the logarithm of the gene expression values (denoted as $\hat{x}_g$) are calculated and then normalized as follows: $\bar{x}_g = (\hat{x}_g - \mu)/\sigma$. Here, $\mu$ and $\sigma$ are the mean and standard deviation of the $\hat{x}_g$ values in the training set only.

First, we apply the multi-modal NSGA-II on 50 genes which were used in another Leukemia study,[11] considering minimization of above three objectives ($f_1$, $f_2$, and $f_3$). For NSGA-II, we choose a population of size 500 and run up to 500 generations with a single-point crossover with a probability of 0.85 and a bit-wise mutation with a probability of $p_m = 0.05$. In this case, we use $\theta = 30\%$ threshold on the prediction strength. All non-dominated solutions found by NSGA-II are shown in Fig. 3. There are a number of important matters to observe from the figure:

(1) The minimum size of a classifier to classify all 72 samples correctly (100% correct with zero mismatches) is five.
(2) There are eight different such five-gene classifiers discovered.

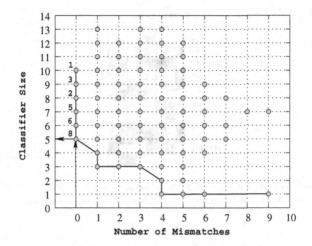

**Fig. 3.** All non-dominated solutions found for the 50-gene Leukemia dataset.

(3) Besides the five-gene classifiers, NSGA-II has found other bigger sized classifiers (six to 10-gene long) capable of making 100% correct classification. Multiple such classifiers are also found. For example, five, seven-gene classifiers are discovered by the proposed method.

(4) Smaller-sized (less than five-gene) classifiers are inadequate to make 100% correct classifications. For example, some three-gene classifiers are enough to correctly classify 71 out of 72 samples. Multiple copies of each such classifier are found, but for clarity we do not mark them in the figure. Although such less-perfect solutions are meaningless to consider when a classifier with 100% correct classification ability exists, the proposed NSGA-II methodology is capable of finding many such interesting solutions. Moreover, an analysis of these high-performing classifiers may reveal interesting gene combinations worth knowing.

(5) Interestingly, no classifier having more than 10 genes is found to produce 100% correct classification. Many genes in a classifier introduce noise in the classification process, thereby lowering the accuracy. In this problem, six to 10 genes are found to be adequate for 100% correct classification.

(6) Interestingly, to achieve a particular accuracy in the classification task, there exist differently sized classifiers and NSGA-II is capable of finding many such solutions in a single simulation run.

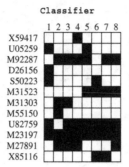

**Fig. 4.** The multi-modal solutions in 50-gene Leukemia dataset with 30% threshold on prediction strength. Each column represents a classifier with marked genes as members of the classifier.

It can be observed from the figure that there exist several five-gene classifiers with zero mismatches in all 72 leukemia samples. Interestingly, the multi-modal NSGA-II has also discovered a number of multi-modal solutions corresponding to each optimal solution. We show all eight such five-gene classifiers making no mismatches in all samples in Fig. 4. Each column in the figure represents a classifier. Figure 4 brings out an interesting aspect. Among eight different five-gene combinations, three genes (accession numbers: M92287, U82759, and M23197) found to appear in more than 75% of the obtained high-performing classifiers. Moreover, the gene M23197 appears in all eight classifiers. Such information about frequently appearing genes in high-performing classifiers is certainly useful to biologists.

### 4.1. Complete leukemia study

Since the above results unveiled the importance of using multi-modal NSGA-II in finding multiple distinct Pareto-optimal solutions, we now apply the same multi-modal NSGA-II on the complete Leukemia dataset to investigate if there exist a smaller gene subset (having less than five genes) with a 100% correct classification accuracy. In the case of the complete Leukemia dataset, we observe that the preprocessing of micro-array data results in 3859 genes from a total of 7129 original genes. Thus, we use 3859 Boolean variables, each representing whether the corresponding gene would be present in the classifier or not. Because of this large string-length requirement, we have chosen a population size of 1000 here and run the multi-modal NSGA-II for 1000 iterations. We use a mutation probability of 0.0005, so that on an average about two bits get mutated in the complete binary string. We consider a classification procedure with $\theta = 0$, 10, 20 and 30% thresholds on the prediction strength on different simulations. Of the 72 samples available for this cancer data, we have used 38 of them as training and 14 of them as testing within NSGA-II. The remaining 20 samples are used for the final testing purpose (which were not used in the optimization process). Table 1 shows the number of genes in the classifiers ($f_1$), number of mismatches in training ($f_2$) and test ($f_3$) samples within NSGA-II simulations, number of mismatches ($\epsilon$) in the final test samples, and the number of multi-modal solutions ($\alpha$) obtained for different values of $\theta$. Results from both steps are shown in the table. It is interesting to observe that for 0% prediction strength, the multi-modal NSGA-II finds 18 different three-gene classifiers providing a 100% correct classification accuracy. In the next step, when the multi-modal NSGA-II is applied for a fixed classifier size of three, as many as 316 different three-gene classifiers are identified. Since the search is restricted to three-gene classifiers only, the NSGA-II is able to achieve a better search and find many such solutions. This trend is observed in other simulations involving higher prediction strength ($\theta$) values, as well.

A randomly generated 1000 three-gene classifiers produce an average of 25.5 mismatches (with a minimum of 5) out of 72 samples with $\theta = 0\%$ and an average of 32.1 mismatches (with a minimum of 9) with $\theta = 30\%$. Interestingly, in all cases our simulations three-gene classifiers are found to

**Table 1.** Results of the proposed approach on Leukemia datasets. LI presents results from Liu and Iba (2002).

| $\theta$ (%) | Multi-modal NSGA-II | | | | | | Fixed-size NSGA-II | | | | |
|---|---|---|---|---|---|---|---|---|---|---|---|
| | $f_1$ | $f_2$ | $f_3$ | $\epsilon$ | Succ. | $\alpha$ | $f_1$ | $f_2$ | $f_3$ | $\epsilon$ | Succ. | $\alpha$ |
| 0 | 3 | 0 | 0 | 0 | 100 | 18 | 3 | 0 | 0 | 0 | 100 | 316 |
| LI | 16 | | | | 97 | | | | | | | |
| 10 | 3 | 0 | 0 | 0 | 100 | 78 | 3 | 0 | 0 | 0 | 100 | 169 |
| 20 | 3 | 0 | 0 | 0 | 100 | 36 | 3 | 0 | 0 | 0 | 100 | 43 |
| 30 | 3 | 0 | 0 | 0 | 100 | 2 | 3 | 0 | 0 | 0 | 100 | 2 |

be adequate for providing a 100% correct classification on all 72 samples. With an increase in prediction strength, the number of classifiers producing 100% correct classification reduce. Compared to an earlier study[13] on the same problem (which required a 16-gene classifier for 97% classification accuracy), our approach finds a smaller-sized classifier and providing a 100% classification accuracy.

To investigate if there are some common genes among all 316 high-performing classifiers at 0% threshold, we mark the gene numbers column-wise from the obtained classifiers in Fig. 5. There are a total of 316 three-gene classifiers, all capable of making a 100% classification of all 72 samples classifier is shown vertically and is comprised of three shaded circles marking corresponding gene numbers in the range [1, 3859]. It is interesting that most of the classifiers have two common genes (gene numbers 1030 and 3718 representing genes M23197 and M31523, respectively). These two genes were also present in 50-gene Leukemia study discussed earlier and were also found to be present in the five-gene classifier shown in Fig. 4. Interestingly, an investigation shows that along with these two genes none of the other 48 genes used in the original 50-gene study can produce a 100% correct classifier on the available 72 samples. A number of other

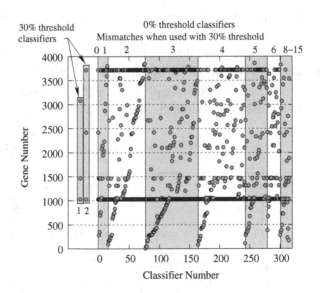

**Fig. 5.** Classifiers producing 100% correct classification in the complete leukemia study.

genes make a better combination with the above two genes. For example, in one of the 316 classifiers producing 100% correct classification the third gene is X64364. This gene was not included in the original 50-gene study.[11] Thus, it can then be argued that the absence of such a gene in our previous study demanded three more genes to be needed to produce a 100% correct classifier.

In the available Leukemia samples, these two genes (M23197 and M31523) show a distinct classification on two classes, as shown in Fig. 6, in which original microarray data is plotted in different shades according to their magnitudes (top ALL, bottom AML). Thus, it is not surprising that these two genes come out to be present in most high-performing classifiers. When each of these genes alone is tested as a classifier, the gene M23197 produces an overall four mismatches and the gene M31523 produces an overall five mismatches. When they together are used as a two-gene classifier, the overall mismatch reduces to one (interestingly, a random set of 1000 two-gene classifiers make an average of 27.6 mismatches). With the help of an appropriate third gene in the classifier, they together can make a 100% correct classification.

The two leftmost classifiers having three genes (shown in Fig. 5) are the only two 100% correct classifiers obtained for 30% prediction strength threshold. Interestingly, one of them (marked 2 in the figure with genes M23197, X64364 and M31523) was also present in the 0% threshold case, but a new three-gene classifier (marked 1 with genes M23197, U05259 and U58046) is also found to provide a 100% correct classification with 30% threshold. When we reevaluate all 316 classifiers obtained in the 0% threshold case for a 30% threshold test, we observe that only one classifier (identical to classifier marked 2 in Fig. 5) produces zero mismatches and it is the same classifier as that obtained in the 30% threshold case. In the figure, all 316 classifiers are grouped according to the extent of their mismatches in

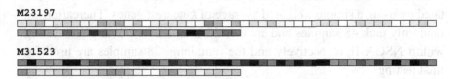

**Fig. 6.** Microarray data on two classes (47 ALL and 25 AML) of Leukemia for two genes shows a clear classification.

the 30% test. The number of mismatched samples in the 30% test are marked group-wise on the top of the figure. Although all 316 three-gene classifiers produce 100% correct classification for the 0% prediction strength case, they produce non-zero mismatches under 30% prediction strength consideration. The appearance of two common genes (M23197 and M31523) in them is a unique feature of this study.

We observe that the gene M23197 appears in all high-performing classifiers, thereby making it the single most important gene in the classification of ALL and AML in a Leukemia sample. This study shows how from original 7129 genes the attention can be focused reliably to three or even to one gene (M23197), achieving more than a thousand-fold advantage. It is worth mentioning that a recent study using an artificial immune system has also stressed the importance of finding multiple high-performing classifiers for Leukemia dataset.[4] That study reported eight classifiers having four to 11 genes capable of making a 100% correct classification. Our result with 316 classifiers requiring only three genes for a 100% correct classification is more significant and reliable.

### 4.2. *Diffuse large B-cell lymphoma dataset*

The diffuse large B-cell lymphoma (DLBCL) dataset[1] contains expression measurements of 96 normal and malignant lymphocyte samples each measured using a specialized cDNA microarray, containing 4026 genes that are preferentially expressed in lymphoid cells or which are of known immunological or oncological importance. There are 42 DLBCL and 54 other cancer disease samples. The expression data was downloaded from http://llmpp.nih.gov/lymphoma/data/figure1.cdt. However, some arrays contain a number of genes with missing expression values. For correcting missing expression values, we have used $k$-nearest neighbor algorithm,[19] in which $k$ genes with similar expression profiles to the gene of interest to impute missing values are selected and the missing values of that particular gene are imputated by using a simple weighted average of $k$ nearest genes. Thereafter, we randomly pick 48 samples and another 20 samples for training and testing within NSGA-II, respectively and the remaining 28 samples are used for final testing.

When we apply the multi-modal NSGA-II with the identical NSGA-II parameters as in the 3859-gene Leukemia case and with a threshold of

**Table 2.** Results of the proposed approach on Lymphoma datasets. LI presents results from Liu and Iba (2002).

| $\theta$ (%) | Multi-modal NSGA-II | | | | | | Fixed-size NSGA-II | | | | | |
|---|---|---|---|---|---|---|---|---|---|---|---|---|
| | $f_1$ | $f_2$ | $f_3$ | $\epsilon$ | Succ. | $\alpha$ | $f_1$ | $f_2$ | $f_3$ | $\epsilon$ | Succ. | $\alpha$ |
| 0 | 3 | 0 | 0 | 1 | 98.9 | 3 | 3 | 0 | 0 | 1 | 98.9 | 190 |
| LI | 18 | | | | 94.0 | 1 | | | | | | |
| 10 | 3 | 0 | 0 | 1 | 98.9 | 5 | 3 | 0 | 0 | 1 | 98.9 | 5 |
| 20 | 5 | 0 | 0 | 1 | 98.9 | 1 | 5 | 0 | 0 | 1 | 98.9 | 684 |
| 30 | 3 | 1 | 0 | 1 | 97.9 | 22 | 3 | 1 | 0 | 1 | 97.9 | 39 |

$\theta = 0\%$, we obtain three, three-gene classifier producing one mismatch in all 96 samples (Table 2). Compared to the earlier study[13] which suggested a 18-gene classifier having 94% successful classification (making six mismatches in 96 samples), our study finds 190, three-gene classifiers providing 98.9% correct classification (just one mismatch in 96 samples). Two genes (1639X and 1720X) are found to be commonly appearing in 190 classifiers. A consideration of 10% prediction strength restricts the search to only five, three-gene classifiers. However, when prediction strength for classification is increased to 20%, a three-gene classifier is found to be inadequate to be the best performer. Instead, a five-gene classifier now makes a zero mismatch in the NSGA-II study and this classifier seems to have made one mismatch in 28 final test samples. Interestingly, the fixed-size NSGA-II simulation finds 684 different five-gene classifiers in one run. When the prediction strength is increased further to 30%, no gene combination is found to provide a zero mismatch in the samples used in the NSGA-II run. However, there are 22, three-gene classifiers which can provide just one mismatch among 48 training samples used in NSGA-II. It is observed that these classifiers also make one more mismatch in the final testing, thereby making a total of two mismatches out of 96 samples. The fixed-size NSGA-II run increases the number of such classifiers to 39. Two genes (1698X and 1643X) appear in all but one classifiers, thereby making the study reliable.

Figure 7 shows all 39 three-gene classifiers and 190 three-gene classifiers found with 30% and 0% prediction strength values, respectively. Interestingly, there is no common gene found between the two sets. Overall, classifiers with 30% prediction strength make two mismatches. When all 190 classifiers obtained using 0% prediction strength are re-evaluated using

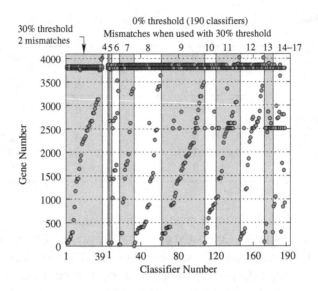

Fig. 7. High-performing classifiers for the lymphoma study.

30% prediction strength threshold, they make four or more mismatches in all 96 samples, as shown in the figure. Thus, all 190 classifiers are inferior to 39 classifiers in the case of 30% prediction strength run. Interestingly, when all 39 classifiers obtained in 30% prediction strength run are re-evaluated using 0% prediction strength threshold, they are also found to make two mismatches, whereas all 190 classifiers made only one mismatch in 0% threshold run. Although there is no common classifiers between the 0% and 30% prediction strength results, the emergence of common genes among high-performing classifiers is striking.

### 4.3. Colon cancer dataset

The Colon gene expression dataset[3] containing expression values of 62 colon biospy samples measured using high density oligonucleotide microarrays containing 2000 genes is available at http://microaaray.princeton.edu/oncology. It contains 22 normal and 40 Colon cancer samples. The dataset is randomly partitioned into three groups, of which randomly picked 31 samples are used for training and other 13 samples are used for testing within NSGA-II. The remaining 18 samples are used for final testing purpose. The gene expression values are log-transformed and then normalized as in the case of Leukemia samples.

Identical NSGA-II parameters to those used in the Leukemia case, except a mutation probability of 0.001, are used. Optimal classifiers and their performances obtained for different $\theta$ values are shown in Table 3. It can be seen from the table that the multi-modal NSGA-II has found four different five-gene classifiers with just one mismatch in all 62 samples without any threshold on the prediction strength. Like in the previous cases, three genes (H08393, M82919 and Z50753) are found to be common among the four classifiers. As we increase the value of $\theta$ to 10%, the mismatches increase. With 20% and 30% prediction strengths, although the classifier size reduces to four, the mismatches increase further. However, it is interesting that even with 30% prediction strength, a four-gene classifier is as good as the 14-gene solution reported the earlier study.[13] Another interesting aspect of the study is that 12 different such four-gene classifiers are discovered.

We present the gene numbers of each of these 12 classifiers in Fig. 8 and once again observe that one gene (H08393) is common to all 12 classifiers and another gene (M36634) is also present in two but all best-performing classifiers. When all four, five-gene classifiers obtained with 0% threshold are re-evaluated using 30% prediction strength threshold, all of them make 19 mismatches, thereby making them inferior classifiers for the 30% prediction strength requirement. However, there is one gene (H08393) found to be common to all 12 best-performing classifiers and all four classifiers found for 0% prediction strength. Interestingly, when all 12 classifier found in 30% prediction strength case are re-evaluated for 0% prediction strength threshold, five mismatches in the training samples and zero

Table 3. Results of the proposed approach on colon datasets. LI presents results from Liu and Iba (2002).

| $\theta$ (%) | Multi-modal NSGA-II | | | | | | Fixed-size NSGA-II | | | | | |
|---|---|---|---|---|---|---|---|---|---|---|---|---|
| | $f_1$ | $f_2$ | $f_3$ | $\epsilon$ | Succ. | $\alpha$ | $f_1$ | $f_2$ | $f_3$ | $\epsilon$ | Succ. | $\alpha$ |
| 0 | 5 | 1 | 0 | 0 | 98.4 | 4 | 5 | 1 | 0 | 0 | 98.4 | 4 |
| LI | 14 | | | | 90.3 | 1 | | | | | | |
| 10 | 5 | 3 | 0 | 1 | 93.5 | 1 | 5 | 2 | 0 | 3 | 91.9 | 2 |
| 20 | 4 | 4 | 0 | 1 | 91.9 | 1 | 4 | 4 | 0 | 1 | 91.9 | 1 |
| 30 | 4 | 5 | 0 | 1 | 90.3 | 5 | 4 | 5 | 0 | 1 | 90.3 | 12 |

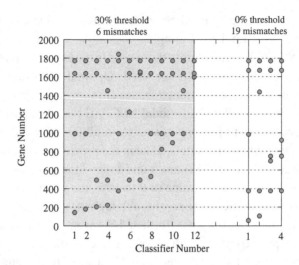

**Fig. 8.** Best-performing classifiers for the complete colon cancer data-sets show at least two common genes. Left-most 12 classifiers produce minimum mismatches in all 62 samples.

mismatches in the testing samples within NSGA-II are observed. According to our definition of dominated solutions (discussed earlier), these classifiers would be dominated by the five-gene classifiers found in the 0% prediction strength case. This amply explains the plausible optimality of the reported classifiers and in turn demonstrates the efficacy of the proposed optimization procedure.

This study shows that the outcome of the classification depends on the chosen prediction strength threshold. Keeping a higher threshold makes a more confident classification, but at the expense of some mismatches, while keeping a low threshold values may make a near 100% classification, but the classification may have been performed with a poor confidence level. Also for a low threshold, there exist more classifiers providing a near 100% correct classification. Compared to past studies, NSGA-II finds better classifiers in all three cancer cases.

### 4.4. *NCI60 multi-class tumor dataset*

Finally, we apply the multi-modal NSGA-II to the NCI60 expression dataset[17] requiring a multi-class classification. In this case, the cDNA microarrays containing 9703 spotted cDNA probes were used to measure the variation in gene expression values among 64 cancer cell lines. The experimental observations were derived from tumors with different sites of origin: 7 breast, 6 CNS, 7 colon, 6 leukemia, 8 melanoma, 9 non-small-cell-lung-carcinoma (NSCLC), 6 ovarian, 8 renal, 4 reproductive, 2 prostate, and 1 unknown cell line. In order to compare with past studies,[10,15] two prostate and one unknown cell line observations are excluded from analysis, leaving a total of 61 samples. The dataset is available from http://genome-www.stanford.edu/sutech/download/nci60/dross_arrays_nci60.tgz. In this study, the gene expression measurements are preprocessed based on the guidelines given in Ref. 15 and 6167 genes are retained for 61 samples. Out of 61 samples, we have used 32 samples as training cases and 17 samples as test cases in the optimization process. The remaining 12 samples are kept reserved for the final testing of the optimized classifiers. In each of these three groups, at least one sample from each of the nine classes is kept.

In this case, we have used a population of size 1000 and run NSGA-II up to 1000 generations with a crossover probability of 0.8 and with a mutation probability of 0.0001. The results obtained with $\theta = 0\%$ and $\theta = 30\%$ are shown in Table 4. The two-stage multi-modal NSGA-II has found a 22-gene classifier making 88.5% classification accuracy without any threshold on prediction strength. This classifier is better than any other previously reported results. The previous study[15] used the same dataset and solved the problem with a GA/MLHD (maximum likelihood) classification approach and reported a 14-gene classifier having 80.3% overall classification accuracy. Despite our classifier being more accurate, a striking difference between our study and the GA/MHLD study is that the latter used a drastically reduced search space. They allowed only 11 to 15 genes to be considered in the classifier, on the other hand, in our first step we

**Table 4.** The results obtained with multi-modal NSGA-II with $\theta = 0\%$ and $\theta = 30\%$ thresholds on a complete dataset of 6167 genes. Parameters $f_1$, $f_2$, $f_3$, $\epsilon$, and $\epsilon_T$ represents gene subset size, mismatches in training samples, mismatches in test samples, and mismatches in the remaining test samples, and the overall mismatches in all samples. Suc. refers to percentage classification success.

| $\theta$ | $f_1$ | $f_2$ | $f_3$ | $\epsilon$ | $\epsilon_T$ | Suc. | $\alpha$ |
|---|---|---|---|---|---|---|---|
| 0 | 22 | 3 | 2 | 2 | 7 | 88.5 | 1 |
| 30 | 16 | 5 | 3 | 3 | 11 | 82.0 | 5 |
| GA/MHLD | 14 | | | | 12 | 80.3 | |

have not considered any such restrictions and the classifier can be of any size between zero and 6167. Table 4 summarizes our results. Our classifier with 22 genes is able to make a better accuracy (7 errors) compared to that found in the GA/MHLD study (12 errors). With a more stringent classification confidence of 30%, 16-gene classifiers are enough to make 82% accurate classification to the available samples. An interesting aspect is that of the 16 genes found in each of the five classifiers, 12 genes are common to all. When a classifier constructed with these 12 genes alone are tested to all 61 samples, a total of 15 errors are observed for each classifier. Interestingly, when 1000 random 12-gene classifiers are constructed and tested on all 61 samples, an average of 50.5 errors (with a minimum of 40 errors) per classifier is observed. Thus, it can be concluded that the 12-gene combination discovered in this study demonstrates a significant classification task.

The 22-gene classifier obtained with $\theta = 0$ makes 11 and 5 errors in training and test samples within NSGA-II when tested with 30% prediction strength. Hence, this classifier is not a reliable one with 30% prediction strength. In fact, NSGA-II could not find any classifier producing an error lesser than 11 in the case of $\theta = 30$. However, among the 22-gene classifier obtained for $\theta = 0$ and all five classifiers obtained for $\theta = 30$, there are four genes in common.

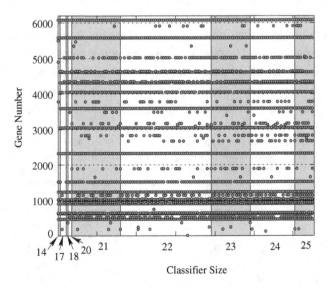

**Fig. 9.** Classifiers producing 86.9% correct classification with $\theta = 0\%$ in the NCI60 study.

Although a 22-gene classifier is found to provide an overall accuracy of 88.5% (with 7 errors) in all 61 samples for threshold $\theta = 0$, the NSGA-II has found 131 different classifiers, having sizes varying from 14 to 25, capable of making a 86.9% accuracy (with 8 errors). Figure 9 plots the gene numbers for these 131 classifiers. It is evident from the figure that about 80% genes are common among these classifiers (as shown by several commonly appearing shaded circles along rows). It is worth emphasizing that the proposed procedure is capable of narrowing down to only about 14 to 25 genes out of 6167 genes for further investigation relating to the differences in nine cancer diseases. To the best of our knowledge, a classification accuracy of 88.5% found in this study is the best reported result to this problem so far.

## 5. Conclusions

The identification of gene subsets responsible for classifying disease samples to fall in one category or another has been dealt in this paper. By treating the resulting optimization problem with three objectives, we have applied a multi-objective evolutionary algorithm (NSGA-II) to find optimal gene-subsets for five different available microarray datasets. The following conclusions can be drawn from the study:

- Compared to past studies involving similar classification of microarray dataset, the proposed optimization algorithm with a multi-objective optimizer finds smaller-sized classifiers, yet resulting in more accurate classifications.
- The proposed approach is capable of finding *multiple* different classifiers (gene combinations) each of the same size and classification accuracy. The knowledge of multiple high-performing classifiers brings out information about commonly appearing genes, thereby providing confidence in the classification task.
- A generic procedure is adopted for tasks involving two or more classes.
- Reliability in classification is ensured in obtained classifiers by using a prediction strength consideration in evaluating a classifier.

Due to the availability of only a few samples compared to the large number of gene expression values in disease samples, it is likely that in such classification problems there may exist multiple gene combinations producing a high classification accuracy. To improve the classification accuracy, a better model than that used in Eq. (1) can also be tried. It is important to highlight that since the results of this study are obtained purely from a computational point of view, more biologically meaningful classifiers can be obtained by using appropriate biological information to the NSGA-II operators. However, besides finding multiple such classifiers in a single simulation run, the present computational approach has revealed another unique aspect of gene subset identification problem. The obtained high-performing multiple classifiers are found to have a number of common genes. It is needless to write that a knowledge of such common genes and their relationships may provide useful insights to biologists for unveiling salient information about the cause and cure of the disease termed as the 'innovization' task [x]). In most cancer disease datasets used here, the study has unveiled such vital information for their further and immediate processing. The procedure suggested in this study can also be used in other more complex classification tasks.

## References

1. A. A. Alizadeh, M. B. Eisen, R. E. Davis, C. Ma, I. S. Losses, A. Rosenwald *et al.*, Distinct types of diffuse large B-cell lymphoma identified by gene expression profiling, *Nature* **403**, 503–511 (2000).
2. E. L. Allwein, R. E. Schapire and Y. Singer, Reducing multiclass to binary: a unifying approach for margin classifiers, *Proc. 17th International Conf. on Machine Learning*, Morgan Kaufmann, San Francisco, CA, pp. 9–16 (2000).
3. U. Alon, N. Barkai, D. A. Notterman, K. Gish, S. Ybarra, D. Mack and A. J. Levine, Broad patterns of gene expression revealed by clustering analysis of tumor and normal colon tissues probed by oligonucleotide arrays, *Proceedings of National Academy of Science, Cell Biology* **96**, 6745–6750 (1999).
4. S. Ando and H. Iba, Artificial immune system for classification of cancer, *Proceedings of the Applications of Evolutionary Computing (LNCS 2611)*, Springer, Berlin, Germany, pp. 1–10 (2003).
5. R. C. Bose and D. K. Ray-Chaudhari, On a class of error correcting binary group codes, *Information Control* **2**, 68–79 (1960).
6. K. Deb, Multi-Objective Optimization using Evolutionary Algorithms, Wiley, Chichester, UK (2001).
7. K. Deb, S. Agrawal, A. Pratap and T. Meyarivan, A fast and elitist multi-objective genetic algorithm: NSGA-II. *IEEE Transactions on Evolutionary Computation* **6**(2), 182–197 (2002).
8. K. Deb and T. Goel, A hybrid multi-objective evolutionary approach to engineering shape design, *Proceedings of the First International Conference on Evolutionary Multi-Criterion Optimization (EMO-01)*, pp. 385–399 (2001).
9. K. Deb and A. R. Reddy, Classification of two-class cancer data reliably using evolutionary algorithms, *BioSystems* **72**(1–2), 111–129 (2003).

10. J. Fridlyand, S. Dudoit and T. P. Speed, Comparison of discrimination methods for the classification of tumors using gene expression data, *Journal of the American Statistical Association* **97**, 77–87 (2002).
11. T. R. Golub, D. K. Slonim, P. Tamayo, C. Huard, M. Gaasenbeek, J. P. Mesirov *et al.*, Molecular classification of cancer: class discovery and class prediction by gene expression monitoring, *Science* **286**, 531–537 (1999).
12. R. Kohavi and G. H. John, Wrappers for feature subset selection, *Artificial Intelligence Journal, Special Issue on Relevance* **97**, 234–271 (1997).
13. J. Liu and H. Iba, Selecting informative genes using a multiobjective evolutionary algorithm, *Proceedings of the World Congress on Computational Intelligence (WCCI-2002)*, pp. 297–302 (2002).
14. J. Liu, H. Iba and M. Ishizuka, Selecting informative genes with parallel genetic algorithms in tissue classification, *Genome Informatics* **12**, 14–23 (2001).
15. C. H. Ooi and P. Tan, Genetic algorithms applied to multi-class prediction for the analysis of gene expression data, *Bioinformatics* **19**(1), 37–44 (2003).
16. S. Ramaswamy, P. Tamayo, R. Rifkin, S. Mukherjee, C. H. Yeang, M. Angelo *et al.*, Multiclass cancer diagnosis using tumor gene expression signatures, *Proceedings of the National Academy of Science* **98**(26), 15149–15154 (2001).
17. D. T. Ross, M. B. Eisen U. Scherf, C. M. Perou, C. Rees, P. Spellman, V. Iyer, S. S. Jeffrey, M. Van de Rijn and M. Waltham *et al.*, Systematic variation in gene expression patterns in human cancer cell lines, *Nature Genetics* **24**, 227–235 (2000).
18. G. Syswerda, Uniform crossover in genetic algorithms, *Proceedings of the Third International Conference on Genetic Algorithms*, pp. 2–9 (1989).
19. O. Troyanskaya, M. Cantor, G. Sherlock, P. Brown, T. Hastie, R. Tibshirani, D. Botstein and B. R. Altman, Missing value estimation methods for DNA microarrays. *Bioinformatics* **17**(6), 520–525 (2001).
20. K. Deb and A. Srinivasan, Innovization: Innovating design principles through optimization. *Proceedings of the Genetic and Evolutionary Computation Conference (GECCO-2006)*, New York: The Association of Computing Machinery (ACM), pp. 1629–1636 (2006).

# CHAPTER 11

# FEATURE SELECTION FOR CANCER CLASSIFICATION USING ANT COLONY OPTIMIZATION AND SUPPORT VECTOR MACHINES

A. Gupta[*], V. K. Jayaraman[†,‡,¶] and B. D. Kulkarni[†,§,¶]

[*]*Summer Trainee, Chemical Engineering Department,*
*IIT Kharagpur, India*
[†]*Chemical Engineering Division*
*National Chemical Laboratory, Pune, India*
[‡]*vk.jayaraman@ncl.res.in*
[§]*bd.kulkarni@ncl.res.in*

Feature selection has been an important preprocessing step in cancer classification using microarray data. It is well known that feature selection improves classification performance considerably. However, a significant problem in feature selection is to traverse the vast search space in order to select the best feature subset, which predicts with the highest classification accuracy. In this chapter, we present a novel feature selection method, using Ant Colony Optimization and Support Vector Machines. Ant Colony Optimization is used to select the most informative feature subsets, which give higher classification accuracy evaluated using Support Vector Machines. The proposed method can efficiently perform feature selection, and improve the classification accuracy significantly. Our approach has been evaluated on three widely used cancer datasets and has shown excellent performance.

## 1. Introduction

Microarray technology today has enabled us to rapidly measure the expression levels of thousands of genes simultaneously in a tissue sample. The gene expression patterns have already shown promising results in wide variety of problems, especially in the field of clinical medicine. One such field

---

[¶] Corresponding authors.

is cancer diagnosis, where gene expression profiling can be used to construct a classifier, which can predict an unknown test sample as malignant or benign, or classify it into its sub types. Classification on the basis on gene expression profiles has been considered by Golub et al.[1] for classification of acute leukemia, Alon et al.[2] for clustering of normal and tumor tissues, Nutt et al.[3] for classifying different kinds of tumors and Alizadeh et al.[4] for diffuse large B-cell lymphoma. The major hurdle in such microarray classification is the non-availability of sufficient number of samples as compared to the huge number of features (genes), which makes the statistical analysis difficult. However, the number of genes actually required for the classification task is usually very small, while a large number of them are usually redundant or act as noise. Identification of these discriminatory genes can, not only decrease the complexity of the task at hand, but also significantly increase the accuracy of classification. Furthermore, this process of gene selection can be used for selection of potential biomarkers for the problem, which can be then used for drug discovery.

Various classification techniques have been applied for microarray data, namely, Support Vector Machines (SVM),[5,6] Logistic Regression,[7] k-NN.[13] These classification techniques are normally used in conjunction with various feature selection methods,[6,8–19] which are usually classified into *Filter*, *Wrapper* and *Embedded* approaches,[8] depending on the usage of the classification algorithm. *Filter* approach is a preprocessing step, which attempts to select the good features from the complete set of features using only intrinsic characteristics of the data. A filter method generally uses statistical correlations to evaluate the relevance of each feature (or feature subset) to the output class label, and thus independent of the classification method to be used. The best features are then used for the classification task.*Wrapper* approach, on the other hand uses a classification algorithm to evaluate a feature subset, and the quality of the subset is given by the performance of the classifier. Since this approach involves the use of a classification algorithm, it is usually computationally more intensive than the filter approach. *Embedded* approach involves the selection of a feature subset during the training of a classification predictor. Embedded and wrapper methods are often closely related to each other. Since the wrapper and embedded methods take the classification method into account, it can be argued that they will perform better than the filter methods.

Feature selection methods can also be classified as *univariate* or *multivariate* approaches, depending on whether an individual feature or a subset of features is evaluated. A *univariate* approach evaluates the relevance or importance of each feature independently, and then ranks them in order of their importance. In contrast the *multivariate* approach evaluates the relevance of a subset of features instead of a single feature, and then tries to find the best subset of features. It can be argued that the classification will depend on the combination of different features, and thus it may so happen that two very informative features when selected together may not be able to classify at all. Also, a combination of seemingly redundant features may classify well. Thus, a multivariate approach in this sense can be more desirable than univariate approaches.

It can be seen that the number of possible combinations of feature subsets is huge even for moderate number of features and the problem of finding an optimal feature subset is NP-hard. For a large number of features, the search space becomes exceedingly large, making an exhaustive search computationally intractable. Thus, we require the use of heuristic search methods that can obtain sufficiently near-optimal solutions within feasible time. Various methods like, Sequential search, Meta Heuristic and Exponential search, have been proposed which search the combinatorial space. Most common meta heuristic technique applied to the field of feature selection is Genetic Algorithms. In Refs. 11–17, Genetic Algorithms have been specifically used for the purpose of gene selection. Here, Genetic algorithms are used for multivariate feature selection using SVM and kNN as classifiers.

The case of finding an optimal feature subset can be considered as a combinatorial optimization problem. Ant Colony optimization (ACO), is another meta-heuristic technique which is known to effectively solve such problems.[20] ACO has recently been used for the purpose of feature selection.[21,22] In this chapter, we present an Ant Colony Based gene selection method using Support Vector Machines as a classifier. This method of feature selection has then been applied to two datasets, and comparable results have been obtained for each.

The rest of the work is organized as follows. A review of Ant Colony Optimization and Support Vector Machine methods is presented in Secs. 2 and 3. The proposed approach is explained in Sec. 4, and Sec. 5 outlines the proposed algorithm. The datasets used and the experimental setup has been described in Sec. 6. The results of the proposed algorithm have been presented in Sec. 7. Finally, we conclude this chapter with Sec. 8.

## 2. Ant Colony Optimization

Ant Colony Optimization is a meta-heuristic approach, first proposed by Dorigo et al.,[23] which can be used to solve various combinatorial problems. ACO simulates the behavior of real ants to find the shortest path between their nests to the food source. Initially the ants search for the food source in a random fashion. When an ant finds a food source, it deposits an odorous chemical called pheromone. When a random ant comes across this pheromone it will follow the pheromone trail with high probability. Thus pheromone acts as a medium of communication between the ants. Such form of indirect communication by modification of environment is called "stigmergy". The probability of a path to be chosen depends on the amount of pheromone on the path. Larger amount of pheromone will increase the probability of the path to get selected, which will again deposit pheromone and thus enforcing the path in an autocatalytic manner. This forms a positive feedback loop, which will attract more ants to follow the high pheromone trail, thus helping the ants to find the shortest path. This behaviour of the ants has been illustrated in Fig. 1. However, the pheromone deposited evaporates over time, thus allowing the ants to explore new paths.

Figure 2 shows the ability of ants to adapt to the changes in environment. Initially the ants follow a direct path from their nests to the food source. When an obstacle is placed in their way, the ants start exploring all the paths around the obstacle to their food source and then back. Gradually, the better path (i.e. the shorter path in this example) receives more pheromone deposition making it even more favorable. The longer path on the other hand receives less amount of pheromone. Owing to the evaporation of pheromone the attractiveness of the longer path decreases further. Eventually, all the ants converge to the shorter path. Thus, pheromone evaporation is required to prevent ants from choosing a previously explored unfavorable path.

ACO has been successfully applied to the combinatorial optimization problems like TSP,[23] sequential ordering,[24] scheduling[25] and process optimization.[26,27] It has recently been used in the context of feature selection.[21,22]

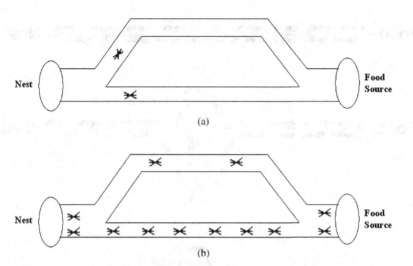

**Fig. 1.** (a) Ants initially explore all the paths with equal probability; (b) Ants follow the better (shorter) path where more pheromone is deposited, however, some ants still explore other paths.

In this chapter, we present an ACO model, which acts as a search algorithm for selecting optimal feature subsets from a large number of possible subsets. We have used SVM as a classifier for the purpose of evaluating the feature subsets.

## 3. Support Vector Machines

Support Vector Machines is a learning algorithm originally developed by Vapnik.[28] It is rigorously based on statistical learning theory and has been used extensively for the purpose of classification in a wide variety of fields.[29,30] It has also been applied in the context of microarray for cancer classification.[5,6,10,14,15]

SVM separates a given set of binary-labeled data by constructing a hyper-plane, which maximizes the margin between the nearest data points of each class. For linearly non-separable problems SVM transforms the input data into a very high-dimensional feature space and then employs a linear hyperplane for classification. Introduction of a feature space creates a computationally intractable problem. SVM handle this by defining appropriate kernels so that the problem can be solved in the input space itself. The

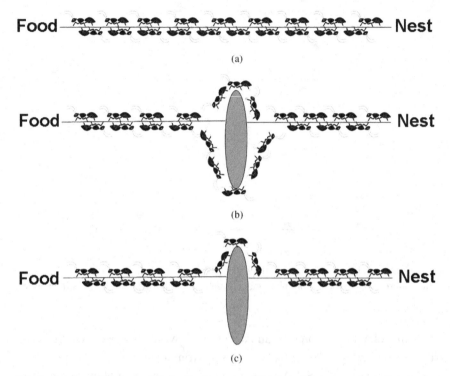

**Fig. 2.** (a) Ants follow a path from food source to their nest (b) Ants explore all the paths when an obstruction is placed in their path (c) Ants converge to the shorter path from the food source to their nest.

problem of maximizing the margin can be reduced to the solution of a convex quadratic optimization problem, which has a unique global minimum. We consider a binary classification training data problem,

$$\{x_i, y_i\}, \quad i = 1, 2, 3, \ldots, L, \tag{1}$$

where $y_i \in \{-1, 1\}$, and $x_i \in \mathbb{R}^d$.

Here, $x_i$ represents the input training vector, and $y_i$ represents the input training label for the $i$th input pattern. The SVM classifier for the case on linearly inseparable data is given by

$$f(x) = \sum_{i=1}^{m} y_i \alpha_i K(x_i, x) + b, \tag{2}$$

where $K$ is the kernel matrix, and $m(< L)$ is the number of input patterns having nonzero values of the Langrangian multipliers ($\alpha_i$), and $b$ is the bias.

In case of microarray, $x_i$ is the $i$th tissue sample, and $y_i$ is the class representing benign or malignant tissue. These $m$ input patterns are called support vectors, and hence the name support vector machines. The Langrangian multipliers ($\alpha_i$) can be obtained by solving the following dual form of the quadratic programming (QP) problem:

$$w(\alpha) = \sum_{i=1}^{L} \alpha_i - \left(\frac{1}{2}\right) \sum_{i,j=1}^{L} \alpha_i \alpha_j y_i y_j K(x_i, x_j). \quad (3)$$

Subject to the constraints,

$$0 \le \alpha_i \le C, \quad i = 1, 2, \ldots, L,$$

$$\sum_{i=1}^{L} \alpha_i y_i = 0,$$

where $C$ is the cost parameter, which controls the number of non-separable points. Increasing $C$ will increase the number of support vectors thus allowing fewer errors, but making the boundary separating the two classes more complex. On the other hand, a low value of $C$ allows more non-separable points, and therefore, has a simpler boundary.

We obtain the values for $\alpha_i$ by solving the QP, we can get the value of bias $b$,

$$b = -\frac{1}{2} \left[ \max_{\{i \mid y_i = -1\}} \left( \sum_{j \in \{SV\}}^{m} y_j \alpha_j K(x_i, x_j) \right) \right.$$
$$\left. + \min_{\{i \mid y_i = +1\}} \left( \sum_{j \in \{SV\}}^{m} y_j \alpha_j K(x_i, x_j) \right) \right]. \quad (4)$$

A kernel matrix is a symmetric positive definite matrix. Various types of kernels can be used depending on the classification problem. Some widely used kernels are:

Linear $\qquad K(x_i, x_j) = x_i^T x_j$
Polynomial $\qquad K(x_i, x_j) = (\gamma x_i^T x_j + r)^d$
Radial Basis Function $\qquad K(x_i, x_j) = \exp\left(-\gamma \left\| x_i - x_j \right\|^2\right).$

In this study however, we have found that the linear kernel performs better than the polynomial or RBF kernel. The SVM experiments done in this work have been done using an implementation of LIBSVM.[31]

## 4. Proposed Ant Algorithm

In this approach to feature selection, the ACO system has been applied to select various feature subsets from the original dataset, which are then evaluated using a SVM classifier. This evaluation is then used by the ACO to produce better feature subsets.

Let $S = \{f_j\}, j = 1, 2, \ldots N$ be the original set of $N$ features, and $s_i \in S$ is a feature subset of subset size $n$, selected by ant $i$. The pheromone value of a feature $f_j$ is represented by $\tau(f_j)$, and $\eta(f_j)$ is a heuristic function value, which represents the measure of quality of the feature $f_j$. Initially the pheromone level of all the features is same, and is initialized as $\tau_0$. However, as the ant colony progresses through the generations, pheromone value changes using the global and the local updating rules. The heuristic function represents the quality of every feature, which can be used with a prior knowledge of relative importance of features. This allows us to selectively favor a feature, thereby improving its chance to get selected in the best feature subset. However, in this case, all the features have been given equal opportunity to get selected in a best feature subset, by keeping the quality of every feature to be same. The ants select the features using the state transition rules, which use the pheromone value as well as the heuristic function value of the features to probabilistically decide whether a feature is to be selected or not. After all the ants have selected their feature subsets, they are evaluated for accuracy, and the pheromone value is updated using the global and local updating rules. The following sections describe each of the processes in detail.

### 4.1. State transition rules

Every ant builds a solution subset in a step-by-step manner, by applying a probabilistic decision policy at every step. Every feature represents a state in the path from the nest to the food source. The ant can be taken as moving through various states as it selects its features. This is done by employing the state transition rules, where the next state (or feature) is decided by its pheromone level and the heuristic value. The state transition rules used in this work are given as:

$$f = \begin{cases} max\{\tau(f)\eta(f)^\beta\} & \text{if } (q < q_0) \text{ Exploitation} \\ \dfrac{\tau(f)\eta(f)^\beta}{\sum_{f \in j_k(r)} \tau(f)\eta(f)^\beta} & \text{otherwise Biased Exploration} \end{cases} \quad (5)$$

A particular ant $i$ selects a feature $f$, using the above rule, where $j_k(r)$ is the set of features which do not belong to the subset of the ant $i$. The parameter $\beta$ represents the relative importance of the heuristic function value over the pheromone level. A value of $\beta = 0$ will not consider the heuristic function for the feature selection process, while a value of $\beta = 1$ will give equal importance to the heuristic function and the pheromone value. $q_0$ is a fixed parameter known as probability exploitation factor between 0 and 1, and $q$ is a random number uniformly distributed over the range [0,1]. The state transition rule selects the features by maximizing the product of the pheromone level and the heuristic function. The pheromone here represents the learned knowledge, which is shared between all the ants, thereby acting as an indirect medium of communication as well. Thus, an ant adds a new feature by considering the previously gained knowledge as well as the heuristic function value. Parameter $q$ decides the relative importance of the two modes of feature selection by ants, namely, Exploitation mode and Biased Exploration mode. In Exploitation mode, the ants select the features having the maximum value of the product of pheromone and heuristic values, thereby exploiting the previously gained knowledge. This rule favors the selection of those features, which are known to be better, either by heuristic function value, or by the knowledge gained by the ants before or by both. In the Biased Exploration mode, the ants probabilistically select the features, which were not selected by it before. The probability of selecting a feature is again, determined by the product of the pheromone and heuristic function values. This mode ensures that ants do not stick to their original paths but use the previously learned knowledge as well as the heuristics to keep exploring new paths, which may provide better solution. The parameter $q_0$ is a user-defined adjustable parameter, which depends on the problem. A high value of $q_0$ will ensure that the known "good" features are selected, keeping the number of newly explored features less. In contrast, a low value of $q_0$ will increase the ants' ability to look for newer solutions. It should be taken care that an already selected feature is not selected again in an ant's subset. Also the scales of the pheromone values and the heuristic function values should be made same, to prevent any of them to dominate over the other.

### 4.2. Evaluation procedure

Every feature subset constructed by the ants in the previous step is then evaluated to determine the ant with the best feature subset. We have considered the leave-one-out-cross-validation (LOOCV) accuracy on the training data as a measure of ranking the feature subsets, using SVM as a classifier. This method is most commonly used methods in the feature selection methods.[5,11,12,15,18,19] In this method, in every iteration one training sample is taken out, and the SVM model is trained on the remaining data. This model is then tested on the left out sample. This is repeated for every training sample, and the average accuracy over all the iterations is found, which is called the LOOCV Accuracy. Ants with a higher LOOCV are considered to have better feature subsets. The ant with the highest LOOCV is taken as the winner ant.

### 4.3. Global updating rule

After all the ants have been evaluated, the winner ant is identified and the global updating rule is applied, to only those features which belong to the winning ants subset. The global updating rule increases the value of the pheromone of these features according to the relation,

$$\tau(f)' = (1 - \kappa) \times \tau(f) + \kappa \times \sigma. \qquad (6)$$

Here, $\sigma$ is the accuracy of the globally best solution obtained for the current subset size, and $\kappa$ is the pheromone decay parameter, in the range of [0,1]. This rule allows the winner ant to deposit more pheromone on its selected path by increasing the pheromone values of its selected features. This increase in the pheromone level will increase the probability of these features to be selected. Thus, the pheromone value of those features, which appear in the winner ant's subsets will be increased in every generation, thus gradually making them more attractive for the other ants. This increased pheromone helps the ants to move to better solutions in every generation, by more number of "good" features and lesser number of "bad" features.

### 4.4. Local updating rule

The local updating rule is applied to all the features, which were not selected in the winner ants subset. The rule decreases the pheromone level of the features, which were selected by the ants but did not win by a very small amount, but keeps the pheromone level of the features not yet explored same as the initial value. This makes the irrelevant features less desirable, while keeping the desirability of the unexplored features at the same level, thus reducing the probability of selecting the irrelevant features. It also enables the ants to keep selecting the unexplored features. The local updating rule is given by:

$$\tau(f)' = (1 - \alpha) \times \tau(f) + \alpha \times \tau_0. \qquad (7)$$

Here, the parameter $\alpha$ lies in the range, $0 < \alpha < 1$, and is called the local pheromone update strength parameter. Algorithm for the above mentioned procedure is given in the next section.

### 5. Algorithm Outline

Each ant $i$, in the Ant Colony System selects a subset $s_i$ of $n$ features from the original dataset $S$ of $N$ features. This subset will correspond to the path of the ant from its nest to the food source. Thus for a total of $r$ ants, we get $r$ subsets of $n$ features each. Each of these subsets will be evaluated using the SVM classifier to give an accuracy value, which will represent the quality of the path used by every ant. Accordingly, the ants will deposit some amount of pheromone that will be higher for the feature subsets yielding higher accuracy values. This is then repeated for a specified number of generations to obtain the best feature subset of size $n$. The number of features in a subset, $n$, is decreased at a constant rate, such that the whole process is repeated for decreasing values of $n$. The decrement rate depends on the search space, however in this work, a constant decrement rate of 1 has been chosen. Also the initial and final subset size are problem dependent. We now give an outline of the overall procedure to be followed (see also flow chart in Fig. 3):

(1) *Initialization*: The pheromone values of all the features are initialized with the same value $\tau_0$. The ants are initialized by randomly selecting

features subsets of size $n = n_i$. However, other initialization rules can also be used.
(2) *Solution Construction*: All the $r$ ants construct their paths by selecting a subset of $n$ features, using the state transition rules (except for the first time when the ants randomly select the features).
(3) *Evaluation*: Feature subsets of all the $r$ ants are used to calculate their qualities in terms of accuracies using the SVM classifier, and the winner ant (the one with the highest classification accuracy) is identified.
(4) *Global Updating*: Global updating rule is applied to the winner ant, which has produced the highest classification accuracy. This is done by increasing the pheromone value of all the features that have been a part of the winning ant subset.
(5) *Local Updating*: Local updating rule is applied to all the ants except the winner ant. This is done by slightly decreasing the value of pheromone of all the features that had not been a part of the winning ant subset.
(6) *Record the best solution of this subset size*: The best solution out of all the solutions in all the generations of the present subset size is recorded.
(7) *Record the overall best solution*: The best solution obtained among all the subset size uptil now is recorded.
(8) *Decrement the subset size*: The subset size is decreased by the specified amount until $n < n_f$ , and same procedure is repeated from Step 2.
(9) Report the Best Solution over all the subset sizes.

## 6. Experiments

### 6.1. *Datasets*

The proposed method has been applied to three microarray datasets from cancer research, Colon Cancer,[2] Brain Cancer[3] and Leukemia.[1]

### 6.1.1. *Colon cancer dataset*

The colon cancer dataset consists of expression levels of 40 tumor and 22 normal colon tissues. The 2000 genes having the highest minimal intensity across the 62 tissue samples were selected by Ref. 2. The dataset is available

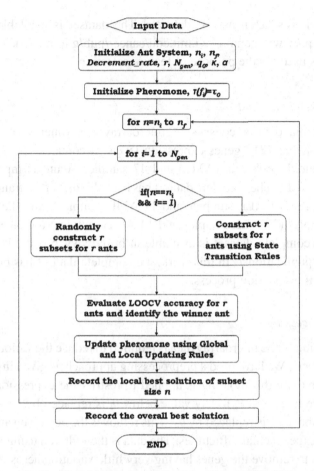

**Fig. 3.** Flowchart for the proposed Ant Colony Algorithm.

at http://microarray.princeton.edu/oncology/. All the 62 samples have been used during the feature selection process.

### 6.1.2. Brain cancer dataset

The Brain cancer dataset consists of gene expression levels of 12 625 genes in a set of 50 gliomas; 28 glioblastomas and 22 anaplastic oligodendrogliomas. A subset of 21 samples (14 glioblastomas and 7 anaplastic oligodendrogliomas) with classic histology was selected as training set, while the rest 29 samples (14 glioblastomas and 15 anaplastic oligodendrogliomas) was taken as test dataset.[3] The dataset is available on the website http://www-genome.wi.mit.edu/cancer/pub/glioma. The complete dataset was used for the purpose of feature selection.

### 6.1.3. *Leukemia dataset*

The Leukemia dataset consists of microarray experiments of 72 samples each having 7219 genes expression levels. It contains 25 samples of Acute Myeloid Leukemia (AML) and 47 samples Acute Lymphoblastic Leukemia (ALL). The complete dataset has been divided into a training data of 38 samples (27 ALL samples and 11 AML samples), and the remaining 34 samples (20 ALL samples and 14 AML samples) are taken as test data.[1] The complete dataset is available at http://www.broad.mit.edu/cgi-bin/cancer/publications/. In this work, the complete dataset has been used for the feature selection process.

## 6.2. *Preprocessing*

Preprocessing of the microarray data is required to reduce the various experimental effects. We have used a preprocessing approach as given in Refs. 32 and 16. Firstly, a thresholding is applied to restrict the expression values between the range $(\theta_l, \theta_h)$. Any value below $\theta_l$ is replaced by $\theta_l$; similarly any value above $\theta_h$ is replaced by $\theta_h$. This removes the negative values arising due to experimental difficulties. The data is then filtered using variation filters so as to remove the genes having very little variation across the samples. Those genes whose maximum and minimum values differ by less than a value of $\Delta$ are removed. Furthermore, those genes with a ratio of maximum to minimum expression levels less than $\Omega$ are removed as well. This ensures that the remaining genes have enough variability to be a discriminatory gene. Natural logarithm of the gene expression values was taken. These logged gene expression profiles can now, either be scaled between 0 and 1 for every gene, or can be standardized to have zero mean and unit standard deviation for a gene across all the samples. Different values of the parameters $\theta_l$, $\theta_h$, $\Delta$, $\Omega$ have been used for every dataset.

For the case of Colon Cancer Dataset, filtered data is already available, and thus no preprocessing was performed. For the Brain Cancer dataset, parameter values of $\theta_l = 20$, $\theta_h = 16,000$, $\Delta = 100$, $\Omega = 3$ reduced the number of genes from 12 625 to 4434. Then the gene expression values were linearly scaled from 0 to 1 after log-transformation. Leukemia dataset was preprocessed using the parameter values of $\theta_l = 20$, $\Delta = 500$, $\Omega = 5$ resulting in only 3870 genes. These values were then log-transformed and standardized to have zero mean and unit standard deviation.

### 6.3. Experimental setup

For the SVM classifier, it was found that linear kernels outperform the Polynomial and RBF Kernels for the above two datasets. Table 1 shows the various ACO parameters chosen for the study. The initial number of features for both the datasets was taken as 100, with a decrement of 1 feature per iteration until the number of features reaches 20. The number of ants chosen is 40. A higher number of ants imply a larger computational time, while small number of ants may not be sufficient to effectively explore all the potential states. Again, a higher number of generations will increase the computational time many fold, while a lesser number of generations may not be able to obtain the best solution in the given subset size. This value has been chosen by trial and error. We have found that 40 generations per subset size is enough for the given problems.

The exploitation probability factor ($q_0$) decides the relative importance of the exploitation and the biased exploration mode. In the exploitation mode, those features, which are known to be good, are selected, however, in the biased exploration mode the ants explore the new features. A high value of $q_0 = 0.8$ has been chosen so that the ants retain the known "good" features, as well as keep exploring the new features simultaneously. The pheromone decay parameter ($\kappa$) and the pheromone update strength parameter ($\alpha$) have been taken as 0.8 and 0.2 respectively.

Table 1. Parameters for the Ant Colony System

| Parameters | Values |
|---|---|
| Initial Feature Subset Size ($n_i$) | 100 |
| Final Feature Subset Size ($n_f$) | 20 |
| Decrement Rate | 1 |
| Number of Ants ($r$) | 40 |
| Number of Generations | 40 |
| Exploitation Probability Factor ($q_0$) | 0.8 |
| Pheromone Decay Parameter ($\kappa$) | 0.8 |
| Pheromone Update Strength ($\alpha$) | 0.2 |

## 7. Results and Discussion

Table 2 shows the results when all the complete datasets have been used as training data, without applying any feature selection. This gives us an estimate of the baseline accuracy of the datasets. After preprocessing, the number of genes for the case of Leukemia was reduced to 3870 genes. The Leave-One-Out Cross Validation (LOOCV) accuracy obtained using all the 3870 genes in this case was 98.61%. No preprocessing was done for the case of Colon Cancer dataset. We obtained an accuracy of 80.64%, by the LOOCV procedure using all the 2000 genes. The Brain Cancer dataset gives us a LOOCV accuracy of 80.00% using all the 4434 genes left after preprocessing.

After this, we applied the proposed algorithm to the above three datasets. Table 3 shows the results of our approach on the three datasets. The highest classification accuracy obtained during the process has been reported along with the number of genes in the subset. In the calculation of these results LOOCV accuracy has been used on both the complete datasets. Perfect classification was achieved for all the three datasets using our approach. It shows that the feature selection process increases the classification accuracy considerably. Also, the number of genes required for the classification purpose also decreases drastically. Only 20 genes were used for the classification purpose. Many different subsets with varying number of genes were

Table 2. LOOCV accuracy on the complete datasets using all the genes.

| Dataset | Number of genes | LOOCV accuracy |
|---|---|---|
| Leukemia | 3870 | 98.61% |
| Colon Cancer | 2000 | 80.64% |
| Brain Cancer | 4434 | 80.00% |

Table 3. Results of ACO/SVM on the complete datasets.

| Dataset | Number of genes | LOOCV accuracy |
|---|---|---|
| Leukemia | 20 | 100% |
| Colon Cancer | 20 | 100% |
| Brain Cancer | 20 | 100% |

found to obtain a LOOCV accuracy of 100%, however, only the smallest subset has been reported here. Further, in order to check the stability of the method, we conducted five independent runs for all the three datasets. Perfect classification was achieved for all the three datasets in all the five runs for various gene subsets. Also, 100% classification accuracy was achieved for all the three datasets using only 20 genes. The results obtained for the case of Leukemia data are comparable to the previous results obtained in Ref. 11, where a GA-based approach has been considered for the purpose of feature selection. Our method obtains a higher accuracy for the case of Colon cancer dataset. We also applied our approach for the case of Brain Cancer dataset, achieving perfect classification for this case as well. This is higher than that obtained by Ref. 16, where a GA-based approach has been used.

For each of the three datasets, gene subsets yielding the highest accuracy were recorded for all subset sizes ranging from 100 to 20. Using these subsets, we find the number of times a particular gene occurs in the subsets. It is expected that a gene which is more highly related with the classification task is likely to have more number of occurrences than others. Tables 4–6 report the 15 most frequently selected genes along with

Table 4. Most frequently selected genes for leukemia.

| Acc. No. | Gene description | No. |
|---|---|---|
| M34344_at | ITGA2B Integrin, alpha 2b (platelet glycoprotein IIb of IIb/IIIa complex, antigen CD41B) | 70 |
| X77737_at | Red cell anion exchanger (EPB3, AE1, Band 3) 3 non-coding region | 61 |
| X68560_at | SP3 Sp3 transcription factor | 61 |
| X74008_at | PPP1CC Protein phosphatase 1, catalytic subunit, gamma isoform | 59 |
| M34539_at | FKBP1 FK506-binding protein 1 (12kD) | 57 |
| L19437_at | TALDO Transaldolase | 55 |
| D61391_at | Phosphoribosypyrophosphate synthetase-associated protein 39 | 50 |
| M16276_at | HLA-DQB1 Major histocompatibility complex, class II, DQ beta 1 | 49 |
| X76104_at | GB DEF = DAP-kinase mRNA | 42 |
| U34877_at | Biliverdin-IXalpha reductase mRNA | 42 |
| D86969_at | KIAA0215 gene | 41 |
| U07695_at | HTK Hepatoma transmembrane kinase | 40 |
| S65738_at | Actin depolymerizing factor [human, fetal brain, mRNA, 1452 nt] | 40 |
| U28014_at | ICH-2 PROTEASE PRECURSOR | 39 |
| U88047_at | DNA binding protein homolog (DRIL) mRNA, partial cds | 39 |

Table 5. Most frequently selected genes for colon cancer.

| Acc. No. | Gene description | No. |
| --- | --- | --- |
| H08393 | COLLAGEN ALPHA 2(XI) CHAIN (Homo sapiens) | 81 |
| H16096 | MITOCHONDRIAL PROCESSING PROTEASE BETA SUBUNIT PRECURSOR (Rattus norvegicus) | 81 |
| T47383 | ALKALINE PHOSPHATASE, PLACENTAL TYPE 1 PRECURSOR (Homo sapiens) | 81 |
| K03474 | Human Mullerian inhibiting substance gene, complete cds | 80 |
| R09138 | DIHYDROPTERIDINE REDUCTASE (Homo sapiens) | 69 |
| X57351 | INTERFERON-INDUCIBLE PROTEIN 1-8D (HUMAN); contains MSR1 repetitive element | 63 |
| X14830 | Human mRNA for muscle acetylcholine receptor beta-subunit | 62 |
| H69872 | PROTEIN KINASE C, DELTA TYPE (Homo sapiens) | 57 |
| U00968 | STEROL REGULATORY ELEMENT BINDING PROTEIN 1 (HUMAN) | 55 |
| R74208 | GENERAL NEGATIVE REGULATOR OF TRANSCRIPTION SUB-UNIT 4 (Saccharomyces cerevisiae) | 53 |
| M31776 | Human brain natriuretic protein (BNP) gene, complete cds | 50 |
| R10066 | PROHIBITIN (Homo sapiens) | 48 |
| T57882 | MYOSIN HEAVY CHAIN, NONMUSCLE TYPE A (Homo sapiens) | 48 |
| R62945 | COMPLEMENT DECAY-ACCELERATING FACTOR 1 PRECURSOR (Homo sapiens) | 47 |
| R92729 | X BOX BINDING PROTEIN-1 (Homo sapiens) | 46 |

their number of occurrences for the three datasets. It can be seen that some genes are selected in the best gene subsets more frequently than others.

In the case of leukemia dataset, Table 6 shows that the most frequently selected gene was a part of the best gene subsets for 70 times out of 81, while others were selected considerably less frequently. Thus, it can be said that this gene is highly important for the purpose of classifying between AML and ALL. This analysis might reveal biological relevance of genes for particular type of diseases, and further help in biomarker discovery. The case of Colon Cancer as given in Table 5 reveals that the top three genes were selected in all the 81 best gene subsets of varying size. The gene H08393 COLLAGEN ALPHA 2(XI) CHAIN (Homo sapiens) has also been selected in the top scoring gene subset in Ref. 8. The results show that, number of occurrences of the top four genes is considerably higher than the rest of the genes. Hence, it is likely that these four genes can more effectively

Table 6. Most frequently selected genes for brain cancer.

| Acc. No. | Gene description | No. |
|---|---|---|
| 891_at | Homo sapiens GLI-Krupple related protein (YY1) mRNA, complete cds | 72 |
| 41193_at | Homo sapiens mRNA for DUSP6, complete cds | 66 |
| 31807_at | Homo sapiens cDNA | 62 |
| 31671_at | Human retropseudogene MSSP-1 DNA, complete cds | 59 |
| 37705_at | Human (ard-1) mRNA, complete cds | 57 |
| 34873_at | Homo sapiens mRNA for nebulette | 57 |
| 38630_at | Homo sapiens mRNA; cDNA DKFZp434B102 (from clone DKFZp434B102) | 56 |
| 31511_at | Human ribosomal protein S9 mRNA, complete cds | 55 |
| 34593_g_at | Human ribosomal protein S17 mRNA, complete cds | 50 |
| 35786_at | Homo sapiens mRNA for KIAA0476 protein, complete cds | 48 |
| 33883_at | Homo sapiens mRNA for Efs1, complete cds | 46 |
| 33862_at | Homo sapiens phosphatidic acid phosphohydrolase homolog (Dri42) mRNA, complete cds | 44 |
| 424_s_at | Homo sapiens N-sam mRNA for fibroblast growth factor receptor | 43 |
| 41223_at | Homo sapiens nuclear-encoded mitochondrial cytochrome c oxidase Va subunit mRNA, complete cds | 40 |
| 36562_at | Homo sapiens KIAA0427 mRNA, complete cds | 40 |

differentiate between the normal and tumour cells than the others. A similar analysis of Brain Cancer dataset is presented in Table 6.

## 8. Conclusions

Feature selection for the purpose of cancer classification improves the classification performance to great extent, while simultaneously reducing the number of genes by a great extent. In this work, we presented a novel approach to cancer classification, employing gene selection through Ant Colony Optimization and Support Vector Machines. The method is essentially a multivariate wrapper approach, which has been used to select a gene subset from the complete set of features, on the basis of its classification ability. The approach is evaluated on three widely used datasets, Leukemia,

Colon Cancer and Brain Cancer. The results show that the method can obtain very high classification accuracy for all the three the datasets. The results are comparable to the existing results in literature.

## Acknowledgments

The financial assistance received from the Department of Biotechnology (DBT), the Government of India, New Delhi is gratefully acknowledged.

## References

1. T. R. Golub, D. K. Slonim, P. Tamayo, C. H. M. Gaasenbeek, J. P. Mesirov, H. Coller, M. L. Loh, J. R. Downing, M. A. Caligiuri, C. D. Bloomfield and E. S. Lander, Molecular classification of cancer: class discovery and class prediction by gene expression monitoring, *Science* **286**, 531–537 (1999).
2. U. Alon, N. Barkai, D. A. Notterman, K. Gish, S. Ybarra, D. Mack and A. J. Levine, Broad patterns of gene expression revealed by clustering of tumor and normal colon tissues probed by oligonucleotide arrays, *Proc. Natl. Acad. Sci. USA* **96**, 6745–6750 (1999).
3. C. L. Nutt, D. R. Mani, R. A. Betensky, P. Tamayo, J. G. Cairncross, C. Ladd, U. Pohl, C. Hartmann, M. E. McLaughlin, T. T. Batchelor, P. M. Black, A. von Deimling, S. L. Pomeroy, T. R. Golub and D. N. Louis, Gene expression-based classification of malignant gliomas correlates better with survival than histological classification, *Cancer Res.* **63** 1602–1607 (2003).
4. A. A. Alizadeh, M. B. Eisen *et al.*, Distinct types of diffuse large B cell lymphoma identified by gene expression profiling, *Nature* **403**, 503–511 (2000).
5. T. S. Furey, N. Cristianini, N. Duffy, D. W. Bednarski, M. Schummer and D. Haussler, Support vector machine classification and validation of cancer tissue samples using microarray expression data, *Bioinformatics* **16**(10), 906–914 (2000).
6. X. Zhang, X. Lu, Q. Shi, X. Q. Xu, H. C. Leung, L. N. Harris, J. D. Iglehart, A. Miron, J. S. Liu and W. H. Wong, Recursive SVM feature selection and sample classification for mass-spectrometry and microarray data, *BMC Bioinformatics* **7**, 197 (2006).
7. Z. Liu, D. Chen and H. Bensmail, Gene expression data classification with kernel principal component analysis, *Journal of Biomedicine and Biotechnology*, pp. 155–159 (2005).
8. I. Inza, P. Larranaga, R. Blanco and A. J. Cerrolaza, Filter versus wrapper gene selection approaches in DNA microarray domains, *Artificial Intelligence in Medicine, Data Mining in Genomics and Proteomics* **31**(2), 91–103 (2004).
9. X. Liu, A. Krishnan and A. Mondry, An entropy-based gene selection method for cancer classification using microarray data, *BMC Bioinformatics* **6**, 76 (2005).
10. E. K. Tang, P. N. Suganthan and X. Yao, Gene selection algorithms for microarray data based on least squares support vector machine, *BMC Bioinformatics* **7**, 95 (2006).

11. E. B. Huerta, B. Duval and J. Hao, A hybrid GA/SVM approach for gene selection and classification of microarray data. *EvoWorkshops 2006*, pp. 34–44 (2006).
12. T. Jirapech-Umpai and S. Aitken, Feature selection and classification for microarray data analysis: evolutionary methods for identifying predictive genes, *BMC Bioinformatics* **6**, 148 (2005).
13. L. Li, C. R. Weinberg, T. A. Darden and L. G. Pedersen, Gene selection for sample classification based on gene expression data: study of sensitivity to choice of parameters of the GA/KNN method, *Bioinformatics* **17**, 1131–1142 (2001).
14. X.-W. Chen, Gene selection for cancer classification using bootstrapped genetic algorithms and support vector machines, *Proceedings of the Computational Systems Bioinformatics (CSB'03)* (2003).
15. S. Peng, Q. Xu, X. B. Ling, X. Peng, W. Du and L. Chen, Molecular classification of cancer types from microarray data using the combination of genetic algorithms and support vector machines, *FEBS Letters* **555**(2), 358–362 (2003).
16. T. K. Paul and H. Iba, Gene selection for classification of cancers using probabilistic model building genetic algorithm, *Biosystems* **82**(3), 208–225 (2005).
17. A. R. Reddy and K. Deb, *Classification of Two-Class Cancer Data Reliably Using Evolutionary Algorithms*. Technical Report, KanGAL (2003).
18. K. Yang, Z. Cai, J. Li and G. Lin, A stable gene selection in microarray data analysis, *BMC Bioinformatics* **27**(1), 228 (2006).
19. A. Antoniadis, S. Lambert-Lacroix and F. Leblanc, Effective dimension reduction methods for tumor classification using gene expression data, *Bioinformatics* **19**(5), 563–570 (2003).
20. M. Dorigo, G. D. Caro and L. M. Gambardella, Ant algorithm for discrete optimization, *Artificial Life* **5**(2), 137–172, 767 (1999).
21. R. K. Sivagaminathan and S. Ramakrishnan, A hybrid approach for feature subset selection using neural networks and ant colony optimization, *Expert Systems with Applications*, Uncorrected Proof, Available online 4 May 2006 (in press).
22. A. Ahmed, Ant colony optimization for feature subset selection, *International Journal of Computational Intelligence* **2**(1), 53–58 (2005).
23. M. Dorigo and L. M. Gambardella, Ant colony system: a cooperative learning approach to the traveling salesman problem, *IEEE Transaction on Evolutionary Computation* **1**(1), 53–66 (1997).
24. L. M. Gambardilla and M. Dorigo, HAS-SOP: an hybrid ant system for the sequential ordering problem, *Technical Report*, Lugano, Switzerland IDSIA, pp. 11–97 (1997).
25. V. K. Jayaraman, B. D. Kulkarni, S. Karale and P. Shelokar, Ant colony framework for optimal design and scheduling of plants, *Computers and Chemical Engineering* **24**(8), 1901–1912 (2000).
26. V. K. Jayaraman, B. D. Kulkarni, K. Gupta, J. Rajesh and H. S. Kusumaker, Dynamic optimization of fed-batch reactors using the ant algorithm, *Biotechnology Progress* **17**, 81–88 (2001).
27. V. S. Summanwar, P. S. Shelokar, V. K. Jayaraman and B. D. Kulkarni, Ant colony framework for process optimization: unconstrained and constrained problems with single and multiple objectives, R. Luus (ed), *Recent Developments in Optimization and Optimal Control in Chemical Engineering*, Research Signpost, Trivandrum, India, pp. 67–87.

28. V. Vapnik, The nature of statistical learning theory, Springer-Verlag, Berlin, Germany (1995).
29. A. J. Kulkarni, V. K. Jayaraman and B. D. Kulkarni, Knowledge incorporated support vector machines to detect faults in Tennessee Eastman Process, *Computers & Chemical Engineering* **29**(10), 2128–2133 (2005).
30. S. Idicula-Thomas, A. J. Kulkarni, B. D. Kulkarni, V. K. Jayaraman and P. V. Balaji, A support vector machine-based method for predicting the propensity of a protein to be soluble or to form inclusion body on over-expression in Escherichia coli, *Bioinformatics* **22**, 278–284 (2006).
31. C. Chang and C. Lin, LIBSVM: a library for support vector machines. Software available at http://www.csie.ntu.edu.tw/cjlin/libsvm (2001).
32. K. Deb and A. R. Reddy, Reliable classification of two-class cancer data using evolutionary algorithms, *Biosystems, Computational Intelligence in Bioinformatics* **72**(1,2), 111–129 (2003).

# CHAPTER 12

# SOPHISTICATED METHODS FOR CANCER CLASSIFICATION USING MICROARRAY DATA

Sung-Bae Cho* and Han-Saem Park[†]

*Department of Computer Science, Yonsei University
134 Shinchon-dong, Sudaemoon-ku, Seoul 120-749, Korea
\*sbcho@cs.yonsei.ac.kr,
†sammy@sclab.yonsei.ac.kr*

The development of microarray technology has supplied a large amount of data to many fields. In particular, it has helped to predict and diagnose cancer. Many machine learning techniques have been developed and applied to produce informative results. These classification methods have some practical limitations because microarray data can be noisy and incomplete, and classification algorithm itself cannot be perfect. This chapter presents three sophisticated methods to solve these problems, using the ensemble approach in common. By using multiple features of data and combining the results of multiple classifiers, more accurate prediction can be obtained, and experiments with lymphoma and colon cancer datasets have shown the usefulness of the presented methods.

## 1. Introduction

The development of microarray technology has supplied a great amount of data to many fields including bioinformatics.[1] In particular, it has helped to predict and diagnose cancer.[2] Many machine learning techniques have been developed and applied to produce informative results since it is important for cancer diagnosis and treatment to classify tumors precisely.[3] However, these conventional classification methods have a few practical limitations. Classification process is usually divided into feature selection and classification parts, each of which has been dealt as an important research issue. Since microarray data can be noisy and incomplete, selected features with feature selection methods can be incomplete. Besides the classifications algorithms cannot be perfect.[4]

One of the alternatives in overcoming these limitations is the classifier ensemble approach. Classifier ensemble is a learning strategy to produce a new classification result through the collection of a finite number of single classifiers that are trained for the same task.[5] Using the classifier ensemble makes it possible to obtain high and stable generalization performance. It is demonstrated theoretically and empirically that the ensemble classifier yields higher performance than a good single classifier,[6,7] leading many research groups to have studied the classifier ensemble methods. Tsymbal et al. used several ensemble methods for integration of simple Bayesian classifiers.[8] Opitz and Maclin compared Bagging to Boosting systematically using decision tree and neural network classifiers as a single classifier.[9] Sboner et al. utilized modified voting schema to diagnose melanoma.[10] Cho et al. have used ensemble classifier of several feature selection and classification methods to classify cancer datasets.[3,4]

In this chapter, we present three sophisticated classification methods based on the ensemble approach. The first method uses negatively correlated gene subsets and combines their results with Bayesian approach. The second one uses combinatorial ensemble approach based on elementary single classifiers, and the last one searches the optimal pair of feature-classifier ensemble with genetic algorithm. Finally, experimental results with colon and lymphoma cancer datasets are demonstrated.

## 2. Backgrounds

### 2.1. *DNA microarray*

#### 2.1.1. *DNA microarray*

DNA microarrays consist of thousands of individual DNA sequences printed in a high-density array on a glass microscope slide using a robotic arrayer. The relative abundance of these spotted DNA sequences in two DNA or RNA samples may be assessed by monitoring the differential hybridization of the two samples to the sequences on the array. For mRNA samples, the two samples are reverse-transcribed into cDNA, labeled using different fluorescent dyes mixed (red-fluorescent dye Cy5 and green-fluorescent dye Cy3). After the hybridization of these samples with the arrayed DNA probes, the slides are imaged using a scanner that makes fluorescence measurements for each dye. Equation (1) shows the log ratio between the two intensities of each dye, which is used as the gene expression data.[11–13]

$$gene\_expression = \log_2 \frac{\text{Int(Cy5)}}{\text{Int(Cy3)}}, \qquad (1)$$

where Int(Cy5) and Int(Cy3) are the intensities of red and green colors, respectively. Since more than hundreds of genes are put on the DNA microarray, we can investigate the genome-wide information in short time.

### 2.1.2. Oligonucleotide microarray

Affymetrix (Inc, Santa Clara, CA) has developed the GeneChip oligonucleotide array. This high-density oligonucleotide DNA probe array technology employs photolithography and solid-phase DNA synthesis techniques.

High-density oligonucleotide chip arrays are made using spatially patterned, light-directed combinatorial chemical synthesis, and they contain up to hundreds of thousands of different oligonucleotides on a small glass surface. Synthetic linkers, modified with a photochemically removable protecting groups, are attached to a glass surface, and light is directed through a photolithographic mask to specific areas on the surface to produce localized deprotection. Specific hydroxyl-protected deoxynucleotides are incubated with the surface, and chemical coupling occurs at those sites that have been illuminated in the preceding step. As the chemical cycle is repeated, each spot on the array contains a short synthetic oligonucleotide, typically 20–25 bases long. The oligonucleotides are designed based on the knowledge of the DNA target sequences, to ensure high-affinity and specificity of each oligonucleotide to a particular gene. This allows cross-hybridization with the other similar sequenced gene and local background to be estimated and subtracted.[14,15]

### 2.2. Feature selection methods

Among thousands of genes whose expression levels are measured, not all of them are needed for classification. Microarray data consist of large number of genes in small samples, so we need to select the informative genes for classification. This process is called gene selection or feature selection in machine learning.[16] There have been various studies that select the informative features for the classification.[16,17] In this chapter, eight feature selection methods explained below have used.

Suppose that we have a gene expression pattern $g_i$ ($i = 1\sim2000$ in colon data, and $i = 1\sim4026$ in lymphoma data). Each $g_i$ is a vector of gene expression levels from $N$ samples, $g_i = (e_1, e_2, \ldots, e_N)$. The first $M$ elements $(e_1, e_2, \ldots, e_M)$ are examples of tumor samples, and the other $N - M(e_{M+1}, e_{M+2}, \ldots, e_N)$ are those from normal samples. An ideal gene pattern that relates to tumor class is defined by $g_{ideal\_tumor} = (1, \ldots, 1, 0, \ldots, 0)$, so that all elements from tumor samples are 1 and the others are 0.

The similarity between $g_{ideal\_tumor}$ and $g_i$ tells us how much likely the $g_i$ is to the tumor class. This ideal gene vector is used for some of feature selection methods based on correlation measures (Pearson correlation coefficients (PC) and Spearman correlation coefficients (SC)) and similarity measures (Euclidean distance (ED) and cosine coefficients (CC)). For the information theoretic methods, the frequencies of gene expression sign (+ or −) that belongs specific class (information gain (IG) and mutual information (MI)) or the means and standard deviations of genes belonging to each class (signal to noise ration (SN)) are used to select informative genes. Table 1 provides the equations of these methods. Inclusive of seven methods explained above and principal component analysis (PCA), widely used statistical data analysis technique,[18] a total of eight feature selection methods have been utilized. The decision of the optimal number of genes is another problem, and we have 25 as an optimal number according to the previous study.[4]

## 2.3. Base classifiers

Classification can be defined as the process of approximating I/O mapping from the given observation to the optimal solution. Several methods have been suggested and applied for more accurate classification. In this chapter, four representative classification methods, multi-layer perceptron,[16,19] k-nearest neighbour (KNN),[20] support vector machine (SVM),[21,22] and structure-adaptive self-organization map,[23] have been used. Here, two different similarity measures (cosine coefficients and Pearson correlation coefficients) have been used for KNN, and two different kernel functions (linear and RBF) have been used for SVM, so a total 6 classifiers have been used as base classifiers.

Table 1. Equations of the feature selection methods.

$$PC(g_i, g_{ideal}) = \frac{\sum g_i g_{ideal} - \frac{\sum g_i \sum g_{ideal}}{N}}{\sqrt{\left(\sum g_i^2 - \frac{(\sum g_i)^2}{N}\right)\left(\sum g_{ideal}^2 - \frac{(\sum g_{ideal})^2}{N}\right)}}, \quad (2)$$

$$SC(g_i, g_{ideal}) = 1 - \frac{6\sum(D_g - D_{ideal})^2}{N(N^2 - 1)}, \quad (3)$$

($D_g$ and $D_{ideal}$ are the rank matrices of $g_i$ and $g_{ideal}$)

$$ED(g_i, g_{ideal}) = \sqrt{\sum(g_i - g_{ideal})^2}, \quad (4)$$

$$CC(g_i, g_{ideal}) = \frac{\sum g_i g_{ideal}}{\sqrt{\sum g_i^2 \sum g_{ideal}^2}}, \quad (5)$$

$$IG(g_i|c_j) = P(g_i|c_j) \log \frac{P(g_i|c_j)}{P(c_j) \cdot P(g_i)} + P(\bar{g}_i|c_j) \log \frac{P(\bar{g}_i|c_j)}{P(c_j) \cdot P(\bar{g}_i)}, \quad (6)$$

$$MI(g_i, c_j) = \log \frac{P(g_i, c_j)}{P(g_i) \cdot P(c_j)}, \quad (7)$$

$$SN(g_i, c_j) = \frac{\mu_{c_j}(g_i) - \mu_{\bar{c}_j}(g_i)}{\sigma_{c_j}(g_i) + \sigma_{\bar{c}_j}(g_i)}. \quad (8)$$

## 2.4. Classifier ensemble methods

In this chapter, three classifier combination methods have been used to combine the results of base classifiers.

- Majority voting (MV): This is a simple ensemble method that selects the class favored by the most base classifiers. Majority voting has some advantages that it does not require any previous knowledge as well as any additional complex computation to decide. When $c_i$ is the class $i$ ($I = 1, \ldots, m$), and $s_i(classifier_j)$ is 1 if the output of the $j$th classifier $classifier_j$ equals to the class $i$ otherwise 0, majority voting is defined as follows:

$$c_{ensemble} = \arg\max_{1 \leq i \leq m} \left\{ \sum_{j=1}^{k} s_i(classifier_j) \right\}. \quad (9)$$

- Weighted voting (WV): Poor classifier can affect the result of the ensemble in majority voting because it gives the same weight to all classifiers. Weighted voting reduces the effect of poor classifier by giving a different weight to a classifier based on the performance of each classifier.

When $w_j$ is the weight of the $j$th classifier, weighted voting is defined as follows:

$$c_{ensemble} = \arg\max_{1 \leq i \leq m} \left\{ \sum_{j=1}^{k} w_j s_i(classifier_j) \right\}. \tag{10}$$

- Bayesian combination (BC): While majority voting method combines classifiers with their results, Bayesian combination makes the error possibility of each classifier affect the final result. The method combines classifiers with different weights by using the previous knowledge of each classifier. When $k$ classifiers are combined, $c(classifier_j)$ is the class of the $j$th classifier, and $w_i$ is *a priori* possibility of the class $c_i$, Bayesian combination is defined as follows:

$$c_{ensemble} = \arg\max_{1 \leq i \leq m} \left\{ w_i \prod_{j=1}^{k} P(c_i | c(classifier_j)) \right\}. \tag{11}$$

## 3. Sophisticated Methods for Cancer Classification

This section describes the sophisticated methods for cancer classification. Figure 1 illustrates an overview of the three ensemble methods.

### 3.1. *Ensemble with negatively correlated features*

We define two ideal feature vectors as the one high in class $A$ and low in class $B$ (1, 1, ..., 1, 0, 0, ..., 0), and the other one low in class $A$ and high in class $B$ (0, 0, ..., 0, 1, 1, ..., 1) as shown in Fig. 2 and we select the sets of informative genes with high similarity to each ideal gene vector. The features are selected from DNA microarray data using two different ideal feature vectors, ideal feature A and ideal feature B. The ideal feature is a good arbitrator to distinguish between Classes A and B. SGS I (significant gene subset I) is defined as the gene set selected on the basis of ideal Gene A and SGS II is defined as the gene set selected on the basis of ideal Gene B. SGS is defined as follows:

$$SGS = \arg\max \left\{ Sim(gene_i, Ideal\ gene\ vector) \right\}, \tag{12}$$

where $Sim(X, Y)$ is the similarity of vectors X and Y, which can be replaced by the feature selection method shown in Table 1.

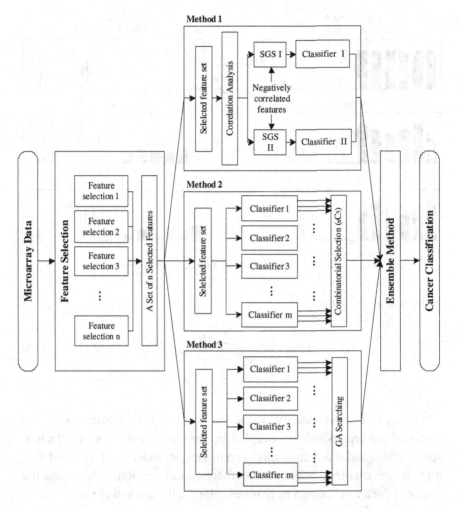

**Fig. 1.** Sophisticated methods for cancer classification — an overview.

Theoretically, the more features we concern, the more effectively the classifier can solve the problems.[3] However, features overlapped in feature space may cause the redundancy of irrelevant information and result in counter effects such as overfitting. When there are $N$ feature selection methods, the set of nonlinear transformation functions that change observation space into feature space is $\phi = \{\varphi_1, \varphi_2, \varphi_3, \ldots, \varphi_N\}$ and $\phi_k \in 2^\phi$, $I(\phi_k)$, the amount of classification information provided by the set of features $\phi_k$,

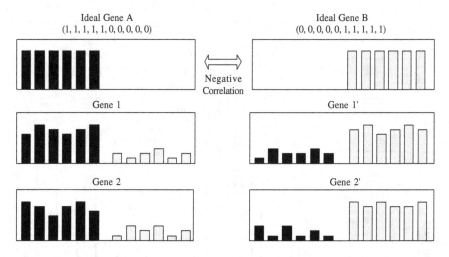

**Fig. 2.** Informative genes selected using negative correlation.

is defined as follows:

$$I(\phi_k) = \frac{a \sum A_i}{\frac{N}{2} \sum_{j=1, j \neq i}^{N} d_{ij}} + b, \tag{13}$$

where $d_{ij}$ is the dependency of the $i$th and the $j$th elements, $A_i$ is the extent of the area which is occupied by the $i$th element from the feature space, and $a$ and $b$ are constants. The higher dependency of a pair of features is, the smaller amount of classification information $I(\phi_k)$ is. As the extent of the area occupied by features is larger, the amount of classification information $I(\phi_k)$ is bigger. If we keep the number of features larger, the numerator of the equation is larger because the extent of the area occupied by features becomes wider. Though the numerator of the equation becomes larger, $I(\phi_k)$ will be mainly decreased without keeping $d_{ij}$ small. Therefore, it is more desirable to use small number of mutually independent features than to unconditionally increase the number of features to enlarge $I(\phi_k)$, the amount of classification information provided to the classifier by the set of features. Correlation between feature sets can be induced from the distribution of feature numbers, or using mathematical analysis based on statistics.

### 3.2. Combinatorial ensemble

Figure 3 shows the examples of good ensemble and bad ensemble. The classifiers of good ensemble have complementary relationship (see Fig. 3(a)), so they generate errors in mutually exclusive way. As a result, the ensemble outperforms the best single classifier. On the contrary, classifiers of bad ensemble, more than half of them generate errors in common space (see Fig. 3(b)), so an error region is enlarged. This can be applied with more classifiers or different combining methods, thereby it is important to find the ensemble which is close to the example shown in Fig. 3(a).

Figure 3 also explains that classifiers should complement one another, but it is difficult to discover the complementary set of classifiers. We try to find the combinatorial ensemble with promising number of single classifiers. From the ensemble of three classifiers, increasing the number to five, seven and more, it is possible to try all the available ensembles of classifiers. However, the time cost gets higher as the possibility to find the optimal ensemble increases. The number of $_nC_k$ is $n!/(n-k)!k!$, and it increases according to $k$ until $k$ reaches a half of $n$. Thus, it takes more time whenever we increase $k$ until $n/2$. Besides, single classifiers, which compose an

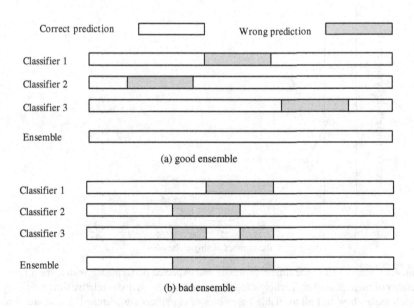

**Fig. 3.** The examples of good and bad ensembles.

ensemble classifier, do not complement efficiently if $k$ is large since the correlation also grows according to $k$. Therefore, considering time cost and performance, the value $k$ should be neither too small nor too large, and the ensemble of five classifiers might be enough to compose a good ensemble.

Three is too small to guarantee an optimal ensemble, and it takes too much time if it is larger than seven. If we use 42 single classifiers, the number of ensembles with seven classifiers is 30 times larger than the one with five classifiers. Figure 4 illustrates the time cost and expected performance in case of using 42 classifiers. The relative time efficiency means the goodness about time cost, i.e. it is better if the value is higher. The expected performance shows the highest score when $k$ is between 5 and 10, and it decreases after $k$ is larger than that because of high correlation.

Here, we introduce a combinatorial ensemble method with multiple feature-classifiers to identify cancer classes. The basic idea of the ensemble classifier scheme is to develop several classifiers trained with multiple features which are selected by various feature selection methods and then

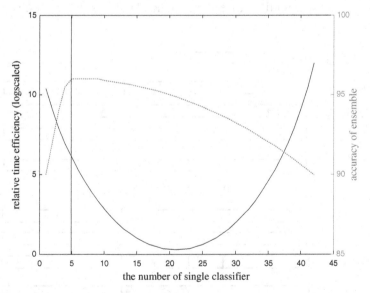

**Fig. 4.** Comparison of the time complexity and expected performance with respect to the number of single classifiers participating at an ensemble. Solid line is relative time efficiency which means how fast all available ensembles of classifiers are searched. The dotted line represents the expected performance by ensemble.

classify a given input pattern by utilizing a combination method. Given $i$ features and $j$ classifiers, there are $i \times j$ feature-classifier combinations. There are $_nC_5$ possible ensemble classifiers when $i \times j$ is $n$ and 5 feature-classifier combinations are selected. In this case, we can find the optimal combination of feature-classifier pair after $n(n-1)(n-2)(n-3)(n-4)/(5 \times 4 \times 3 \times 2 \times 1)(= nC_5)$ trials.

### 3.3. Searching optimal ensemble with GA

We have 48 feature-classifiers from 8 feature selection methods and 6 classifiers that yield $2^{48}$ (about $2.5 \times 10^{11}$) combinations of the different ensemble classifiers. The optimal ensemble can be searched by enumerating and comparing all the combinations, but it requires so much cost that we cannot try all the possible ensembles. Cost is a little reduced if we use the constraints such as the number of classifiers provided in Sec. 3.2, but it is still too high. Hence, an efficient method to find the optimal ensemble is needed and we can exploit the genetic algorithm (GA) to work out this problem.

GA is a stochastic search method that has been successfully applied in many search, optimization, and machine learning problems.[26] GA maintains a population of encoded candidate solutions that are competitively manipulated by applying some variation operators to find a global optimum. A population comprises of many chromosomes that correspond to candidate solutions, which represent specific status or values.

GA proceeds in an iterative way by generating new populations of strings from old ones. Every string is the encoded version of a candidate solution. An evaluation function assigns fitness measure to every string, and it indicates the fitness to the problem. The standard GA applies genetic operators such as selection, crossover, and mutation to an initially random population in order to compute the next population of new strings.

GA can be applied effectively to solve combinatorial optimization problems. Figure 5 illustrates the structure of chromosome and what it represents. Each chromosome is composed of 48 bits string, each of which indicates whether the corresponding feature-classifier pair is joined to the ensemble or not. Each bit corresponds to a specific feature-classifier pair, say the first bit to CC-MLP, the second bit to ED-MLP, and so on. Figure 5 also provides an ensemble of the second (ED-MLP), third (IG-MLP), and sixth (PCA-MLP) feature-classifier pairs. If new feature-classifier pairs are added, the

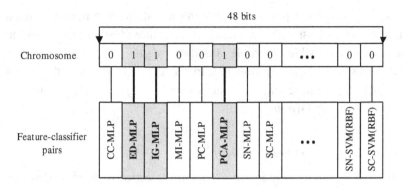

Fig. 5. Structure of one chromosome and its phenotype.

same process is repeated after changing the chromosome structure with the number of individual feature-classifier pairs.

The fitness is evaluated by accuracy of the ensemble on the validation dataset.

$$fitness = \frac{\text{\# of correctly classified samples}}{\text{\# of total validation samples}}. \qquad (14)$$

Figure 6 summarizes the GA procedure. First, the chromosomes are selected randomly as many as the population size. Each of them is evaluated with fitness function by the Eq. (14). Genetic operations are performed with chromosomes. Good chromosomes have higher chance as being selected than poor ones. We have used two different strategies for selection: roulette wheel and rank-based. In a roulette wheel selection, each chromosome has a probability to be selected in proportion to its fitness. In a rank-based selection, it has a probability to be selected according to its order of fitness. After selecting pairs of chromosomes, the information of chromosome for each part is crossed over. Some bits are mutated according to the mutation rate. Through the genetic operations, the chromosomes evolve to the optimal ensembles. Chromosomes are evaluated once more, and the selection, genetic operations and fitness evaluation steps are repeated until the solution is satisfied with the given condition.

We have utilized majority voting and weighted voting methods described in Sec. 2.4 to combine feature-classifier pairs. For weighted voting, the accuracy of the corresponding single feature-classifier pair for validation dataset is used as weight.

| GA procedure |
|---|
| Input:<br>    Basic feature selection methods and classifiers, *FeatureSelection* and *Classifiers*.<br>Output:<br>    Optimal ensemble of one feature selection method and one classifier,*Optimal Ensemble*.<br>Function Description:<br>    **InitPopulation** (*pSize, FeatureSelection, Classifiers*): A function that makes initial chromosomes of size '*pSize*' randomly.<br>    **EvaluateFitness** (*Chromosomes*): A function that evaluates the fitness of chromosomes in terms of classification accuracy.<br>    **Select** (*Chromosomes*): A function that selects superior chromosomes based on the fitnesses.<br>    **CrossoverNMutate** (*Chromosomes*): A function that performs GA operations such as crossover and mutation. |
| {<br>    *Chromosomes* = **InitPopulation** (*pSize, FeatureSelection, Classifiers*);<br>    **EvaluateFitness** (*Chromosomes*);<br>    **while** (*OptimalSolutionIsFound*) {<br>        **Select** (*Chromosomes*);<br>        **CrossoverNMutate** (*Chromosomes*);<br>        **EvaluateFitness** (*Chromosomes*);<br>    }<br>    **return** *OptimalEnsemble*;<br>} |

Fig. 6. The GA procedure.

## 4. Experiments

### 4.1. Datasets

For experiments, colon and lymphoma cancer datasets have been used.

- Colon cancer dataset: Colon dataset consists of 62 samples of colon epithelial cells taken from colon-cancer patients. Each sample contains 2000 gene expression levels. Even though an original data consist of 6000 gene expression levels, 4000 out of 6000 were removed based on the confidence in the measured expression levels. 40 of 62 samples are colon cancer samples and the remaining are normal samples. Each sample was taken from tumors and normal healthy parts of the colons of the same patients and measured using high density oligonucleotide arrays.[24] 31 out of 62 samples have been used as training data, and the remaining 31 samples have been used as test data (available at: http://www.sph.uth.tmc.edu:8052/hgc/default.asp).

- Lymphoma cancer dataset: B cell diffuse large cell lymphoma (B-DLCL) is a heterogeneous group of tumors, based on significant variations in morphology, clinical presentation, and response to treatment. Gene expression profiling has revealed two distinct tumor subtypes of B-DLCL: germinal center B cell-like DLCL and activated B cell-like DLCL.[27] Lymphoma dataset consists of 24 samples of GC B-like and 23 samples of activated B-like. 22 out of 47 samples have been used as training data and the remaining have been used as test data (available at: http://genome-www.stanford.edu/lymphoma).

### 4.2. Ensemble with negative correlated features

Four feature selection methods of PC, SC, ED and CC have been used, and multi-layer perceptron has been used as a classifier. As explained in Sec. 3.1, MLP I is defined as the result of MLP trained by the SGS I and MLP II is defined as the result of MLP trained by the SGS II. Bayesian combination has been used for an ensemble method.

Table 2 shows the recognition rate of the base classifiers in each dataset. In Colon dataset, MLP I with CC produces the best recognition rate of 83.9%. In lymphoma dataset, MLP II with SC produces the best recognition rate of 88.0%. MLP II outperforms MLP I in both datasets.

The best recognition rate of ensemble classifier is 87.1% in colon dataset and 92.0% in lymphoma dataset. The performance of the ensemble classifier is superior to the base classifiers in all benchmark datasets. Compared with

Table 2. Recognition rate with features and classifiers (%).

|     | Colon | | Lymphoma | |
| --- | --- | --- | --- | --- |
|     | MLP I | MLP II | MLP I | MLP II |
| PC  | 74.2 | 77.4 | 64.0 | 72.0 |
| SC  | 58.1 | 64.5 | 60.0 | 88.0 |
| ED  | 67.8 | 77.4 | 56.0 | 72.0 |
| CC  | 83.9 | 77.4 | 68.0 | 76.0 |
| Avg. | 71.0 | 74.2 | 62.0 | 77.0 |

the best recognition rates of base classifiers, the performance of ensemble is superior.

Compared the results of MLP I with the ones of MLP II, the negatively correlated features set (SGS I + SGS II) does not outperform in the average recognition rate, but outperforms in the best recognition rate. While the best recognition rate of the ensemble of MLP I and MLP II decrease as the number of combined classifiers increases, the best recognition rate of the ensemble of the negatively correlated coefficient feature set increases.

Table 3 shows the confusion matrix for the best ensemble of 4 classifiers. The result of negatively correlated feature set (SGS I + SGS II) is the best in both datasets. In colon dataset, the *false positive rate*, the proportion of negatives that were incorrectly classified as positive, is reduced in negatively correlated feature set compared with the result of MLP I or MLP II. In lymphoma dataset, the *false negative rate*, the proportion of positives that were incorrectly classified as negative, is reduced much in negatively correlated feature set compared with the result of MLP I or MLP II.

These experiments show that the ensemble classifier works and we can improve the classification performance by combining independent sets of classifiers learned from negatively correlated features.

Table 3. Confusion matrix for the best ensemble of 4 classifiers.

Colon

| | | MLP I + II | | | MLP I | | | MLP II | |
|---|---|---|---|---|---|---|---|---|---|
| | | Predicted | | | Predicted | | | Predicted | |
| | | 0 | 1 | | 0 | 1 | | 0 | 1 |
| Actual | 0 | 8 | 3 | 0 | 6 | 5 | 0 | 5 | 6 |
| | 1 | 1 | 19 | 1 | 1 | 19 | 1 | 1 | 19 |

Lymphoma

| | | MLP I + II | | | MLP I | | | MLP II | |
|---|---|---|---|---|---|---|---|---|---|
| | | Predicted | | | Predicted | | | Predicted | |
| | | 0 | 1 | | 0 | 1 | | 0 | 1 |
| Actual | 0 | 13 | 1 | 0 | 12 | 2 | 0 | 11 | 3 |
| | 1 | 1 | 10 | 1 | 7 | 4 | 1 | 3 | 8 |

## 4.3. Combinatorial ensemble

All six classifiers explained in Sec. 2.3, seven feature selection methods except PCA explained in Sec. 2.2, and all combination methods have been used for experiments. We have conducted all $_{42}C_m$ ($m = 3, 5, 7,$ and $42$) combinations of ensemble, and have investigated the best recognition rate and average recognition rate. We have chosen an odd number of $m$ such as 3, 5, and 7 to avoid the case of tie.

The results on the test data are as shown in Tables 4 and 5. KNN and MLP produce the best recognition rate among the classifiers on the average. Using KNN classifier, performance of PC is better than CC as a similarity measure.

Recognition rates by ensemble classifiers are shown in Table 6. The numbers next to the ensemble methods are the numbers of single classifiers

**Table 4.** Recognition rate with features and classifiers in colon dataset (%).

|      | MLP  | SASOM | SVM (L) | SVM (R) | KNN (C) | KNN (P) |
|------|------|-------|---------|---------|---------|---------|
| PC   | 74.2 | 74.2  | 64.5    | 64.5    | 71.0    | 77.4    |
| SC   | 58.1 | 45.2  | 64.5    | 64.5    | 61.3    | 67.7    |
| ED   | 67.8 | 67.6  | 64.5    | 64.5    | 83.9    | 83.9    |
| CC   | 83.9 | 64.5  | 64.5    | 64.5    | 80.7    | 80.7    |
| IG   | 71.0 | 71.0  | 71.0    | 71.0    | 74.2    | 80.7    |
| MI   | 71.0 | 71.0  | 71.0    | 71.0    | 74.2    | 80.7    |
| SN   | 64.5 | 45.2  | 64.5    | 64.5    | 64.5    | 71.0    |
| PCA  | 87.1 | 74.2  | **87.7**| 67.1    | 80.0    | 78.1    |
| Avg. | 72.2 | 64.1  | 69.0    | 66.5    | 73.7    | 77.5    |

**Table 5.** Recognition rate with features and classifiers in lymphoma dataset (%).

|      | MLP  | SASOM | SVM (L) | SVM (R) | KNN (C) | KNN (P) |
|------|------|-------|---------|---------|---------|---------|
| PC   | 64.0 | 48.0  | 56.0    | 60.0    | 60.0    | 76.0    |
| SC   | 60.0 | 68.0  | 44.0    | 44.0    | 60.0    | 60.0    |
| ED   | 56.0 | 52.0  | 56.0    | 56.0    | 56.0    | 68.0    |
| CC   | 68.0 | 52.0  | 56.0    | 56.0    | 60.0    | 72.0    |
| IG   | **92.0** | 84.0 | **92.0** | **92.0** | **92.0** | **92.0** |
| MI   | 72.0 | 64.0  | 64.0    | 64.0    | 80.0    | 64.0    |
| SN   | 76.0 | 76.0  | 72.0    | 76.0    | 76.0    | 80.0    |
| PCA  | 87.2 | 84.0  | 88.4    | 58.4    | 86.0    | 86.4    |
| Avg. | 71.9 | 66.0  | 66.1    | 63.3    | 71.3    | 74.8    |

Table 6. The best recognition rate by ensemble classifier.

|  |  | Colon | Lymphoma |
|---|---|---|---|
| MV | 3 | 93.5 | 96.0 |
|  | 5 | 93.5 | 100.0 |
|  | 7 | 93.5 | 100.0 |
|  | All | 71.0 | 80.0 |
| WV | 3 | 93.5 | 96.0 |
|  | 5 | 93.5 | 100.0 |
|  | 7 | 93.5 | 100.0 |
|  | All | 71.0 | 88.0 |
| BC | 3 | 93.5 | 96.0 |
|  | 5 | 93.5 | 100.0 |
|  | 7 | 93.5 | 100.0 |
|  | All | 74.2 | 92.0 |

combined, and the best recognition rate of ensemble classifier is 93.5% in colon dataset, and 100.0% in lymphoma dataset, which appeared when five or seven feature-classifier pairs form an ensemble. However, we can conclude $_nC_5$ combination method is better in terms of efficiency. A previous study has also demonstrated that five is quite a feasible number for the optimal ensemble.[28] Compared with the best recognition rates of base classifier, 83.9%, and 92.0% on each dataset in Tables 4 and 5, the performance of ensemble is at least equal to or superior to any other single classifiers. In both datasets, the performance of ensemble classifier outperforms the best single classifier, and ensemble classifier with all feature-classifier combinations is inferior to one with five feature-classifier combinations.

While there is little difference in the best recognition rate of the ensemble classifier according to the ensemble method or the number of combined classifiers, there is a difference in average recognition rate of the ensemble classifier. As the number of combined classifiers is increased, the average recognition rate of the ensemble classifier is also increased in lymphoma dataset but slightly decreasing in colon dataset. Figure 7 shows the average recognition rate of the ensemble classifier on benchmark dataset. For both datasets, Bayesian combination is the best among three ensemble methods, and weighted voting is superior to majority voting when the number of combined classifiers is increased.

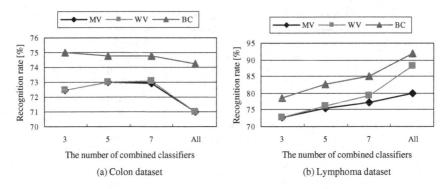

**Fig. 7.** Average recognition rate of the ensemble.

### 4.4. *Optimal ensemble with GA*

All feature selection and classification methods and two ensemble methods of majority voting and weighted voting described in Sec. 2 have been used for experiments. For GA, we have used roulette wheel and rank-based methods for selection. In the rank-based selection, we give higher rank to the chromosomes whose number of 1 is smaller than the others for tie. Because preliminary results showed that it sometimes converged to a local minimum when the GA was run with less than 100 chromosomes, we set the chromosome size as greater than 100. Experiments are conducted with different crossover rates of 0.3, 0.5, 0.7 and 0.9, and mutation rates of 0.01 and 0.05. GA stops when it finds the perfect ensemble on the validation dataset, or when the generation exceeds 10 000.

We have tried to find the optimal ensemble with validation dataset and GA has practically found the outstanding ensembles close to 100% of classification accuracy. In colon cancer dataset, GA has found interesting ensembles that yield higher classification performance than any other individual feature-classifier pairs. The best accuracy of individual feature-classifier pairs is 87.7%, but the ensembles searched by GA yield 90.0–96.7% accuracy. Two of the best ensembles are presented in Table 7. Although the accuracy of individual pairs is of about 71.0–83.9%, the ensemble accuracy reaches 96.7% by majority voting. Similar performance has been obtained in the weighted voting. We can see that some general feature-classifiers which produce 67.7–80.7% of accuracy yield 96.7% of the ensemble result in a complementary way.

Table 7. Two cases of the optimal ensembles on colon cancer dataset.

| Combining methods | Feature-classifier pair | Accuracy (%) |
|---|---|---|
| Majority voting | CC-MLP | 83.9 |
| | MI-MLP | 71.0 |
| | PC-MLP | 74.2 |
| | ED-KNN(C) | 83.9 |
| | ED-KNN(P) | 83.9 |
| | PC-KNN(P) | 77.4 |
| | SN-KNN(C) | 64.5 |
| | PC-SASOM | 74.2 |
| | Ensemble | 96.7 |
| Weighted voting | MI-MLP | 71.0 |
| | PC-MLP | 74.2 |
| | CC-KNN(P) | 80.7 |
| | PC-KNN(C) | 71.0 |
| | SC-KNN(P) | 67.7 |
| | MI-SASOM | 71.0 |
| | Ensemble | 96.7 |

In lymphoma cancer dataset, the average accuracy of individual feature-classifiers is 44.0–92.0%. In every case, the excellent ensemble that has more than 90% of classification accuracy is found. Table 8 shows some optimal ensembles that classify samples correctly. SC-SASOM pair has 68.0% of accuracy, but it plays an important role in the ensemble in majority voting method. In the same manner, CC-SASOM pair yields only 52.0% of accuracy, but it takes its part most effectively in the weighted voting.

It is important to show the generality since the test on the validation dataset has the possibility of overfitting. The LOOCV on colon cancer dataset, we could also get the excellent ensemble that is superior to any individual feature-classifiers. The best ensemble has misclassified 6 out of 62 samples whose ids are 2, 8, 15, 18, 20 and 22. The classification accuracy of this ensemble is 90.3%. In lymphoma cancer the dataset misclassified only 3 out of 47 samples. This leads to 93.6% accuracy, which is better than any individual feature-classifier pairs. The misclassified samples are the 4th, 26th and 32nd that are misclassified more than half of total experiments by the best ensemble on the validation dataset. These results are summarized in Fig. 8.

Table 8. Two cases of the optimal ensembles on Lymphoma cancer dataset.

| Combining methods | Feature-classifier pair | Accuracy (%) |
|---|---|---|
| Majority voting | CC-KNN(P) | 72.0 |
|  | MI-KNN(C) | 80.0 |
|  | SN-KNN(C) | 76.0 |
|  | SC-SASOM | 68.0 |
|  | IG-SVM(L) | 92.0 |
|  | Ensemble | 100.0 |
| Weighted voting | IG-KNN(C) | 92.0 |
|  | MI-KNN(C) | 80.0 |
|  | SN-KNN(C) | 76.0 |
|  | SN-KNN(P) | 80.0 |
|  | CC-SASOM | 52.0 |
|  | IG-SASOM | 84.0 |
|  | PC-SVM(R) | 60.0 |
|  | Ensemble | 100.0 |

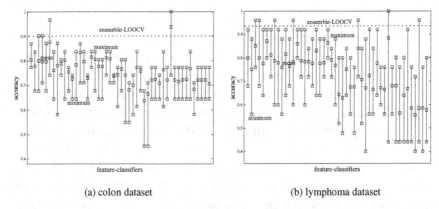

(a) colon dataset  (b) lymphoma dataset

**Fig. 8.** Summary of the classification results with (a) colon dataset and (b) Lymphoma dataset. Lines are the accuracy of individual feature-classifiers. The dotted line indicates the result of leave-one-out cross validation.

## 5. Conclusions

This chapter has introduced three sophisticated classification methods based on classifier ensemble approach. They are supposed to search the informative feature-classifier sets and combine them to overcome the limitations of the basic classification methods. The first method used ensemble classifier with negatively correlated genes, and the second one has used combinatorial ensemble method. The last one searches an optimal ensemble with GA. Finally, we have shown their usefulness with experiments using colon and lymphoma cancer datasets.

**Acknowledgement**

This work was supported by the Korea Science and Engineering Foundation (KOSEF) through the Biometric Engineering Research Center (BERC) at Yonsei University.

**References**

1. A. V. Antonov, I. V. Tetko, M. T. Mader, J. Budczies and H. W. Mewes, Optimization models for cancer classification: extracting gene interaction information from microarray expression data, *Bioinformatics* **20**(5), 644–652 (2004).
2. D. V. Nguyen and D. M. Rocke, Multi-class cancer classification via partial least squares with gene expression profiles, *Bioinformatics* **18**(9), 1216–1226 (2002).
3. S.-B. Cho and J. Ryu, Classifying gene expression data of cancer using classifier ensemble with mutually exclusive features, *Proceedings of the IEEE* **90**(11), 1744–1753 (2002).
4. S.-B. Cho and H.-H. Won, Data mining for gene expression profiles from DNA microarray, *International Journal of Software Engineering* **13**(6), 593–608 (2003).
5. Z.-H. Zhou, J. Wu and W. Tang, Ensembling neural networks: many could be better than all, *Artificial Intelligence* **137**(1–2), 239–263 (2002).
6. E. Bauer and R. Kohavi, An empirical comparison of voting classification algorithms: bagging, boosting and variants, *Machine Learning* **36**(1–2), 105–139 (1999).
7. L. K. Hansen and P. Salamon, Neural network ensembles, *IEEE Transactions on Pattern Analysis and Machine Intelligence* **12**(10), 993–1001 (1990).
8. A. Tsymbal, S. Puuronen and D. W. Patterson, Ensemble feature election with the simple Bayesian classification, *Information Fusion* **4**, 87–100 (2003).
9. D. Opitz and R. Maclin, Popular ensemble methods: an empirical study, *Journal of Artificial Intelligence Research* **11**, 169–198 (1999).
10. A. Sboner, C. Eccher, E. Blanzieri, P. Bauer, M. Cristofolini, G. Zumiani and S. Forti, A multiple classifier system for early melonama diagnosis, *Artificial Intelligence in Medicine* **27**, 29–44 (2003).
11. D. Lashkari, J. Derisi, J. McCusker, A. Namath, C. Gentile, S. Hwang, P. Brown and R. Davis, Yeast microarrays for genome wide parallel genetic and gene expression analysis, *Proc. Nat. Acad. Sci. USA* **94**, 13057–13062 (1997).
12. J. Derisi, V. Iyer and P. Brosn, Exploring the metabolic and genetic control of gene expression on a genomic scale, *Science* **278**, 680–686 (1997).

13. M. Eisen, P. Spellman, P. Brown and D. Bostein, Cluster analysis and display of genome-wide expression patterns, *Proc. Nat. Acad. Sci. USA* **95**, 14863–14868 (1998).
14. R. J. Lipshutz, S. P. A. Fodor, T. R. Gingeras and D. J. Lockhart, High density synthetic oligonucleotide arrays, *Nature Genetics* **21**, 20–24 (1999).
15. R. Shamir and R. Sharan, Algorithmic approaches to clustering gene expression data, *Current Topics in Computational Biology*, T. Jiang, T. Smith, Y. Xu, M. Q. Zhang (eds), MIT Press (2001).
16. J. Khan, J. S. Wei, M. Ringner, L. H. Saal, M. Ladanyi, F. Westermann, F. Berthold, M. Schwab, C. R. Antonescu, C. Peterson and P. S. Meltzer, Classification and diagnostic prediction of cancers using gene expression profiling and artificial neural networks, *Nature Medicine* **7**, 673–679 (2001).
17. J. Liu, H. Iba and M. Ishizuka, Selecting informative genes with parallel genetic algorithms in tissue classification, *Genome Informatics* **12**, 14–23 (2001).
18. I. T. Jolliffe, *Principal Compoentn Analysis*, Springer, New York (1986).
19. R. O. Duda, P. E. Hart and D. G. Stork, *Pattern Classification*, 2nd Edition, Wiley Interscience (2001).
20. J. Li and L. Wong, Identifying good diagnostic gene groups from gene expression profiles using the concept of emerging patterns, *Bioinformatics* **18**(5), 725–734 (2002).
21. M. P. S. Brown, W. N. Grundy, D. Lin, N. Cristianini, C. W. Sugnet, T. S. Furey, M. Jr. Ares and D. Haussler, Knowledge-based analysis of microarray gene expression data by using support vector machines, *Proc. Nat. Acad. Sci. USA* **97**, 262–267 (2000).
22. T. S. Furey, N. Cristianini, N. Duffy, D. W. Bednarski, M. Schummer and D. Haussler, Support vector machine classification and validation of cancer tissue samples using microarray expression data, *Bioinformatics* **16**(10), 906–914 (2000).
23. S.-B. Cho, Self-organizing map with dynamical node splitting: application to handwritten digit recognition, *Neural Computation* **9**, 1345–1355 (1997).
24. A. Ben-Dor, L. Bruhn, N. Friedman, I. Nachman, M. Schummer and N. Yakhini, Tissue classification with gene expression profiles, *Journal of Computational Biology* **7**, 559–584 (2000).
25. M. Dettling and P. Buhlmann, *How to Use Boosting for Tumor Classification with Gene Expression Data*, Technical Report, Department of Statistics, ETH Zurich (2002).
26. T. M. Mitchell, *Machine Learning*, Carnegie Mellon University (1997).
27. I. S. Lossos, A. A. Alizadeh, M. B. Eisen, W. C. Chan, P. O. Brown, D. Bostein, L. M. Staudt and R. Levy, Ongoing immunoglobulin somatic mutation in germinal center B cell-like but not in activated B cell-like diffuse large cell lymphomas, *Proc. Nat. Acad. Sci. USA* **97**(18), 10209–10213 (2000).
28. C. Park and S.-B. Cho, Evolutionary ensemble classifier for lymphoma and colon cancer classification, *Proc. Congress on Evolutionary Computation*, pp. 2378–2385 (2003).

# CHAPTER 13

# MULTIOBJECTIVE EVOLUTIONARY APPROACH TO FUZZY CLUSTERING OF MICROARRAY DATA

Anirban Mukhopadhyay

*Department of Computer Science and Engineering*
*University of Kalyani, Kalyani-741235, India*
*anirbanbuba@yahoo.com*

Ujjwal Maulik

*Department of Computer Science and Engineering*
*Jadavpur University, Kolkata-700032, India*
*drumaulik@cse.jdvu.ac.in*

Sanghamitra Bandyopadhyay

*Machine Intelligence Unit*
*Indian Statistical Institute, Kolkata-700108, India*
*sanghami@isical.ac.in*

Recent advancements in microarray technology allows simultaneous monitoring of the expression levels of a large number of genes over different time points. Clustering is one of the important data mining tools for analyzing such microarray data. This article considers the problem of fuzzy partitioning as one of searching for a suitable set of cluster centers such that some fuzzy cluster validity (goodness) measure is optimized. Genetic algorithm, a popular search and optimization tool, is utilized to perform the clustering, taking the validity measure as the fitness value. However, no single validity measure works uniformly well for different kinds of data sets. Hence, simultaneous optimization of a number of validity criteria is considered here. In this article, a multiobjective optimization algorithm is utilized to address the fuzzy clustering problem where a number of fuzzy cluster validity indices are simultaneously optimized. The resultant near-Pareto-optimal set of solutions consists of a number of non-dominated solutions, from which the user can choose the most promising one according to the problem specifications. Real-coded encoding of the cluster centers is used for this purpose. The performance of the multiobjective fuzzy clustering algorithm has been compared with other well known clustering techniques (partitional and hierarchical), that are widely used for clustering gene expression data, to prove its effectiveness on a variety of publicly available data such as Yeast Sporulation and Human Fibroblasts Serum data. The performance has been analyzed both quantitatively and by visualization. Biological interpretations are also made for the resulting solutions.

## 1. Introduction

Classical approach to genomic research was based on the local study and collection of data on single genes. The advent of microarray technology has now made it possible to have a global and simultaneous view of the expression levels for many thousands of genes over different time points during some biological processes. Advances in microarray technology in recent years have major impacts in many fields such as medical diagnosis, characterizing various gene functions, understanding different molecular biological processes etc.[2,10,11,16,19]

New application opportunities have been created for data mining methodologies[7,20] due to the development of microarray technologies. However, microarray chips gather expression levels of large number of genes, hence producing huge amount of data to handle. A primary approach to analyze such large amount of data is clustering,[23,33] which is a popular unsupervised pattern recognition technique. Clustering techniques aim to discover natural grouping of objects in a data set based on some similarity/dissimilarity measures. Several clustering methods have been adopted in clustering gene expression data over the past few years such as K-means, K-medoids, Fuzzy C-means and hierarchical clustering methodologies.

A lot of uncertainty and imprecisions are associated with gene expression data as many genes can exhibit similar characteristics. Hence a particular gene may not be assigned to a particular class with reasonable amount of certainty. Hence it is natural to apply the fuzzy clustering algorithm for partitioning microarray data sets.

In recent years, some Genetic Algorithm based clustering tehniques[26,27] have been developed and they have been proved to be efficient in clustering many real life data sets.[4,27] Genetic Algorithms (GAs)[14,18,28] are randomized search and optimization techniques guided by the principles of evolution and natural genetics, and have a large amount of implicit parallelism.

They provide near-optimal solutions of an objective or fitness function in complex, large and multimodal landscapes. In GAs, the parameters of the search space are encoded in the form of *strings* (or, *chromosomes*). A fitness function is associated with each *string* that represents the degree of goodness of the solution encoded in it. Biologically inspired operators like *selection*, *crossover* and *mutation* are used over a number of generations for generating potentially better strings. Genetic and other evolutionary algorithms have been earlier used for pattern classification including clustering of data.[5,6,26,27] Traditional GA based clustering techniques use a single cluster validity measure to reflect the goodness of an encoded clustering. This measure is used as the fitness function in the GA. However, a single cluster validity measure is seldom equally applicable for different kinds of data sets. A wrong choice of the validity measure may lead to poor clustering results.

This article demonstrates an application of a recently developed multiobjective genetic clustering technique,[8] which simultaneously optimizes more than one objective function, for automatically partitioning microarray data set. In multiobjective optimization (MOO),[12,15] search is performed over a number of, often conflicting, objective functions. In contrast to single objective optimization, which yields a single best solution, in MOO the final solution set contains a number of Pareto-optimal solutions, none of which can be further improved on any one objective without degrading it in another. Real-coded multiobjective genetic algorithms (MOGAs) are used in this regard in order to determine the appropriate cluster centers and the corresponding partition matrix. NSGAII,[12,13,15] a popular elitist MOGA, is used as the underlying optimization strategy. The Xie-Beni (XB) index[34] and the FCM[9] measure ($J_m$) are used as the objective functions. Note that any other and any number of objective functions could be used instead of the above mentioned two.

A number of gene expression data sets are used for experiments such as *Yeast Sporulation* and *Human Fibroblasts Serum* data. Clustering results are reported in terms of a cluster validity measure (*Silhouette index*) and verified visually using some microarray visualization tools. Performance of the multiobjective technique has been compared with that of the single objective version, K-means, K-medoids, fuzzy C-means and hierarchical algorithms.

## 2. Structure of Gene Expression Data Sets

A microarray experiment typically measures the expression levels of large number of genes across different time points. A gene expression data consisting of $n$ genes and $m$ time points are usually expressed as a real valued $n \times m$ matrix $M = [g_{ij}]$, $i = 1, 2, \ldots, n$, $j = 1, 2, \ldots, m$. Here each element $g_{ij}$ represents the expression level of the $i$th gene at the $j$th time point.

$$M = \begin{bmatrix} g_{11} & g_{12} & \cdots & g_{1m} \\ g_{21} & g_{22} & \cdots & g_{2m} \\ \cdots & \cdots & \cdots & \cdots \\ g_{n1} & g_{n2} & \cdots & g_{nm} \end{bmatrix}.$$

The raw gene expression data consists of noise, some variations arising from biological experiments and missing values. Hence before applying any clustering algorithm, the data is preprocessed. Two widely used preprocessing techniques are missing value estimation and standardization. Standardization is a statistical tool for transforming data into a format that can be used for meaningful cluster analysis.[31] Normalization is a useful standardization process by which each row of the matrix $M$ is standardized to have mean 0 and variance 1. Following are the preprocessing techniques that have been used in this article. First some filtering is applied on the raw data to filter out those genes whose expression levels do not change significantly over different time points. Next the expression values are log transformed and each row is normalized to have mean 0 and variance 1.

## 3. Cluster Analysis

Clustering[3,17,21,23,33] is a popular data mining tool that partitions the input space into $K$ groups $C_1, C_2, \ldots, C_K$ depending on some similarity/dissimilarity metric where the value of $K$ may or may not be known *a priori*. The main objective of any clustering technique is to evolve a partition matrix $U(X)$ of the given data set $X$ (consisting of, say, $n$ patterns,

$X = \{x_1, x_2, \ldots, x_n\}$) such that

$$\sum_{j=1}^{n} u_{kj} \geq 1 \quad \text{for } k = 1, \ldots, K,$$

$$\sum_{k=1}^{K} u_{kj} = 1 \quad \text{for } j = 1, \ldots, n \text{ and,}$$

$$\sum_{k=1}^{K} \sum_{j=1}^{n} u_{kj} = n.$$

The partition matrix $U(X)$ of size $K \times n$ can be represented as $U = [u_{kj}]$, $k = 1, \ldots, K$ and $j = 1, \ldots, n$, where $u_{kj}$ denotes the membership of pattern $x_j$ to cluster $C_k$. In crisp clustering $u_{kj} = 1$ if $x_j \in C_k$, otherwise $u_{kj} = 0$. On the other hand, for fuzzy partitioning of the patterns, $0 < u_{kj} < 1$ and the following conditions hold on $U$ (representing non-degenerate clustering): $0 < \sum_{j=1}^{n} u_{kj} < n$, $\sum_{k=1}^{K} u_{kj} = 1$, and $\sum_{k=1}^{K} \sum_{j=1}^{n} u_{kj} = n$.

Here a brief discussion on some popular clustering techniques used for microarray data clustering is presented.

### 3.1. K-means

K-means[25,30] clustering algorithm is one of the most fundamental partitional clustering approaches. Given a set of $n$ points $X = \{x_1, x_2, \ldots, x_n\}$ to be clustered into $K$ groups $C_1, C_2, \ldots, C_K$, K-means algorithm begins with a set of random $K$ cluster centroids (chosen from $n$ points) $Z = \{z_1, z_2, \ldots, z_K\}$. Then each data point $x_i$, $i = 1, 2, \ldots, n$, is assigned to its nearest cluster center to form $K$ initial clusters. The centroids are subsequently updated by taking means of the points assigned to each cluster. When the new cluster centers have been computed, the points are again assigned to their nearest centers. This process continues until centroid positions do not change in two consecutive iterations. Overall objective of the algorithm is to increase the global compactness of the clustering by minimizing the following metric.

$$\mathcal{U}(C_1, C_2, \ldots, C_K) = \sum_{i=1}^{K} \sum_{x_j \in C_i} D^2(x_j, z_i). \quad (1)$$

Here $D(.)$ is a distance function such as Euclidean or Pearson Correlation metrics and $D(x_j, z_i)$ denotes the distance of pattern $x_j$ from $i$th cluster center. K-means algorithm often gets stuck into local optimum and the final solution depends on initial choice of cluster centers. Also this method does not scale well with large data sets.

### 3.2. K-medoids

Partitioning around medoids (PAM), also called K-medoids clustering,[24] is a variation of K-means with the objective to minimize within the cluster variance W($K$):

$$W(K) = \sum_{i=1}^{K} \sum_{p \in C_i} D(p, m_i), \qquad (2)$$

where $m_i$ is the medoid (most centrally located point) of cluster $C_i$ and $D(.)$ denotes a distance function. The resulting clustering of the data set $X$ is usually only a local minimum of $W(K)$. The idea of PAM is to select $K$ representative points, or medoids, in $X$ and assign the rest of the data points to the cluster identified by the closest medoid. Initially medoids are chosen randomly. Then, each point in $X$ is assigned to its nearest medoid. In each iteration, a new medoid is determined for each cluster by finding the data point with minimum total dissimilarity to all other points of the cluster. Subsequently, all the points in $X$ are reassigned to their clusters in accordance with the new set of medoids. The algorithm iterates until $W(K)$ does not change any more.

### 3.3. Fuzzy C-means

Fuzzy C-means[9] algorithm is the fuzzy version of K-means algorithm. It uses the principles of fuzzy sets to evolve a partition matrix $U(X)$ while minimizing the measure

$$J_m = \sum_{j=1}^{n} \sum_{k=1}^{K} u_{kj}^m D^2(z_k, x_j), \qquad (3)$$

where $n$ is the number of data objects, $K$ represents the number of clusters, $U = [u_{kj}]$ is the fuzzy membership matrix (partition matrix) and $m$ denotes the fuzzy exponent. Here $x_j$ is the $j$th data point and $z_k$ is the center of the $k$th cluster. $D(z_k, x_j)$ denotes the distance of point $x_j$ from the center of the $k$th cluster.

FCM algorithm is based on an alternating optimizing strategy. This involves iteratively estimating the partition matrix followed by computation of new cluster centers. It starts with random initial $K$ cluster centers, and then at every iteration it finds the fuzzy membership of each data point to every cluster using the following equation:[9]

$$u_{ik} = \frac{1}{\sum_{j=1}^{K}\left(\frac{D(z_i, x_k)}{D(z_j, x_k)}\right)^{\frac{2}{m-1}}}, \quad \text{for } 1 \leq i \leq K; \ 1 \leq k \leq n, \quad (4)$$

where $D(z_i, x_k)$ and $D(z_j, x_k)$ are the distances between $x_k$ and $z_i$, and $x_k$ and $z_j$, respectively. $m$ is the weighting coefficient. (Note that while computing $u_{ik}$ using Eq. (4), if $D(z_j, x_k)$ is equal to zero for some $j$, then $u_{ik}$ is set to zero for all $i = 1, \ldots, K, i \neq j$, while $u_{jk}$ is set equal to one.)

Based on the membership values, the cluster centers are recomputed using the following equation:[9]

$$z_i = \frac{\sum_{k=1}^{n}(u_{ik})^m x_k}{\sum_{k=1}^{n}(u_{ik})^m}, \quad 1 \leq i \leq K. \quad (5)$$

The algorithm terminates when there is no more movement in the cluster centers. Finally, each data point is assigned to the cluster to which it has maximum membership.

### 3.4. *Hierarchical agglomerative clustering*

Agglomerative clustering techniques[33] begin with singleton clusters, and combine two least distant clusters at every iteration. Thus in each iteration two clusters are merged, and hence the number of clusters reduces by one. This proceeds iteratively in a hierarchy, providing a possible partitioning of the data at every level. When the target number of clusters ($K$) is achieved, the algorithms terminate. Single, average and complete linkage agglomerative algorithms differ only in the linkage metric used. For the single linkage algorithm, the distance between two clusters $C_i$ and $C_j$ is computed as the smallest distance between all possible pairs of data points $p$ and $q$, where $p \in C_i$ and $q \in C_j$. For average and complete linkage algorithms, the linkage metrics are taken as the average and largest distances respectively.

## 4. Multiobjective Genetic Algorithms

In many real world situations there may be several objectives that must be optimized simultaneously in order to solve a certain problem. This is in contrast to the problems tackled by conventional GAs which involve optimization of just a single criterion. The main difficulty in considering multiobjective optimization is that there is no accepted definition of optimum in this case, and therefore it is difficult to compare one solution with another. In general, these problems admit multiple solutions, each of which is considered acceptable and equivalent when the relative importance of the objectives is unknown. The best solution is subjective and depends on the need of the designer or decision maker.

Traditional search and optimization methods such as gradient descent search, and other non-conventional ones such as simulated annealing are difficult to extend as they are to multiobjective case, since their basic design precludes the consideration of multiple solutions. On the contrary, population based methods like evolutionary algorithms are well suited for handling such situations. The multiobjective optimization can be formally stated as in Ref. 12.

Find the vector $\overline{x}^* = [x_1^*, x_2^*, \ldots, x_n^*]^T$ of decision variables which will satisfy the $m$ inequality constraints:

$$g_i(\overline{x}) \geq 0, \quad i = 1, 2, \ldots, m, \quad (6)$$

the $p$ equality constraints

$$h_i(\overline{x}) = 0, \quad i = 1, 2, \ldots, p, \quad (7)$$

and optimizes the vector function

$$\overline{f}(\overline{x}) = [f_1(\overline{x}), f_2(\overline{x}), \ldots, f_k(\overline{x})]^T. \quad (8)$$

The constraints given in Eqs. (6) and (7) define the feasible region $\mathcal{F}$ which contains all the admissible solutions. Any solution outside this region is inadmissible since it violates one or more constraints. The vector $\overline{x}^*$ denotes an optimal solution in $\mathcal{F}$. In the context of multiobjective optimization, the difficulty lies in the definition of optimality, since it is only rare that we will find a situation where a single vector $\bar{x}^*$ represents the optimum solution to all the objective functions.

The concept of *Pareto optimality* comes handy in the domain of multiobjective optimization. A formal definition of Pareto optimality from the viewpoint of minimization problem may be given as follows.

A decision vector $\bar{x}^*$ is called Pareto optimal if and only if there is no $\bar{x}$ that dominates $\bar{x}^*$, i.e. there is no $\bar{x}$ such that

$$\forall i \in \{1, 2, \ldots, k\}, \quad f_i(\bar{x}) \leq f_i(\bar{x}^*),$$

and

$$\exists i \in \{1, 2, \ldots, k\}, \quad f_i(\bar{x}) < f_i(\bar{x}^*).$$

In other words, $\bar{x}^*$ is Pareto optimal if there exists no feasible vector $\bar{x}$ which causes a reduction on some criterion without a simultaneous increase in at least one other. In this context, two other notions, *weakly non-dominated* and *strongly non-dominated* solutions are defined.[12] A point $\bar{x}^*$ is a weakly non-dominated solution if there exists no $\bar{x}$ such that $f_i(\bar{x}) < f_i(\bar{x}^*)$, for $i = 1, 2, \ldots, k$. A point $\bar{x}^*$ is a strongly non-dominated solution if there exists no $\bar{x}$ such that $f_i(\bar{x}) \leq f_i(\bar{x}^*)$, for $i = 1, 2, \ldots, k$, and for at least one $i$, $f_i(\bar{x}) < f_i(\bar{x}^*)$. In general, Pareto optimum usually admits a set of solutions called *non-dominated* solutions.

There are different approaches to solving multiobjective optimization problems,[12,15] e.g. aggregating, population based non-Pareto and Pareto-based techniques. In aggregating techniques, the different objectives are generally combined into one using weighting or goal based method. Vector Evaluated Genetic Algorithm (VEGA) is a technique in the population based non-Pareto approach in which different subpopulations are used for the different objectives. Multiple Objective Genetic Algorithm (MOGA), Non-dominated Sorting Genetic Algorithm (NSGA), Niched Pareto Genetic Algorithm (NPGA) constitute a number of techniques under the Pareto based approaches. However, all these techniques[15] were essentially non-elitist in nature. Non-dominated Sorting Genetic Algorithm II (NSGAII),[15] Strength Pareto Evolutionary Algorithm (SPEA)[36] and Strength Pareto Evolutionary Algorithm-2 (SPEA2)[35] are some recently developed multiobjective elitist techniques. The present article uses NSGAII as underlying multiobjective framework for developing the fuzzy clustering algorithm.

## 5. The Multiobjective Fuzzy Clustering Technique

In this section, we describe the use of NSGAII for evolving a set of near-Pareto-optimal non-degenerate fuzzy partition matrices. Xie-Beni index[34] and the FCM[9] measure are considered as the objective functions that must be minimized simultaneously. The technique is described below in detail.

### 5.1. *Chromosome representation and population initialization*

In the NSGAII based fuzzy clustering, the *chromosomes* are made up of real numbers which represent the coordinates of the centers of the partitions. The length of each *chromosome* is $d \times K$, where $d$ is the number of features and $K$ is the number of clusters. For $K$ clusters, the centers encoded in a *chromosome* in the initial population are randomly selected $K$ distinct points from the data set.

### 5.2. *Computation of objective functions*

In this article, the Xie-Beni (XB) index[34] and $J_m$[9] measure are taken as the two objectives that need to be simultaneously optimized. For computing the measures, the centers encoded in a *chromosome* are first extracted. Let the set of the cluster centers is denoted as $Z = \{z_1, z_2, \ldots, z_K\}$. The membership values $u_{ik}, i = 1, 2, \ldots, K$ and $k = 1, 2, \ldots, n$, are computed as follows:[9]

$$u_{ik} = \frac{1}{\sum_{j=1}^{K} \left( \frac{D(z_i, x_k)}{D(z_j, x_k)} \right)^{\frac{2}{m-1}}}, \quad \text{for } 1 \leq i \leq K; \; 1 \leq k \leq n, \quad (9)$$

where $D(z_i, x_k)$ and $D(z_j, x_k)$ are as described earlier. $m$ is the weighting coefficient. (Note that while computing $u_{ik}$ using Eq. (9), if $D(z_j, x_k)$ is equal to zero for some $j$, then $u_{ik}$ is set to zero for all $i = 1, \ldots, K$, $i \neq j$, while $u_{jk}$ is set equal to one.) Subsequently, the centers encoded in a chromosome are updated using the following equation:[9]

$$z_i = \frac{\sum_{k=1}^{n} (u_{ik})^m x_k}{\sum_{k=1}^{n} (u_{ik})^m}, \quad 1 \leq i \leq K, \quad (10)$$

and the cluster membership values are recomputed.

The XB index is defined as a function of the ratio of the total variation $\sigma$ to the minimum separation $sep$ of the clusters. Here $\sigma$ and $sep$ can be written as

$$\sigma(U, Z; X) = \sum_{i=1}^{K} \sum_{k=1}^{n} u_{ik}^2 D^2(z_i, x_k), \qquad (11)$$

and

$$sep(Z) = \min_{i \neq j}\{D^2(z_i, z_j)\}. \qquad (12)$$

The XB index is then written as

$$\text{XB}(U, Z; X) = \frac{\sigma(U, Z; X)}{n \times sep(Z)} = \frac{\sum_{i=1}^{K} \sum_{k=1}^{n} u_{ik}^2 D^2(z_i, x_k)}{n \times \min_{i \neq j}\{D^2(z_i, z_j)\}}. \qquad (13)$$

Note that when the partitioning is compact and good, value of $\sigma$ should be low while $sep$ should be high, thereby yielding lower values of the Xie-Beni (XB) index. The objective is therefore to minimize the XB index for achieving proper clustering.

The FCM measure $J_m$ is defined as follows:

$$J_m = \sum_{j=1}^{n} \sum_{k=1}^{K} u_{kj}^m D^2(x_j, z_k), \quad 1 \leq m \leq \infty, \qquad (14)$$

where $m$ is the fuzzy exponent.

### 5.3. Selection, crossover and mutation

The popularly used genetic operations are *selection*, *crossover* and *mutation*. The *selection* operation used here is the crowded binary tournament selection used in NSGAII. After *selection*, the selected *chromosomes* are put in the mating pool. *Crossover* and *mutation* are performed on the *chromosomes* selected in the mating pool. The most characteristic part of NSGAII is its elitism operation, where the non-dominated solutions among the parent and child populations are propagated to the next generation. For details on the different genetic processes, the reader may refer to Ref. 15. The near-Pareto-optimal strings of the last generation provide the different solutions to the clustering problem.

## 5.4. Choice of objectives

As multiobjective clustering deals with simultaneous optimization of more than one clustering objective, hence its performance depends highly on the choice of these objectives. Careful choice of objectives can produce remarkable results, whereas arbitrary or unintelligent objective selection may lead to bad situations. The selection of objectives should be such so that they can balance each other critically and are contradictory in nature. Contradiction in the objective functions is beneficial since it guides to global optimum solution. It also ensures that no single clustering objective is optimized leaving other probable significant objectives unnoticed.

In this article XB and $J_m$ validity indices have been chosen as the two objectives to be optimized. From Eq. (14), it can be noted that $J_m$ calculates the global cluster variance, i.e. it considers within the cluster variance summed up over all the clusters. Lower value of $J_m$ indicates better compactness of clusters. On the other hand, XB index (Eq. (13)) is a combination of global (numerator) and particular (denominator) situations. The numerator is similar to $J_m$, but the denominator has a factor that gives the separation between two minimum distant clusters. Hence, this factor only considers the worst case, i.e. which two clusters are closest to each other and forgets about the other partitions. Here, greater value of the denominator (lower value of the whole index) signifies better solution. Hence it is evident that both $J_m$ and XB indices should be minimized in order to get good solutions. Although the numerator of XB index ($\sigma$ in Eq. (13)) is similar to $J_m$, the denominator contains an additional term ($sep$) representing the separation between the two nearest clusters. Therefore, XB index is minimized by minimizing $\sigma$ (or $J_2$), and by maximizing the $sep$. These two terms may not attain their best values for the same partitioning when the data has complex and overlapping clusters, such as microarray data.

## 5.5. Distance measures

In the context of gene expression data clustering, the choice of effective distance metrics plays a key role. Two popular distance measures viz. Euclidean and Pearson Correlation based distance measures are defined below.

*Euclidean distance*: This is probably the most popular distance measure used in clustering. In a $p$-dimensional space, the Euclidean distance

$d(g_i, g_j)$ between two feature vectors $g_i = (g_{i1}, g_{i2}, \ldots, g_{ip})$ and $g_j = (g_{j1}, g_{j2}, \ldots, g_{jp})$ is defined as:

$$d(g_i, g_j) = \sqrt{\sum_{l=1}^{p}(g_{il} - g_{jl})^2}. \quad (15)$$

The use of Euclidean distance measure is very common in clustering gene expression profiles. This distance measure generally works well if the feature vectors are normalized to have mean 0 and variance 1.

*Pearson Correlation*: Given two feature vectors, $g_i$ and $g_j$, Pearson correlation coefficient $Cor(g_i, g_j)$ between them is computed as

$$Cor(g_i, g_j) = \frac{\sum_{l=1}^{p}(g_{il} - \mu_{g_i})(g_{jl} - \mu_{g_j})}{\sqrt{\sum_{l=1}^{p}(g_{il} - \mu_{g_i})^2} \sqrt{\sum_{l=1}^{p}(g_{jl} - \mu_{g_j})^2}}. \quad (16)$$

Here $\mu_{g_i}$ and $\mu_{g_j}$ represent the arithmetic means of the components of the feature vectors $g_i$ and $g_j$ respectively. Pearson correlation coefficient defined in Eq. (16) is a measure of similarity between two objects in the feature space. The distance between two objects $g_i$ and $g_j$ is computed as $1 - Cor(g_i, g_j)$, which represents the dissimilarity between those two objects.

In this article we have used Pearson Correlation based distance measure as it is proved to be more suitable for microarray data sets.

## 6. Experimental Results

This section provides clustering results obtained by applying different clustering algorithms on two publicly available gene expression data sets: *Yeast Sporulation data* and *Human Fibroblasts Serum data*. These data sets are described below.

### 6.1. *Yeast sporulation data*

Microarray data on the transcriptional program of sporulation in budding yeast collected and analyzed[11] has been considered here. The sporulation data set is publicly available at the website http://cmgm. stanford.edu/pbrown/sporulation. DNA microarray containing 97% of the known and predicted genes involved, 6118 in total, is used. During the sporulation process, the mRNA levels were obtained at seven time points 0, 0.5, 2, 5, 7, 9 and 11.5 hours. The ratio of each gene's mRNA level (expression) to its mRNA level in vegetative cells before transfer to the sporulation medium is measured. Consequently, the ratio data are then log-transformed. Among the 6118 genes, the genes whose expression levels did not change significantly during harvesting have been ignored from further analysis. This is determined with a threshold level of 1.6 for the root mean squares of the log2-transformed ratios. The resulting set consists of 474 genes. The resulting data is then normalized so that each row has mean zero and variance one. The number of clusters chosen as 7 as in Ref. 11.

### 6.2. Human fibroblasts serum data

This dataset[22] contains the expression levels of 8613 human genes. The data set is obtained as follows: first, human fibroblasts were deprived of serum for 48 hours and then stimulated by addition of serum. After the stimulation, expression levels of the genes were computed over twelve time points and an additional data point was obtained from a separate unsynchronized sample. Hence the data set has 13 dimensions. A subset of 517 genes whose expression levels changed substantially across the time points have been chosen[16]). The data is then log2-transformed and subsequently, the expression vectors are normalized to have mean zero and variance one. Serum data set is clustered into 10 clusters as in Ref. 22. This data set can be downloaded from http://www.sciencemag.org/feature/data/984559.shl.

### 6.3. Performance validation

Performances of the clustering algorithms are examined both quantitatively and visually. A cluster validity index called Silhouette index[29] has been used for quantitative analysis. For visual demonstration, two cluster visualization tools Eisen plot and Cluster Profile plot are utilized. These are described below.

*Silhouette Index*: Silhouette width reflects the compactness and separation of the clusters. Given a set of $n$ genes $G = \{g_1, g_2, \ldots, g_n\}$ and a clustering of the genes $C = \{C_1, C_2, \ldots, C_K\}$, the silhouette width $s(g_i)$ for each gene $g_i$ belonging to cluster $C_j$ denotes a confidence measure of belongingness, and it is defined as follows:

$$s(g_i) = \frac{b(g_i) - a(g_i)}{max(a(g_i), b(g_i))}. \qquad (17)$$

Here $a(g_i)$ denotes the average distance of gene $g_i$ from the other genes of the cluster to which gene $g_i$ is assigned (i.e. cluster $C_j$), and $b(g_i)$ represents the minimum of the average distances of gene $g_i$ from the genes of the clusters $C_l, l = 1, 2, \ldots, K$, and $l \neq j$. The value of $s(g_i)$ lies between $-1$ and 1. Large value of $s(g_i)$ (near to 1) indicates that the gene $g_i$ is well clustered. A small value of $s(g_i)$ (around 0) means that the gene lies between two clusters and a negative value of $s(g_i)$ refers that gene $g_i$ is probably placed in a wrong cluster. Overall silhouette index $s(C)$ of a clustering $C$ is defined as mean silhouette width of all the genes (Eq. (18)).

$$s(C) = \frac{1}{n}\sum_{i=1}^{n} s(g_i). \qquad (18)$$

Greater value of $s(C)$ (near to 1) indicates that most of the genes are correctly clustered and this in turn reflects better clustering solution. Silhouette index can be used with any kind of distance metrics.

*Eisen plot*: Eisen plot (see Fig. 1, for example) was introduced by M. B. Eisen.[16] The matrix $M$ representing the expression data (see Sec. 2) is plotted as a table where the element $g_{ij}$ (i.e. the expression value of $i$th gene at $j$th time point) is denoted by coloring the corresponding cell of the table with a color similar to the original color of its spot on the microarray. The dark gray and light gray colors represent high and low expression values respectively. A black color denotes a zero expression value which indicates absence of differential expression. The rows are ordered to keep similar genes together. In our representation the cluster boundaries are identified by white colored blank rows.

*Cluster profile plot*: The cluster profile plot (see Fig. 2, for example) shows for each cluster the normalized gene expression values (light gray) of the genes of that cluster with respect to the time points. Along with this, average expression values of the genes of the cluster over different time points are shown as a black line together with the standard deviation within the cluster at each time point. Cluster profile plot demonstrates how the nature of the plots of the expression values of the genes across the time points differ for each cluster. Also, the genes within a cluster are similar in nature in terms of the expression profile.

### 6.4. Input parameter values

The parameters for the GA based clustering techniques (both for single objective and multiobjective) are fixed as follows: number of generations = 100, population size = 50, crossover probability = 0.8, mutation probability = $\frac{1}{length\ of\ chromosome}$. The number of iterations for K-means, K-medoids and Fuzzy C-means algorithms is taken as 200 unless they converge before that. All the algorithms are run for several times and only the best solutions obtained in terms of the silhouette index are considered here.

### 6.5. Quantitative assessments

The Silhouette index values for all the algorithms for the two data sets are reported in Table 1. The algorithms are differentiated in three fundamental categories viz. partitional, hierarchical and GA-based algorithm. K-means, K-medoids and FCM come under the first category. Three hierarchical agglomerative linkage clustering algorithms (single, average and complete linkage) fall under the second category. The GA based algorithms may be single objective (with objective functions either $J_m$ or XB), or they

Table 1. Silhouette index values of clustering solutions obtained by different algorithms for Yeast Sporulation and Human Fibroblasts Serum data sets.

| Clustering type | Algorithm | Data sets | |
|---|---|---|---|
| | | Sporulation | Serum |
| Partitional | K-means | 0.5733 | 0.3245 |
| | K-medoids | 0.5936 | 0.2609 |
| | Fuzzy C-means | 0.5879 | 0.3304 |
| Hierarchical | Single linkage | −0.4913 | −0.3278 |
| | Average linkage | 0.5007 | 0.2977 |
| | Complete linkage | 0.4388 | 0.2776 |
| GA based | Single objective GA ($J_m$) | 0.5886 | 0.3434 |
| | Single objective GA (XB) | 0.5837 | 0.3532 |
| | Multiobjective GA | 0.6465 | 0.4135 |

can be multiobjective (which optimizes both $J_m$ and XB simultaneously). As the multiobjective algorithm generates a set of Pareto optimal solutions, the solution producing the best Silhouette index value is chosen and reported here.

As seen from the table, for both the data sets, the multiobjective clustering algorithm produces best Silhouette index values. For Yeast Sporulation data, K-medoids algorithm is the best among the partitional methods producing a Silhouette index value of 0.5936. For hierarchical methods, average linkage has the largest Silhouette index (0.5007). In case of GA based methods, the multiobjective technique gives a Silhouette value of 0.6465, whereas the single objective methods with objectives $J_m$ and XB provide Silhouette values of 0.5886 and 0.5837 respectively. For the Fibroblasts Serum data, the multiobjective GA clustering technique produces a Silhouette value of 0.4135, which is better than all other methods. For this data set, FCM is the best partitional method (Silhouette value: 0.3304). Average linkage algorithm again provides the best Silhouette value (0.2977) among the hierarchical methods. Hence it is evident from the table that the multiobjective algorithm outperforms all other clustering methods.

Tables 2 and 3 show some interesting observations for the Yeast Sporulation data and Serum data, respectively. For example, Table 2 reports the $J_m$ and XB index values for the GA based clustering techniques for Sporulation

**Table 2.** Silhouette index and objective function values of different GA based algorithms for Yeast Sporulation data.

| Clustering algorithm | $J_m$ | XB | Sil |
|---|---|---|---|
| Single objective GA ($J_m$) | 2.2839 | 0.0178 | 0.5886 |
| Single objective GA (XB) | 2.3945 | 0.0148 | 0.5837 |
| Multiobjective GA | 2.3667 | 0.0164 | 0.6465 |

**Table 3.** Silhouette index and objective function values of different GA based algorithms for Human Fibroblasts Serum data.

| Clustering algorithm | $J_m$ | XB | Sil |
|---|---|---|---|
| Single objective GA ($J_m$) | 7.8041 | 0.0516 | 0.3434 |
| Single objective GA (XB) | 8.0438 | 0.0498 | 0.3532 |
| Multiobjective GA | 7.9921 | 0.0500 | 0.4135 |

data. It can be noticed that the single objective GA that minimizes $J_m$ index produces the best final $J_m$ value of 2.2839, whereas the multiobjective algorithm obtains a $J_m$ value of 2.3667. Similarly, the XB index produced by the single objective algorithm (minimizing XB) is 0.0148, but the multiobjective technique provides a slightly larger XB index value of 0.0164. However, the Silhouette index value (that no algorithm optimizes directly) for multiobjective method (0.6465) is better than that produced by the single objective method optimizing $J_m$ (0.5886) and single objective algorithm optimizing XB (0.5837). Note that similar results are obtained for the Serum data set also. These results imply that neither $J_m$ nor XB alone are sufficient for generating good clustering solutions. Both the indices should be optimized simultaneously in order to obtain superior clustering results.

## 6.6. *Visualization of results*

In this section the clustering solutions provided by different clustering algorithms are visualized using Eisen plot and cluster profile plots. The best clustering results obtained by the algorithms in each category of clustering methods (partitional, hierarchical and GA based methods) for both the data sets are visualized. Figure 1 shows the Eisen plots of clustering solutions

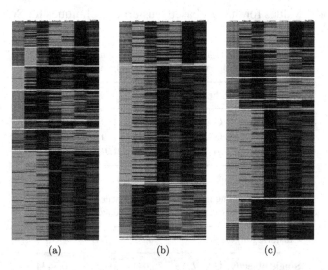

**Fig. 1.** Clustering results obtained on Yeast Sporulation data using (a) K-medoids algorithm, (b) average linkage algorithm, (c) multiobjective genetic algorithm optimizing $J_m$ and XB.

obtained for Yeast Sporulation data set for K-medoids, Average linkage and multiobjective GA clustering algorithms respectively.

It is evident from Fig. 1 that average linkage algorithm performs poorly than the other two algorithms. In fact this algorithm produces two singleton clusters (clusters 2 and 4) and a cluster with only two genes (cluster 7). Also the expression pattern within a cluster is not similar across different genes. The K-medoids algorithm produces clusters that have corresponding clusters for multiobjective technique. For example, cluster 7 of K-medoids algorithm corresponds to cluster 5 of the multiobjective method. However, the expression patterns are more prominent for multiobjective algorithm and with no doubt, it provides the best solution.

For the purpose of illustration, the cluster profile plots for the clusters produced by the multiobjective algorithm for Yeast Sporulation data are shown in Fig. 2. The cluster profile plots demonstrate how the gene

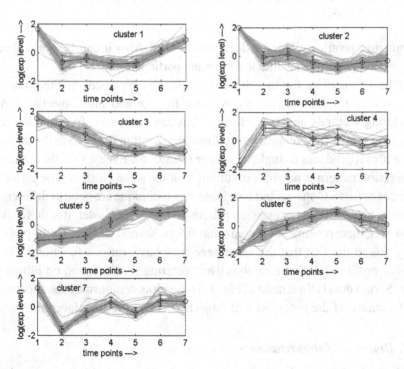

**Fig. 2.** Cluster profile plots for clustering solution obtained by multiobjective GA clustering algorithm for Yeast Sporulation data.

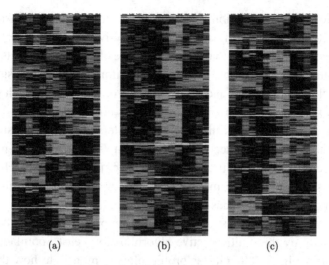

**Fig. 3.** Clustering results obtained on Human Fibroblasts Serum data using (a) fuzzy C-means algorithm, (b) average linkage algorithm (c) multiobjective genetic algorithm optimizing $J_m$ and XB.

expression profiles differ for different clusters. Also it can be noted that expression patterns are similar within any particular cluster.

The Eisen plots for the clustering solutions obtained on Fibroblasts Serum data for fuzzy C-means, Average linkage and multiobjective GA clustering methods are shown in Fig. 3. It can be noticed from the figure that the performance of the average linkage algorithm is not up to the mark. Here also it produces a singleton cluster (cluster 6). Except this, the overall expression pattern is not uniform throughout the same clusters. The performances of fuzzy C-means and multiobjective technique are good as different clusters have different expression pattern and within a cluster, it is uniform. However, there remains some confusion among several classes. This is probably due to the fact that the same gene may have various functions. The cluster profile plots for multiobjective clustering results found on Fibroblasts Serum data is illustrated in Fig. 4. These plots demonstrate the superior performance of the proposed multiobjective GA clustering algorithm.

## 6.7. *Biological interpretation*

Biological meaning of a cluster of genes can be assessed by studying the functional annotation of the genes within that cluster. It is expected that

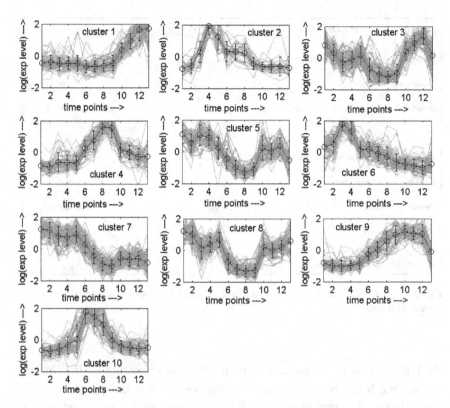

**Fig. 4.** Cluster profile plots for clustering solution obtained by multiobjective GA clustering algorithm for Human Fibroblasts Serum data.

genes within a cluster are expressed similarly as they have similar functionality or they are involved in the same biological processes. In this article, biological meanings of the clusters are found by using Gene Ontology (GO)[32] terms. A widely used web based tool *FatiGO*[1] is used for this purpose (www.fatigo.org). FatiGO extracts the Gene Ontology terms for a query and a reference set of genes and further computes various statistics for the query set. In our experiment, a query is the set of genes of a cluster and union of the genes from the other clusters is taken as the reference set. The GO level is fixed at three.

It is not possible to evaluate each cluster of the clustering solutions produced by all the algorithms here. So two interesting clusters from the clustering results obtained on Yeast Sporulation data set by multiobjective and average linkage algorithms are examined. Figure 5 shows a part of

```
Molecular function. Level: 3         0   20   40   60   80   100      p-values(*)
                                         6.98%
transcriptional activator activity     | 1.48%                         0.05697  1
                                         4.65%
isomerase activity                     | 0.74%                         0.09240  1
                                         9.30%
structural constituent of ribosome      3.70%                          0.11000  1
                                         6.98%
helicase activity                       2.22%                          0.11209  1
                                         2.33%
channel or pore class transporter activity  | 0 %                      0.13738  1
                                         2.33%
carbohydrate transporter activity      | 0.37%                         0.25627  1
                                         2.33%
chromatin binding                      | 0.37%                         0.25627  1

                                         0   20   40   60   80   100
                                                  Cluster Query
                                                  Cluster reference
(*) Unadjusted p-value; FDR(indep.)adjusted p-value;
```

Fig. 5. Part of the FatiGO result of cluster 2 of multiobjective clustering on the Yeast Sporulation data.

the FatiGO results of cluster 2 of multiobjective clustering on the Sporulation data. It can be noted that percentage of genes in the query cluster is considerably different from that of the reference cluster in almost all the functionalities. This implies that the correct genes are selected to be grouped in that particular cluster. A part of the FatiGO results of cluster 3 of Average linkage clustering on the Sporulation data is shown in Fig. 6. It can be observed that for most of the functionalities, percentage of genes within the query cluster is almost equal to that of the reference clusters. It means that the cluster 3 is not well clustered as genes with various functionalities are incorrectly grouped in the same cluster.

For illustration purpose, Table 4 shows some genes of cluster 2 of the clustering result obtained on the Sporulation data using multiobjective technique. The biological processes in which each gene is involved are also shown. It can be observed that all the genes are involved in some common biological processes and hence they are correctly clustered within the same group.

Considering all the experimental results, it can be concluded that the said multiobjective clustering algorithm outperforms all the partitional and

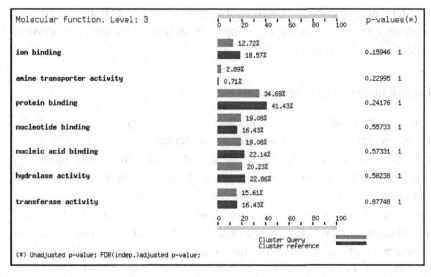

**Fig. 6.** Part of the FatiGO result of cluster 3 of average linkage clustering on the Yeast Sporulation data.

**Table 4.** Some genes of cluster 2 of multiobjective clustering on Yeast Sporulation data and corresponding biological processes.

| Gene | Biological process |
| --- | --- |
| YBR189W | cellular physiological process, metabolism, regulation of physiological process, regulation of cellular process |
| YBL072C | cellular physiological process, metabolism |
| YBR050C | cellular physiological process, metabolism, regulation of physiological process, regulation of cellular process, negative regulation of biological process |
| YBL023C | cellular physiological process, metabolism, regulation of physiological process, regulation of physiological process, regulation of cellular process |
| YBR173C | cellular physiological process, metabolism, response to stress, response to endogenous stimulus |
| YBR066C | cellular physiological process, metabolism, regulation of physiological process, regulation of cellular process, filamentous growth |
| YBR020W | cellular physiological process, metabolism |
| YBR062C | cellular physiological process, metabolism |
| YBL021C | cellular physiological process, metabolism, regulation of physiological process, regulation of cellular process |

hierarchical algorithms, as well as its single objective counterpart. It also produces clusters consisting of biologically similar genes whose expression patterns are similar to each other.

## 7. Conclusions and Discussions

This article demostrates an application of a recently proposed multiobjective genetic algorithm based fuzzy clustering technique for clustering microarray data sets. The clustering algorithm simultaneously optimizes two cluster validity indices, namely, $J_m$ and XB. The algorithm is designed on the framework of NSGAII multiobjective GA. The clustering algorithm has been applied on two microarray gene expression data sets viz., Yeast Sporulation data and Human Fibroblasts Serum data. Its performance has been evaluated on the basis of Silhouette index and compared with several other well known clustering techniques. Two cluster visualization tools namely Eisen plot and cluster profile plots are used for visualizing the clustering solutions. The biological interpretations of the clustering solution have been given based on gene annotation using a web based Gene Ontology tool (FatiGO). The superiority of the multiobjective technique has been demonstrated and use of multiple objectives rather than single objective has been justified. As a part of future work, use of some other validity indices, may be more than two, should be studied. The authors are currently working in these directions.

## References

1. F. Al-Shahrour, R. Diaz-Uriarte and J. Dopazo, Fatigo: a web tool for finding significant associations of gene ontology terms with groups of genes, *Bioinformatics* **20**(4), 578–580 (2004).
2. A. A. Alizadeh, M. B. Eisen, R. Davis, C. Ma, I. Lossos, A. Rosenwald, J. Boldrick, R. Warnke, R. Levy, W. Wilson, M. Grever, J. Byrd, D. Botstein, P. O. Brown and L. M. Straudt, Distinct types of diffuse large b-cell lymphomas identified by gene expression profiling, *Nature* **403**, 503–511 (2000).
3. M. R. Anderberg, *Cluster Analysis for Application*, Academic Press (1973).
4. S. Bandyopadhyay, An efficient technique for superfamily classification of amino acid sequences: feature extraction, fuzzy clustering and prototype selection. *Fuzzy Sets and Systems* **152**, 5–16 (2005).
5. S. Bandyopadhyay, Simulated annealing using reversible jump Markov chain Monte Carlo algorithm for fuzzy clustering, *IEEE Transactions on Knowledge and Data Engineering* **17**(4), 479–490 (2005).

6. S. Bandyopadhyay and U. Maulik, Genetic clustering for automatic evolution of clusters and application to image classification, *Pattern Recognition* (accepted).
7. S. Bandyopadhyay, U. Maulik, L. Holder and D. J. Cook (eds), *Advanced Methods for Knowledge Discovery from Complex Data*, Springer, UK (2005).
8. S. Bandyopadhyay, U. Maulik and A. Mukhopadhyay, Multiobjective genetic clustering for pixel classification in remote sensing imagery, *IEEE Transactions on Geoscience and Remote Sensing* **45** (5), 1506–1511 (2007).
9. J. C. Bezdek, *Pattern Recognition with Fuzzy Objective Function Algorithms*, Plenum, New York (1981).
10. R. J. Cho, M. J. Campbell, E. A. Winzeler, L. Steinmetz, A. Conway, L. Wodica and T. G. Wolfsberg et al., A genome-wide transcriptional analysis of mitotic cell cycle, *Mol. Cell.* **2**, 65–73 (1998).
11. S. Chu, J. DeRisi, M. Eisen, J. Mulholl and D. Botstein, P. O. Brown and I. Herskowitz, The transcriptional program of sporulation in budding yeast, *Science* **282**, 699–705 (1998).
12. C. A. Coello Coello, A comprehensive survey of evolutionary-based multiobjective optimization techniques, *Knowledge and Information Systems* **1**(3), 129–156 (1999).
13. C. A. Coello Coello, A. Carlos and A. D. Christiansen, An approach to multiobjective optimization using genetic algorithms, *Intelligent Engineering Systems Through Artificial Neural Networks*, ASME Press, St. Louis Missouri, USA **5**, 411–416 (1995).
14. L. Davis (ed), *Handbook of Genetic Algorithms*, Van Nostrand Reinhold, New York (1991).
15. K. Deb, *Multi-objective Optimization Using Evolutionary Algorithms*, John Wiley and Sons, Ltd, England (2001).
16. M. B. Eisen, P. T. Spellman, P. O. Brown and D. Botstein, Cluster analysis and display of genome-wide expression patterns, *Proc. Nat. Acad. Sci.*, USA, pp. 14863–14868 (1998).
17. B. S. Everitt, *Cluster Analysis*, Halsted Press, Third Edition (1993).
18. D. E. Goldberg, *Genetic Algorithms in Search, Optimization and Machine Learning*, Addison-Wesley, New York (1989).
19. T. R. Golub, D. K. Slonim, P. Tamayo, C. Huard, M. Gassenbeek, J. P. Mesirov, H. Coller, M. L. Loh, J. R. Downing, M. A. Caligiuri, D. D. Bloomeld and E. S. Lander, Molecular classification of cancer: class discovery and class prediction by gene expression monitoring, *Science* **286**, 531–537 (1999).
20. J. Han and M. Kamber, *Data Mining: Concepts and Techniques*, Morgan Kaufmann Publishers, San Francisco, USA (2000).
21. J. A. Hartigan, *Clustering Algorithms*, Wiley (1975).
22. V. R. Iyer, M. B. Eisen, D. T. Ross, G. Schuler, T. Moore, J. C. F. Lee, J. M. Trent, L. M. Staudt, Jr., J. Hudson, M. S. Boguski, D. Lashkari, D. Shalon, D. Botstein and P. O. Brown, The transcriptional program in the response of the human fibroblasts to serum, *Science* **283**, 83–87 (1999).
23. A. K. Jain and R. C. Dubes, *Algorithms for Clustering Data*, Prentice-Hall, Englewood Cliffs, NJ (1988).
24. L. Kaufman and P. J. Roussenw, *Finding Groups in Data: An Introduction to Cluster Analysis*, John Wiley & Sons, NY, US (1990).

25. J. B. MacQueen, Some methods for classification and analysis of multivariate observations, *Proc. 5th Berkley Symposium on Mathematical Statistics and Probability, Statistics* **1**, 281–297 (1967).
26. U. Maulik and S. Bandyopadhyay, Genetic algorithm based clustering technique, *Pattern Recognition* **33**, 1455–1465 (2000).
27. U. Maulik and S. Bandyopadhyay, Fuzzy partitioning using a real-coded variable-length genetic algorithm for pixel classification, *IEEE Transactions on Geoscience and Remote Sensing* **41**(5), 1075–1081 (2003).
28. Z. Michalewicz, *Genetic Algorithms + Data Structures = Evolution Programs*, Springer-Verlag, New York (1992).
29. P. J. Rousseeuw, Silhouettes: a graphical aid to the interpretation and validation of cluster analysis, *J. Comp. Appl. Math.* **20**, 53–65 (1987).
30. S. Z. Selim and M. A. Ismail, K-means type algorithms: a generalized convergence theorem and characterization of local optimality, *IEEE Transactions on Pattern Analysis and Machine Intelligence* **6**, 81–87 (1984).
31. W. Shannon, R. Culverhouse and J. Duncan, Analyzing microarray data using cluster analysis, *Pharmacogenomics* **4**(1), 41–51 (2003).
32. The Gene Ontology Consortium, Gene ontology: tool for the unification of biology, *Nature Genetics* **25**, 25–29 (2000).
33. J. T. Tou and R. C. Gonzalez, *Pattern Recognition Principles*, Addison-Wesley, Reading (1974).
34. X. L. Xie and G. Beni, A validity measure for fuzzy clustering, *IEEE Transactions on Pattern Analysis and Machine Intelligence* **13**, 841–847 (1991).
35. E. Zitzler, M. Laumanns and L. Thiele, *SPEA2: Improving the Strength Pareto Evolutionary Algorithm*, Technical Report 103, Gloriastrasse 35, CH-8092, Zurich, Switzerland (2001).
36. E. Zitzler and L. Thiele, *An Evolutionary Algorithm for Multiobjective Optimization: The Strength Pareto Approach*, Technical Report 43, Gloriastrasse 35, CH-8092, Zurich, Switzerland (1998).

# INDEX

$\alpha$-helices, 123, 125, 169
$\beta$-sheets, 123, 125
$\beta$-strands, 123, 169
*Evolutionary algorithms*, 26
*neural network models*, 26
1SMT, 194
3D reconstruction algorithm, 169

ab initio, 154, 155
active site, 135, 137, 140, 142, 147, 149, 194
allergen motif, 120
allergenicity, 118, 119
alpha carbon, 170
amino acid, 110, 122, 124, 134, 156, 175
analogy-based learning, 26
ant colony optimization, 262
ant colony systems (ACS), 51
artificial immune systems, 46
artificial intelligence (AI), 25
artificial life, 46

base calling, 8
belief networks, 39
bi-gram, 106
binding, 136
biochemical system theory, 208
bioinformatics, 109
biomarker, 18
BLAST, 111
bond length, 170
breadth-first traversal, 69

cancer classification, 263
cDNA arrays, 17
chaos theory, 50
chromosome, 190
class prediction, 233
classification, 235

classifier ensemble, 282, 285, 300
ClustalW, 7
cluster profile plot, 317
cluster validity index, 305
clustering, 304, 306
   agglomerative, 309
   fuzzy C-means, 304, 308, 312
   hierarchical, 304, 309
      average linkage, 310
      complete linkage, 310
      single linkage, 310
   K-means, 304, 307
   K-medoids, 304, 308
codon usage, 9
colon dataset, 250
combinatorial ensemble, 282, 289, 290, 296, 301
comparative genomics, 15–17
compensatory mutations, 12
competitive learning, 35
complete link algorithm, 78, 79
contact density, 154, 157, 167
contact map, 157, 158, 167, 168, 170
continuous wavelet transforms, 123
copy number polymorphism, 18
cross-entropy error, 165

data driven approach, 4
data mining, 114, 304
data standardization, 306
decoupled S-system model, 210
depth-first traversal, 69
discrete Fourier transform (DFT), 115
differential evolution, 213
directed acyclic graphs, 160
discrete wavelet transform, 115, 124
distance measures, 314
   Euclidean, 314
   Pearson Correlation, 314

## Index

DNA sequencing, 8
DNA microarray, 17, 232, 282, 283, 286
docking, 136
double-barreled sequencing, 8
discrete wavelet transform (DWT), 115
dynamic programming, 111
dynamic programming algorithm, 6, 11, 13

Eisen plot, 317
electrostatic potential, 138, 139
elitism, 313
epidemiological studies, 59
evolution, 5, 6
evolutionary algorithms, 64, 305, 310
exon, 10
expectation-maximization algorithm, 13
expressed sequence tag, 10

FatiGO, 323
feature, 103, 106
feature selection, 261
   embedded approach, 260
   filter approach, 260
   multivariate, 261
   univariate, 261
   wrapper approach, 260
feature vector, 103
filtering, 134, 162, 166
frequency domain, 115
fuzzy logic, 28
fuzzy membership, 308
fuzzy-GA synergism, 45

GA-belief network synergism, 45
gapped seed, 7
Gaussian curvature, 139
Gaussian RBF kernel, 103
gene annotation, 10
gene circuit, 206
gene context, 16
gene expression, 209, 304, 306
gene expression profiling, 18
gene fusion, 16
gene network, 207
gene ontology, 323

gene regulation, 206
gene regulatory network, 206, 209
gene subset identification, 233
genes, 304
genetic algorithm (GA), 65, 184, 304, 311
   chromosome, 312
   crossover, 305, 313
   multiobjective, 305, 310
   mutation, 305, 313
   selection, 305, 313
   single objective, 305
genome, 206
genome alignment, 15
genome rearrangement, 16
genome tiling array, 18
genomics, 4
global alignment, 111
global alignment problem, 6
gene ontology (GO), 323
gradient-descent, 160
granular computing, 46

helices, 155
hidden chains, 161
hidden Markov model, 10
HIV-I capsid, 194
HIV-I Integrase, 194
HIV-I Nef, 194
human fibroblasts serum, 315
hybrid algorithm, 69, 78
hybrid evolutionary algorithm, 242
hydrogen bonds, 155
hydrophobic cores, 122
hydrophobic strength, 138, 139
hydrophobicity, 125
hypothetical proteins problem, 17

information theory, 103
InsightII, 193
intron, 10

kernel, 87, 103, 106
kernel matrix, 265
kinetic orders, 209, 212

Lagrangian multipliers, 265

## Index

LBU, 188
leave-one-out-cross-validation, 233, 268
leukemia dataset, 242
ligand, 137, 143
ligand design, 185
local alignment problem, 6
local alignments, 111
local search, 217
longest common sequence, 174
lymphoma dataset, 248

machine learning, 154, 156, 232
marginalized kernel, 97
Markov model, 9
mass spectrometry, 126
mass spectroscopy, 19
maximal exact match, 7
maximal unique match, 7
mean curvature, 139
MEME, 13
memetic algorithms, 59, 69
microarray, 206, 259, 304
   data, 260
   datasets, 270
   preprocessing, 272
microarray visualization, 305
minimum description length, 103
mining, 134
molecular biology, 59
Morlet wavelet transforms, 125
motifs, 114, 118, 134
multi-class classification, 235
multi-modal NSGA-II, 238
multi-objective optimization, 233
multi-valued logic, 29
multiple sequence alignment, 59, 162

NCI60 dataset, 252
negatively correlated features, 286, 295
neural networks, 122, 160, 162
neuro-belief network synergism, 45
neuro-fuzzy synergism, 44
neuro-fuzzy-GA synergism, 46
neuro-GA synergism, 44
niched pareto genetic algorithm (NPGA), 311

non-dominated, 311
non-dominated sorting genetic algorithm (NSGA), 311
non-dominated sorting genetic algorithm II (NSGA-II), 238, 311
non-monotonic logic, 29
normal vectors, 139
normalization, 306
NP-hardness, 59

oligonucleotide array, 17
open reading frames, 9
optimal ensemble, 289–292, 297–299, 301
orthologous genes, 16
orthonormal, 112
over-fitting, 103

partitioning around medoids (PAM), 308
paralogous genes, 16
Pareto optimality, 311
Pareto-optimal, 238
partial matching, 30
particle swarm optimization (PSO), 47
pathway, 15, 17
peptide database searching tools, 19
peptide sequencing, 19
peptides, 125
pharmacophore, 185, 187, 194
phrap, 8
phylogenetic inference, 79
phylogenetic tree, 7, 59, 60
phylogenetics profile, 16
post-translational modification, 19
prediction strength, 235
primary sequence, 154, 155
primary structure, 138
principal eigenvector, 157, 167
principal eigenvalue, 167
probabilistic reasoning, 28
procedure Evolutionary-computation, 37
progressive alignment, 7
protein contact density, 167
protein folding, 157
protein repeat motifs, 122
protein secondary structure, 125, 155
protein sequence, 110, 114

protein structural motifs, 137
protein structure, 59, 135, 154
protein structure prediction, 13, 14
proteins folding, 122
proteomics, 18, 126
PSI-BLAST, 156, 165, 168

quadratic programming, 265

random walk, 97
rational drug design, 184
recurrent neural networks, 161
recurring substring, 103, 106
reinforcement learning, 36
relative solvent accessibility, 154
repeats, 8
residue contact maps, 154
resonance recognition model, 124
Rfam database, 102
RNA classification, 87
RNA secondary structure, 11
root mean square deviation, 174
rough sets, 49

S-system model, 208
SCFG likelihood, 94
scoring schemes for sequence alignment, 6
sdiscover, 103
secondary structure, 106, 154
secondary structure prediction, 166
seed-and-extend strategy, 7
sensitivity, 104
sequence alignment, 111
sequence analysis, 110
sequence by synthesize, 9
sequence motif, 12, 137
sequence profile, 7, 13
sequential minimal optimization, 104
shotgun approach, 8
signal processing, 109, 112
silhouette index, 305, 316
silhouette width, 316
simulated annealing, 171, 310
single nucleotide polymorphism, 18
single point crossover, 39

soft computing, 27
solvent accessibility prediction, 156
spatial resolutions, 124
strength pareto evolutionary algorithm (SPEA), 311
strength pareto evolutionary algorithm-2 (SPEA2), 311
specificity, 104
splicing, 10
statistical learning, 156
strands, 155
supervised learning, 35
support vector, 103, 265
surface motifs, 134, 141, 142
support vector machine (SVM), 85, 260, 263
SVM classifier, 103
synaptic inhibition, 32
synergism in soft computing, 44

T-Coffee, 7
threading, 14
time series data, 218
transcriptional factors, 12
tree, 187
trigonometric differential evolution, 213
truncation selection, 192
two hybrid technique, 20
two-points crossover, 39

ultrametric tree, 62, 66, 67
unsupervised learning, 35

van der Waals, 187
variable string length genetic algorithm (VGA), 185, 187
vector evaluated genetic algorithm (VEGA), 311

wavelet coefficients, 113
wavelet transform, 109, 111, 112

Xie-Beni index, 312

yeast sporulation, 315